aus der Reihe:

Innovationen mit Mikrowellen und Licht

Forschungsberichte aus dem Ferdinand-Braun-Institut, Leibniz-Institut für Höchstfrequenztechnik

Band 63

Pietro della Casa

Two-step MOVPE, in-situ etching and buried implantation: applications to the realization of GaAs laser diodes

Herausgeber: Prof. Dr. Günther Tränkle, Prof. Dr.-Ing. Wolfgang Heinrich

Ferdinand-Braun-Institut
Leibniz-Institut
für Höchstfrequenztechnik (FBH)
Gustav-Kirchhoff-Straße 4
12489 Berlin

Tel. +49.30.6392-2600
Fax +49.30.6392-2602

E-Mail fbh@fbh-berlin.de
Web www.fbh-berlin.de

Innovations with Microwaves and Light

Research Reports from the Ferdinand-Braun-Institut, Leibniz-Institut für Höchstfrequenztechnik

Preface of the Editors

Research-based ideas, developments, and concepts are the basis of scientific progress and competitiveness, expanding human knowledge and being expressed technologically as inventions. The resulting innovative products and services eventually find their way into public life.

Accordingly, the *"Research Reports from the Ferdinand-Braun-Institut, Leibniz-Institut für Höchstfrequenztechnik"* series compile the institute's latest research and developments. We would like to make our results broadly accessible and to stimulate further discussions, not least to enable as many of our developments as possible to enhance everyday life.

Broadly tunable laser diodes for measurement applications require multiple growth steps with intermediate patterned processing This is particularly challenging in the presence of Al in GaAs-based device heterostructures. Exposure of AlGaAs to atmosphere is avoided by protective GaAs or GaInP layers. In-situ etching inside of the growth reactor removes oxides as well as these protective layers and thus enables growth on a clean surface. Also, broad area laser diodes can benefit from such processes, leading to enhanced device efficiency. This thesis presents a toolbox of different processes, yielding buried heterostructures with excellent device performance for different applications. Hence, this modular approach paves the way for innovative GaAs-based laser diode concepts.

We wish you an informative and inspiring reading

Prof. Dr. Günther Tränkle Scientific Director	Prof. Dr.-Ing. Wolfgang Heinrich Scientific Deputy Director

The Ferdinand-Braun-Institut

The Ferdinand-Braun-Institut researches electronic and optical components, modules and systems based on compound semiconductors. These devices are key enablers that address the needs of today's society in fields like communications, energy, health and mobility. Specifically, FBH develops light sources from the visible to the ultra-violet spectral range: high-power diode lasers with excellent beam quality, UV light sources and hybrid laser systems. Applications range from medical technology, high-precision metrology and sensors to optical communications in space and integrated quantum technology. In the field of microwaves, FBH develops high-efficiency multi-functional power amplifiers and millimeter wave frontends targeting energy-efficient mobile communications as well as car safety systems. In addition, compact atmospheric microwave plasma sources that operate with economic low-voltage drivers are fabricated for use in a variety of applications, such as the treatment of skin diseases.

The FBH is a competence center for III-V compound semiconductors and has a strong international reputation. FBH competence covers the full range of capabilities, from design to fabrication to device characterization.

In close cooperation with industry, its research results lead to cutting-edge products. The institute also successfully turns innovative product ideas into spin-off companies. Thus, working in strategic partnerships with industry, FBH assures Germany's technological excellence in microwave and optoelectronic research.

Two-step MOVPE, in-situ etching and buried implantation: applications to the realization of GaAs laser diodes

vorgelegt von
dott. mag. (M. Sc.)
Pietro Della Casa
ORCID: 0000-0001-8393-4498

an der Fakultät II - Mathematik und Naturwissenschaften
der Technischen Universität Berlin
zur Erlangung des akademischen Grades

Doktor der Naturwissenschaften
Dr. rer. nat.

genehmigte Dissertation

Promotionsausschuss:

Vorsitzende: Prof. Dr. Kathy Lüdge
Gutachter: Prof. Dr. Michael Kneissl
Gutachter: Dr. Michael Jetter
Gutachter: Prof. Dr. Markus Weyers

Tag der wissenschaftlichen Aussprache: 25.01.2021

Berlin 2021

Bibliografische Information der Deutschen Nationalbibliothek
Die Deutsche Nationalbibliothek verzeichnet diese Publikation in der
Deutschen Nationalbibliografie; detaillierte bibliographische Daten sind im Internet
über http://dnb.d-nb.de abrufbar.
1. Aufl. - Göttingen: Cuvillier, 2021
Zugl.: (TU) Berlin, Univ., Diss., 2021

Nonnenstieg 8, 37075 Göttingen
Telefon: 0551-54724-0
Telefax: 0551-54724-21
www.cuvillier.de

1. Auflage, 2021
Gedruckt auf umweltfreundlichem, säurefreiem Papier aus nachhaltiger Forstwirtschaft.

ISBN 978-3-7369-7397-8
eISBN 978-3-7369-6397-9

Abstract

This work concerns the use of two-steps epitaxial growth, realized with metalorganic vapor-phase epitaxy (MOVPE), combined with in-situ etching and buried ion implantation, for the realization of GaAs-based edge emitting laser diodes, emitting in the near infrared region around 1 μm wavelength. The fabricated devices fall into two categories: tunable lasers (milliwatt range, monomodal) and high-power lasers (watt range, multimodal).

In the first case, the use of multi-step epitaxy is a *requisite*, and the task is that of making now this kind of process feasible with GaAs-based materials, similarly to what has been realized in the past with InP-based materials in the 1.3-1.5 μm wavelength range.

In the second case, a multi-step epitaxy is only an *option*, which can be exploited to introduce additional elements – as buried electrical and optical confinement structures – in the devices; a comparison of their performance and reliability with those of lasers fabricated with single epitaxy is then conducted to assess the effectiveness of this alternative approach.

Common for both cases is the requirement that surface contamination - particularly that due to oxygen - is removed before regrowth. This is more challenging than in the case of InP/GaInAsP structures, because here the ternary AlGaAs is used as cladding and waveguide material, and aluminum forms very strong bonds with oxygen. In turn, oxygen forms deep levels inside the bandgap which are effective non-radiative recombination centers.

Thus, in-situ etching with carbon tetrabromide (CBr_4) is first studied: the experimental results are presented, including kinetic data, the effects of different etching conditions and of the substrate characteristics on obtained morphology. Simple models are proposed to interpret the etching mechanism. Moreover, the utilization of the in-situ etching is discussed in relation to the reduction of residual surface contamination and enabling in-situ pattern transfer.

These investigations pave the way to devices based on 2-step epitaxy, combined with in-situ etching. The successful fabrication of thermally-tuned SG-DBR lasers operating around 975 nm is first described: a tuning range of 21 nm is demonstrated.

The possibility of using electronic injection for the tuning (which would allow a faster wavelength shift) is then explored, and the related issues discussed. It is suggested that fundamental material limitations confine the usable wavelength range to values lying near GaAs fundamental absorption edge at ≈ 870 nm. For wavelengths beyond 900 nm, the change in the refractive index that can be obtained in GaAs is too small, and there is no material option available which can be grown pseudomorphic to GaAs but with smaller bandgap.

High-power broad-area lasers have also been realized, using a two-step epitaxy combined with ex-situ + in-situ etching, to create a buried, shallow "mesa" containing the active zone. This approach allows introducing lateral electrical and optical confinement, and – simultaneously - non-absorbing mirrors at the laser facets.

This is reflected in performance and reliability improvements, but at the current state of development, the energy efficiency is still penalized by the two-step process with respect to a single epitaxy, because of additional non-radiative recombination paths. This is due to some extent to the additional etch-stop layers, needed for selective wet etching, that have to be added to the basic (and optimized) layer structure grown without intermediate patterning. Presently, the advantage of the lateral confinement appears to overcome the extra losses when the width of the lasers is below approximately 30 μm. Optimization of the heterostructure may further reduce the losses, making this strategy interesting even for larger devices.

Finally, a different strategy to create deep lateral current confinement in broad-area lasers, based on ion implantation *followed* by epitaxial regrowth, is presented. Two variants, differing in the position of ion implantation with respect to the active zone, are compared. While implantation into the waveguide introduces issues related to increased non-radiative recombination and optical losses, implantation into the p-cladding allows the improvement of device performance and the simultaneous introduction of non-absorbing mirrors at the facets, with corresponding improvement of the reliability. Here no pre-regrowth patterning through selective wet-etching is required and no etch-stop layers are inserted, so that the optimized layer sequence can be used. Thus, this approach represents a straightforward means to improve the performance of high-power broad-area devices.

In summary, this work has thus evaluated a "tool box" allowing different approaches towards the realization - within GaAs-based structures - of buried functional elements: *electrical* (current injection aperture) or *optical* (gratings, lateral optical confinement, butt-coupled active/passive sections, non-absorbing mirrors). Based on this knowledge, new devices are enabled (*GaAs* widely-tunable SG-DBRs) or new approaches towards improved device characteristics (high-power lasers) are opened up. However, the advantageous features that can be introduced with processes based on two epitaxial growth steps coexist with penalties associated to residual regrowth interface defectivity and to the modifications introduced in the vertical structure in order to allow the intermediate ex-situ processing. A favorable trade-off between these two aspects has to be sought.

Particularly for buried implantation such a good compromise has already been found. A reduction of threshold current by ≈12% and ≈15% increase of slope efficiency with respect to standard lasers with comparable processing and design has been obtained by implantation of oxygen, with a dose of 10^{15} cm^{-2} at 250 keV into the AlGaAs p-cladding layer and subsequent regrowth of the p-GaAs contact layer.

Kurzfassung

Diese Arbeit behandelt die Verwendung von zweistufigen Epitaxieprozessen, realisiert mit metallorganischer Gasphasenepitaxie (MOVPE), kombiniert mit in-situ-Ätzen und vergrabener Ionenimplantation, für die Realisierung von GaAs-basierten kantenemittierenden Laserdioden, die im nahen Infrarotbereich um ≈1 µm Wellenlänge emittieren. Die hergestellten Bauelemente lassen sich in zwei Kategorien einteilen: abstimmbare Laser (Milliwatt-Bereich, monomodal) und Hochleistungslaser (Watt-Bereich, multimodal).

Im ersten Fall ist der Einsatz von mehrstufiger-Epitaxie eine *Voraussetzung* für die Realisierung der entsprechenden Bauelemente. Die Aufgabe besteht darin, solche Prozesse auch mit GaAs-basierten Materialien möglich zu machen, ähnlich wie sie bisher mit InP-basierten Materialien im Wellenlängenbereich von 1,3-1,5 µm realisiert werden.

Im zweiten Fall ist eine mehrstufige Epitaxie nur eine *Option*, die genutzt werden kann, um zusätzliche Elemente – wie vergrabene elektrische und optische Einschlussstrukturen – in die Bauelemente einzubringen. Um die Effektivität dieses alternativen Ansatzes zu ermitteln, ist ein Vergleich der Leistung und Zuverlässigkeit solcher Bauelemente mit denen von in nur einem Epitaxieschritt hergestellten Lasern erforderlich.

Gemeinsam ist in beiden Fällen die Forderung, dass die Kontamination der Oberfläche - insbesondere durch Sauerstoff - vor dem Wachstum entfernt werden muss. Dies ist anspruchsvoller als bei InP/GaInAsP-Strukturen, da hier das ternäre AlGaAs als Material für Wellenleiter- und Mantelschicht verwendet wird und Aluminium sehr starke Bindungen mit Sauerstoff bildet; Sauerstoff wiederum bildet im Halbleiter tiefe Störstellen innerhalb der Bandlücke, die effektive nicht-strahlende Rekombinationszentren sind.

Deswegen wird zunächst das in-situ Ätzen mit Tetrabromkohlenstoff (CBr_4) untersucht: die experimentellen Ergebnisse werden vorgestellt, einschließlich kinetischer Daten und der Auswirkungen verschiedener Ätzbedingungen und der Substrateigenschaften auf die erhaltene Morphologie. Einfache Modelle werden vorgeschlagen, um den Ätzmechanismus zu interpretieren. Darüber hinaus wird der Einsatz des in-situ Ätzens im Zusammenhang mit der Verringerung der Oberflächen Restkontamination und der Ermöglichung der in-situ Musterübertragung diskutiert.

Die Untersuchungen zum in-situ Ätzen ebnen den Weg zu Bauelementen, die auf 2-stufiger Epitaxie, kombiniert mit in-situ Ätzen, basieren. Die erfolgreiche Herstellung von thermisch abgestimmten SG-DBR Lasern mit einem Arbeitsbereich um 975 nm wird zunächst beschrieben: ein Abstimmbereich von 21 nm wird demonstriert.

Anschließend wird die Möglichkeit der Verwendung elektronischer Injektion für die Abstimmung (was eine schnellere Wellenlängenverschiebung ermöglichen würde) untersucht. Es stellte sich heraus, dass grundlegende Materialbeschränkungen den nutzbaren Wellenlängenbereich auf Werte beschränken, die nahe der fundamentalen Absorptionskante von GaAs bei ≈ 870 nm liegen. Für Wellenlängen jenseits von 900 nm ist die in GaAs über Ladungsträgerinjektion erzielbare Änderung des Brechungsindex zu gering. und es gibt keine Materialoption, die pseudomorph zu GaAs ist, aber eine kleinere Bandlücke hat.

Hochleistung-Breitstreifenlasern wurden ebenfalls realisiert, wobei eine zweistufige Epitaxie in Verbindung mit ex-situ + in-situ Ätzen verwendet wurde, um eine vergrabene, flache «Mesa» zu erzeugen, die die aktive Zone umfasst. Dieser Ansatz ermöglicht die Einführung eines lateralen elektrischen und optischen Einschlusses und gleichzeitig nicht-absorbierenden Spiegeln an den Laserfacetten.

Dies spiegelt sich in Leistungs- und Zuverlässigkeitsverbesserungen wider, aber beim aktuellen Entwicklungsstand wird die Energieeffizienz immer noch durch das zweistufige Verfahren im Vergleich zu Wachstum in einem einzigen Schritt aufgrund zusätzlicher nichtstrahlender Rekombinationswege beeinträchtigt. Dies liegt zum Teil an den zusätzlichen Ätzstopp-Schichten, die für das selektive Nassätzverfahren benötigt werden, die zur (optimierten) ursprünglichen Schicht-Struktur ohne

Zwischenprozessierung hinzugefügt werden müssen. Gegenwärtig scheint der Vorteil des lateralen Einschlusses die zusätzlichen Verluste zu überwiegen, wenn die Breite der Laser unterhalb von ca. 30 µm liegt; eine Optimierung der Heterostruktur kann die Verluste weiter verringern, was diese Strategie auch für breitere Bauelemente interessant macht.

Schließlich wird eine andere Strategie zur Schaffung eines tiefen lateralen Stromeinschlusses in Breitstreifenlasern vorgestellt, die auf Ionenimplantation mit anschließendem epitaktischem Überwachsen basiert. Es werden zwei Varianten verglichen, die sich hinsichtlich der Position der Ionenimplantation in Bezug auf die aktive Zone unterscheiden: während die Implantation in die Wellenleiterschicht Probleme im Zusammenhang mit erhöhter nicht-strahlender Rekombination und optischer Verluste aufwirft, ermöglicht die Implantation in die p-Mantelschicht eine Verbesserung der Ausgangsleistung und die gleichzeitige Einführung nicht-absorbierender Spiegel an den Facetten mit einer entsprechenden Erhöhung der Zuverlässigkeit. Hier ist keine Vorwachstumsstrukturierung durch selektives Nassätzen erforderlich und es werden keine Ätzstopp Schichten eingefügt, sodass die optimierten Schichtenfolgen genutzt werden können. Dieser Ansatz stellt somit einen einfachen Weg dar, die Leistungsfähigkeit von Hochleistungs-Breitstreifenlasern zu verbessern.

Im Rahmen dieser Arbeit wurde damit ein «Werkzeugkasten» evaluiert, der unterschiedliche Ansätze zur Realisierung von vergrabenen *elektrischen* (Apertur für die Strominjektion) oder *optischen* (Gitter, lateraler optischer Einschluss, butt-coupled aktive/passive Sektionen, nicht-absorbierende Spiegel) *Funktionselementen* innerhalb von GaAs-Strukturen erlaubt. Basierend auf diesen Erkenntnissen werden neue Lasern (weit-abstimmbare *GaAs* SG-DBR) oder neue Ansätze zur Verbesserung der Lasereigenschaften (Hochleistungslasern) ermöglicht.

Den vorteilhaften Funktionen, die mit den entwickelten, auf zwei epitaktischen Wachstumsschritten basierenden Verfahren eingeführt werden können, , stehen jedoch Nachteile gegenüber, die mit an der überwachsenen Grenzfläche und den für die ex-situ Zwischenprozessierung nötigen Modifikationen der vertikalen Struktur verbunden sind. Daher muss immer ein geeigneter Kompromiss zwischen diesen beiden Aspekten gefunden werden.

Insbesondere für die vergrabene Implantation wurde ein solcher Kompromiss bereits gefunden. Durch die Implantation von Sauerstoff mit einer Dosis von 10^{15} cm^{-2} bei 250 keV in die AlGaAs p Mantelschicht und anschließendem Überwachsen mit der p-GaAs-Kontaktschicht wurde eine Reduktion des Schwellenstroms um ≈12% und eine Steigerung der differentiellen Effizienz (Steilheit) von ≈15% gegenüber Standardlasern mit vergleichbarem vertikalen Aufbau und gleicher Prozessierung erreicht.

Acknowledgments

The opportunity to work in the stimulating environment of the Ferdinand-Braun-Institut has been given to me by Prof. Dr. Günther Tränkle, who I wholeheartedly thank for his trust.

I also thank the head of the material technology department, Dr. Markus Weyers, who has invited me to join his team and has been always an interested and insightful point of reference for my work.

The investigations related to the devices presented in this thesis have been conducted with the involvement of different departments and laboratories. My thanks go to the former head of the optoelectronics department, Dr. Götz Erbert, and to the head of the sensor laboratory, Dr. Bernd Sumpf, both of whom were involved in the realization of the tunable lasers (chapter 5). For the part of this work which is related to buried-mesa lasers (chapter 6) I am particularly indebted to the head of the high-power lasers laboratory, Dr. Paul Crump, and for the part related to lasers with a buried-implanted current aperture (chapter 7) to the head of the optoelectronics department, Dr. Andrea Knigge.

I cannot recollect a single member of the material technology team to whom I do not owe some help, suggestion or constructive discussion; my thanks also go to all of them.

Dr. Frank Bugge has been a precious source of information about the epitaxial growth of arsenide compounds. Dr. Andre Maaßdorf has been a great source of help in countless scientific and technical aspects of my work, and a deeply supportive colleague. Daniel Schauer has been of invaluable help in keeping the MOVPE reactor running.

Dr. Ute Zeimer, Dr. Anna Mogilatenko and Helen Lawrenz have taken care of SEM, CL and EDX characterizations, Dr. Carsten Netzel of low-temperature PL; I am indebted to them not only for their competent work but even for many helpful discussions. Dr. Sylvia Hagedorn and Dr. Sebastian Walde have kindly provided me several AFM measurements.

SIMS measurements have been performed by Dr. Peter Jörchel, Dr. Peter Gehrmann and Dr. Peter Helm from RTG Microanalyse GmbH: to them my thanks for their highly proactive attitude.

For the cleanroom processing of the devices I thank collectively the colleagues of the process technology department, and in particular Christine Münnich, who has taken care of the pre-regrowth surface preparations. I also thank the responsible for the ion implantation, Dr. Andreas Thies, for his energetic cooperation, and Dr. Jörg Fricke for the time he has spent introducing me to several aspects of GaAs process technology.

I am particularly indebted to Dr. Olaf Brox and Dr. Dominik Martin; they have drawn the mask layouts, organized the device processing and part of the device measurements. Dr. Brox has also had a major role in the design and analysis of the tunable lasers - to which also Dr. Mahmoud Tawfieq did contribute.

Dr. Karl Häusler has contributed to the investigation concerning the buried-implanted lasers, coordinating the characterization of mounted devices and the reliability tests. Tim Adam and Christian Goerke have analyzed some of these devices in the framework of their Master's theses, and selected results from the latter concerning beam quality are summarized in chapter 7 (section 7.6.2).

The characterization of the mounted buried-mesa lasers has been supervised by Dr. Jonathan Decker and Dr. Martin Winterfeldt, and has been carried out with the aid of Alexander Putz and Lukas Hadasch.

I am grateful to Dipl. Ing. Armin Liero, who has eagerly accepted to help me with the lifetime measurements based on the reverse recovery-time technique.

Finally, Dr. Hans Wenzel has provided the WIAS-TeSCA simulations, and has been in general a reference point for all the discussions regarding the device physics.

Contents

1	**Introduction**	1
2	**Zincblende III-V semiconductors**	3
2.1	Chapter introduction	3
2.2	Zincblende crystal structure	3
2.3	Point defects in III-V semiconductors	6
2.4	III-V semiconductors and optoelectronics	7
3	**MOVPE growth of III-V compounds**	9
3.1	Introductory remarks on MOVPE technique	9
3.2	Planetary reactors AIX2400G3 and AIX2800G4	10
3.2.1	Reactor chamber	10
3.2.2	MOVPE gas mixing system and exhaust	13
3.3	Precursors selected for the experimental work	15
3.4	Dopants and impurities incorporation	16
3.4.1	Intrinsic/unintentional	16
3.4.2	Dopants from precursors	16
4	**In-situ etching with CBr_4**	17
4.1	Motivation for in-situ etching	17
4.2	Pre-existing research on in-situ etching	18
4.2.1	Chlorine compounds	18
4.2.2	CBr_4	19
4.3	Investigation of CBr_4 etching of GaAs	20
4.3.1	General experimental details	20
4.3.2	GaAs etching: surface morphology	23
4.3.3	GaAs etching: kinetics	29
4.4	Investigation of CBr_4 etching of GaAs assisted with TMGa and TMAl	39
4.4.1	Experimental details	39
4.4.2	Assisted etching: kinetics	39
4.4.3	Assisted etching: morphology	44
4.5	CBr_4 etching of AlGaAs and GaInP	46
4.5.1	AlGaAs	46
4.5.2	GaInP	47

4.6 Regrowth and interface contamination 47

5 SG-DBR tunable lasers 55
5.1 Chapter introduction 55
5.2 SG-DBR lasers 55
5.3 Thermally tuned SG-DBR lasers 57
5.3.1 Structure and process 57
5.3.2 MOVPE and intermediate pattern-definition process 62
5.3.3 Device results 66
5.4 Investigation of electronic tuning 68

6 Buried-mesa broad-area lasers 73
6.1 Chapter introduction 73
6.2 High-power lasers 73
6.2.1 Brightness and beam quality 74
6.2.2 Reliability and maximum output power 75
6.2.3 Design aspects 76
6.2.4 Strategies to introduce lateral confinement 77
6.3 Structure and process 77
6.3.1 The vertical structure 78
6.3.2 The process with two-step epitaxy 79
6.4 Results and discussion 82
6.4.1 Comparison of VS0 and VS2 vertical structures 82
6.4.2 Material characterization of the two-step epitaxy 87
6.4.3 Characterization of two-step epitaxy lasers: as-cleaved devices 89
6.4.4 Characterization of two-step epitaxy lasers: coated and mounted devices 91
6.4.5 Electrical overstress test 93
6.5 Chapter summary and conclusions 95

7 Lasers with buried implantation 97
7.1 Chapter introduction 97
7.2 Ion implantation 98
7.2.1 Interactions in the keV range and implantation profiles 98
7.2.2 Damage, damage removal, damage-isolation and doping 99
7.2.3 Oxygen in GaAs-AlGaAs 100
7.2.4 Silicon in GaAs-AlGaAs 101
7.2.5 Quantum-well intermixing effects of implantation 101

7.3 Device description and fabrication procedure 103
7.3.1 Vertical structure 103
7.3.2 Process with 2-step epitaxy and intermediate implantation 104
7.3.3 Device types 106
7.4 Material characterization 107
7.4.1 Residual implantation damage after regrowth 107
7.4.2 Electrical effects of the implantation 110
7.4.3 Surface morphology and regrowth interface 112
7.5 Characterization of as-cleaved devices 113
7.5.1 Comparison of 2-step and single-step growth 113
7.5.2 PI curves of LBI and STD lasers 114
7.5.3 Effects of implantation on I_{th} and leakage current 115
7.5.4 Effects of implantation on the slope efficiency 116
7.5.5 Effects of implantation on optical absorption 117
7.6 Characterization of coated and mounted devices 119
7.6.1 PI curves 119
7.6.2 Near-field and far-field 120
7.7 Step-stress tests 123
7.8 Chapter summary and conclusions 124

8 Summary and outlook 125

A1 Zincblende III-V semiconductors and related properties 129
A1.1 Appendix content 129
A1.2 Composition, bonding and related properties 129
A1.3 Crystal structure 131
A1.3.1 Zincblende and Wurtzite crystalline structures 131
A1.3.2 Zincblende crystal facets: thermodynamics and surface reconstructions 133
A1.4 Ternary and higher order compounds 135
A1.5 Epitaxial multilayers: mismatch, strain, relaxation 137
A1.6 Defects 139
A1.6.1 Point defects 139
A1.6.2 Electrical characteristics of point defects 139
A1.6.3 Structure of deep defects; an example: silicon DX center 140
A1.6.4 Extended defects 142
A1.7 Electronic structure and related properties 145
A1.7.1 Band structure of crystals 145

A1.7.2 Carrier statistics and semiconductors bands 146
A1.7.3 Interfaces and band offsets 151
A1.7.4 Band structure of III-V Zincblende semiconductors 153
A1.8 Carrier transport 157
A1.8.1 Semiclassical equations of motion – effective mass 157
A1.8.2 Intraband scattering and relaxation time 158
A1.9 Interband transitions 161
A1.9.1 Carrier populations away from equilibrium 161
A1.9.2 Non-radiative recombination at deep centers (SRH) 162
A1.9.3 Surface and interface recombination 163
A1.9.4 Auger 165
A1.9.5 Interband radiative transitions 167
A1.10 Optical properties in the transparency region 170

A2 Some general aspects of III-V MOVPE 173
A2.1 Different III-V epitaxy techniques 173
A2.2 General considerations about MOVPE reactors 174
A2.2.1 MOVPE reactor chamber and overall process: a simplified description 174
A2.2.2 Control of reagent flows 179
A2.3 Precursors for the growth of arsenides and phosphides 181
A2.3.1 General precursors requirements 181
A2.3.2 Molecular structure of the precursors and gas-phase diffusivity 181
A2.3.3 Pyrolysis of the precursors 182
A2.4 Surface processes 187
A2.4.1 Growth modes 187
A2.4.2 Surface chemistry and growth rate 189
A2.5 Stoichiometry, composition and impurity control in MOVPE 193
A2.5.1 V/III ratio, condensation phenomena and stoichiometry 193
A2.5.2 Composition control of multinary alloys 194
A2.5.3 Dopant and impurities incorporation 198

A3 Justification of the equations used in modeling the CBr4+TMAl etch 204

A4 Model for the calculation of α_{ip} in the implanted sections 207

References 215

1 Introduction

This work concerns the use of two-steps epitaxial growth, realized with metalorganic vapor-phase epitaxy (MOVPE), combined with in-situ etching and buried ion implantation, for the realization of GaAs-based edge emitting laser diodes. The fabricated devices fall into two categories: high-power lasers (watt range, multimodal) and tunable lasers (milliwatt range, monomodal).

The experimental work encompasses different investigations, which were in part aimed to get a better understanding of potentials and limitations of the manufacturing technologies - ideally defining some transferrable "building blocks" for device fabrication - and in part more focused on characteristics and performance of the specific devices.

Although a particular emphasis is given to the material-related topics, different aspects – processing, device design and characterization – are often combined in the exposition, as they were in the actual project developments.

The intent of the author is that of presenting the more significant challenges and questions arisen in the various part of the work as "open problems", and to discuss the possible answers, in the belief that the merit and interest of the exposed material lies essentially in this critical process of understanding.

Chapters content

Chapter 2 contains a short, introductory description of III-V Zincblende semiconductors: their main crystalline characteristics, the effect of added or unwanted impurities, their relevance for the realization of optoelectronic devices. A broader – although still very condensed – description of the properties of III-V semiconductors is provided as foundation material appendix 1.

Chapter 3 presents the MOVPE technology, focusing on the epitaxial reactors and the set of reagents used in this work. A more in-depth discussion of several aspects of the MOVPE process can be found in appendix 2.

In **chapter 4** the in-situ etching with carbon tetrabromide (CBr_4) is studied: the experimental results are presented, including kinetic data, the effects of different etching conditions and of the substrate characteristics. Simple models are proposed to interpret the etching mechanism. Moreover, the possible usefulness of the in-situ etching is discussed, in particular in relation to the problems of reducing surface contamination and enabling in-situ pattern transfer. The in-situ etching has been used in the investigations presented in the following three chapters.

Chapter 5 deals with the realization of widely-tunable sampled-grating distributed Bragg reflector lasers (SG-DBR). After a general description of the working principle of these devices, the technological aspects of their realization – in particular in relation to the 2-step epitaxy - are presented. Two approaches to the tuning are compared, thermal and electronic. Only the first has led to the realization of working devices, and the reasons behind the difficulties encountered with the second approach are discussed. Fig. 1.1 represents schematically the different sections of the realized SG-DBRs.

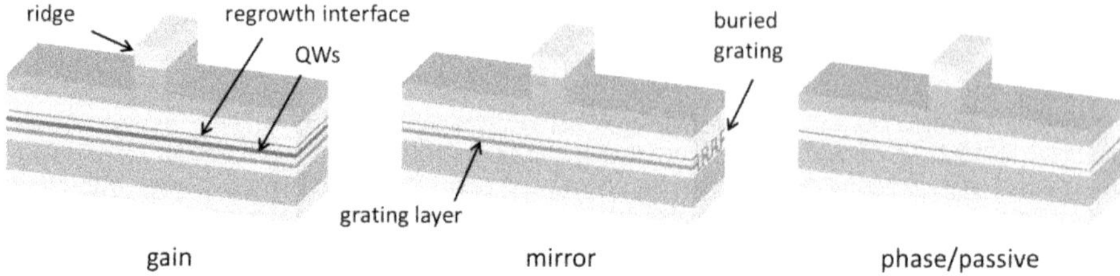

Figure 1.1 Schematic of the different sections of SG-DBR lasers of chapter 5.

Chapter 6 describes the realization of high-power broad-area lasers (BAL), using a two-step epitaxy process to create a buried, shallow "mesa", containing the active zone; the approach allows introducing lateral electrical and optical confinement, and simultaneously non-absorbing mirrors (NAM) at the laser facets (Fig. 1.2). Device results are presented and process limitations are discussed.

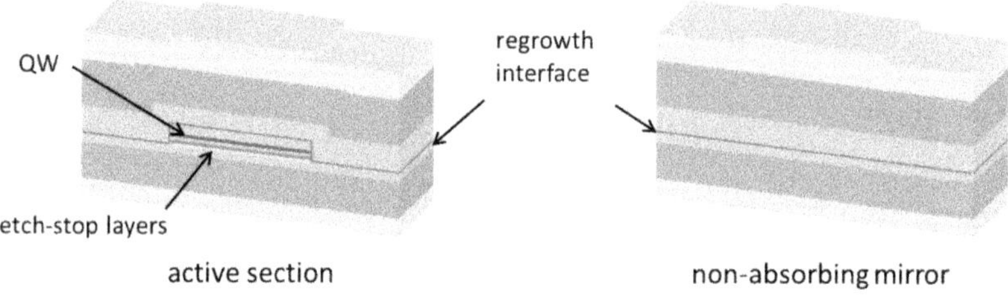

Figure 1.2 Active and passive (NAM) sections of the buried-mesa BALs of chapter 6.

Chapter 7 presents a different strategy to create deep lateral current confinement in BALs, based on ion implantation *followed* by epitaxial regrowth. Two approaches are compared, one more conservative, where the implantation is done in the upper cladding and the regrowth interface is in a "safe" position (it cannot cause non-radiative recombination) and one more challenging, with the implantation and the regrowth interface both in the waveguide layers (Fig. 1.3). While the first approach has given positive results in term of device performance, the second has proved more problematic; results and possible lessons learned are discussed.

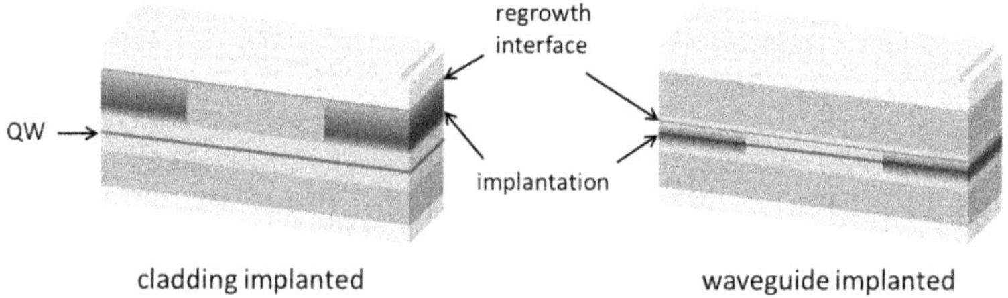

Figure 1.3 Two different positions of implantation and regrowth interface in BALs of chapter 7.

2 Zincblende III-V semiconductors

2.1 Chapter introduction

This chapter briefly introduces III-V semiconductors, with focus on those that have the same crystal structure of GaAs (Zincblende); only a very few aspects are rapidly touched, selected in an effort to provide a minimal material-related background relevant to the experimental part, without encumbering the narrative with an excessive amount of literature-based information.

Nonetheless, a quite broader – although still extremely condensed - discussion of III-V semiconductors' structural, electrical and optical properties, especially those that are more relevant to the realization of optoelectronic devices, is presented as "foundation material" - and possibly useful reference - in Appendix 1.

2.2 Zincblende crystal structure

III-V compounds having As, P and Sb as group V elements crystallize preferably in the Zincblende structure [1].

Within the crystal, each atom has four nearest neighbors of the other species, arranged in a tetrahedron, and the bonding can be interpreted in terms of valence bond theory assuming sp^3 hybridization of both species with formation of four localized sigma bonds, having an ideal angle of 109.47° between each pair.

In Zincblende there are 3 lowest-index high-symmetry families of planes: {100}, {110} and {111}. Figure 2.1 shows a schematic of the corresponding crystal facets and views of atoms and bonds in the crystal, each taken from a direction normal to one of the facets.

The significance of these planes is not only related to symmetry properties but even to chemical reactivity and thermodynamic stability.

Impurity incorporation probability – during the epitaxial growth - can be different between different surfaces. Also the different bonding on different adjacent surfaces can result in different surface diffusion of the atomic species; this is the case for Al and Ga during the growth of the ternary AlGaAs over a patterned surface, resulting in compositional modulation, with zones having a higher or lower Al content. This is generally an unwanted effect in many practical applications, as for example when a ternary or quaternary compound is used for the regrowth over the mesa structure of buried-mesa laser devices, or over the etched Bragg grating in distributed feedback lasers (DFB) and distributed Bragg reflector lasers (DBR), because the compositional perturbation is difficult to control and can even facilitate the formation of extended defects at the conjunction of two growth fronts.

Under surface-reaction-limited conditions, growth and etch rates exhibit selectivity with respect to crystal facets. In the case of growth, the slowest-growing facets emerge over convex substrate geometries, and the fastest-growing facets in the concave; in the case of etch, the situation is reversed, the slowest-etching facets are revealed in concave geometries, the fastest-etching facets over convex geometries. The kinetically more inert faces (slow growth or slow etch) are typically parallel to (111) planes.

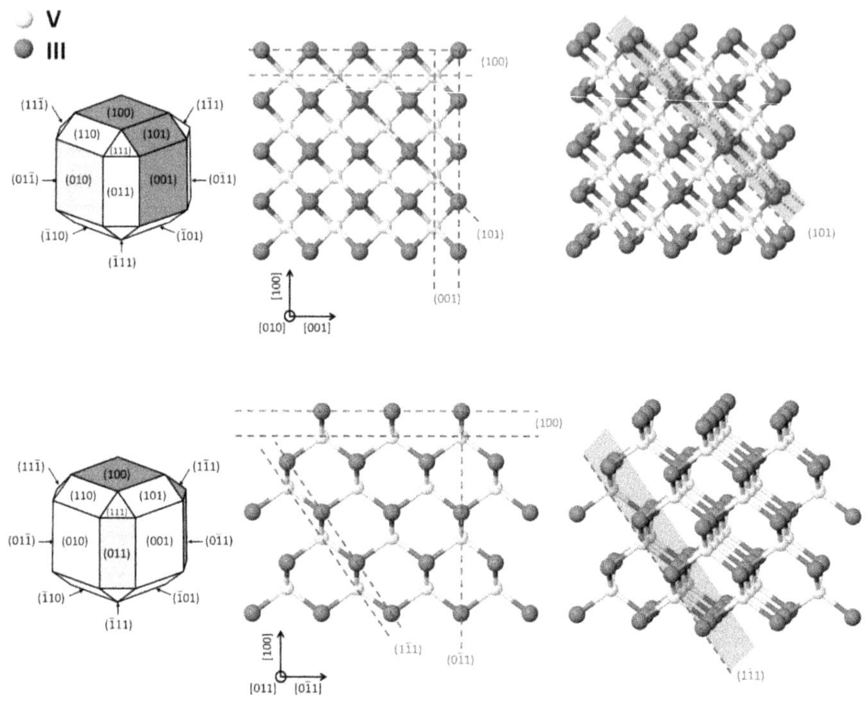

Figure 2.1 Zincblende lowest-index facets and corresponding views of the crystal. The facets in grey are normal to the observation direction, those in red are parallel. The dashed red lines indicate crystal planes perpendicular to the direction of observation. Rightmost, a small twist has been added with respect to the normal view.

The Zincblende Bravais lattice belongs to space group $F\bar{4}3m$ in Hermann-Mauguin notation (T_d^2 Schoenflies); it is symmorphic[1] and contains 24 point symmetry operations [2]. It does not include an inversion center, a characteristic that allows the onset of an electric field in consequence of the application of strain, because this can shift anions and cations in opposite directions (piezoelectricity). Although Zincblende contains directions, as the <111> and <100> along which anions and cations are alternated, the unstrained bulk Zincblende crystal has no net polarization because of its overall symmetry.

The {100} planes are perpendicular to the three 4-fold rotoreflection axes and three 2-fold rotation axis of the space group. Along a [100] direction, there is an alternation of group III and of group V layers: cleavage of the crystal parallel to one (100) plane leaves a facet terminated either with group III or group V atoms, in both cases with 2 dangling bonds for each atom. Addition or removal of a single atom

---

[1] A space group is symmorphic when there is a point such that all symmetry operations are the product of a symmetry operation which keeps this point fixed and a translation.

to/from a (100) surface involves no change in the number of dangling bonds. Most commonly, the semiconductor wafers used for device fabrication have a (100) upper surface.

The {110} planes are parallel to the six reflection planes of the space group, the ideal dihedral angle with {100} planes are 45° and 90°. Cleavage parallel to a (110) plane leaves on the surface group III and group V atoms in equal number, with one dangling bond for each atom; within the plane, the atoms are arranged in zigzag chains of consecutively bonded atoms. The cleavage along a (110) plane is facilitated by the absence of polarity in the [110] direction, a characteristic useful for the separation of discrete devices fabricated on (100) oriented wafers, especially when the facets must have optical quality as in the case of edge emitting lasers. Another characteristic, interesting in relation to edge emitter lasers reliability, is that - at least in the case of GaAs - no midgap surface states are expected to be present in ideally cleaved, relaxed (110) surfaces in absence of any oxidation or contamination [3].

The {111} planes are perpendicular to the four 3-fold rotation axes of the space group, the ideal dihedral angle between {111} and {100} planes is 54.74°, ½ of the tetrahedral bond angle. The {111} planes can be seen as a arranged in double layers, each consisting of two closely spaced planes, one containing only group III and the other only group V atoms, with each atom of one layer bonded to 3 of the other layer. These planes are designated with the letters A and B or with the atomic symbols, for example in the case of GaAs $\{111\}_A \equiv \{111\}_{Ga}$ and $\{111\}_B \equiv \{111\}_{As}$. A cleavage parallel to {111} between the A and B planes of a closely-spaced double layer would leave on the surfaces 3 dangling bonds for each atom, while a cleavage parallel to {111} between A and B planes belonging to consecutive double layers would leave only a single dangling bond for each atom.

When the *facet* of a crystal corresponds to a (111) plane, it is designated A or B according to which kind of atom layer *can* terminate the crystal with only *one* dangling bond per atom, assuming the more energetically stable termination: note that in this case, if each atom retains its own electrons, the group V atoms on a type-B facet will have two electrons per dangling bond while the group III atoms on a type-A facet will have no electrons in the dangling bonds[2]. The $\{111\}_A$ facets are usually the most kinetically stable with respect to chemical etching. The different facets are depicted in Fig. 2.2 along with a representation of a (100) wafer with orientation flats, according to the European-Japanese (E-J) convention.

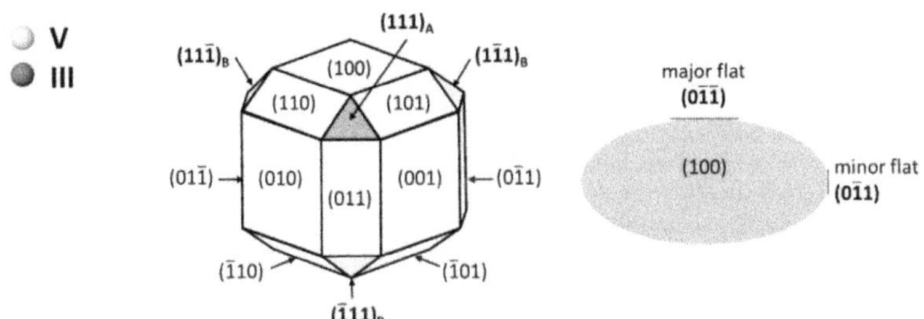

Figure 2.2 Left: $(111)_A$ and $(111)_B$ facets of Zincblende crystal; right:, wafer surface and flat orientation, according to E-J convention.

[2] This is not entirely accurate, because the dangling bonds of a non-reconstructed, neutral (111) surface are expected to bear a fractional charge (Appendix 1).

2.3 Point defects in III-V semiconductors

Defects related to the presence of foreign atoms are referred to as extrinsic defects. Defects that represent deviations from the regular arrangement of the lattice but are not associated to foreign atoms are called intrinsic.

Among the possible criteria for defects classification, one is based on dimensionality, and divides them into zero-dimensional - or point - defects and extended (1- 2- 3- dimensional). The presence of extended defects is in general unwanted in practical applications; their description goes beyond the scope of this chapter, but basic information has been included in Appendix 1. Point defects [4, 5] are not necessarily detrimental: they contribute in a fundamental way in determining the electrical and optical functionality of semiconductor materials and, moreover, they play a key role in diffusion processes and in device degradation processes.

In semiconductors, the electrons primarily involved in the optical and electrical processes relevant to optoelectronic devices, are those belonging to two separate sets of energetically finely-spaced levels, the valence band (VB) and the conduction band (CB); the difference between the lowest energy level of the CB and the highest energy level of the VB is called bandgap. Electrons in the CB are responsible for n-type conductivity (n-carriers), while empty electronic states (or holes) in the VB are responsible for p-type conductivity (p-carriers).

Electrically active defects introduce one or more energy levels within the bandgap, whose associated electronic wavefunctions are localized. Electrons can be exchanged between these localized levels and the bands.

Shallow electronic levels are those whose energy lies near the bottom of the CB or the top of the VB; usually this is intended in the sense that the difference between the level and the nearest band edge is less than the thermal energy k_BT (26 meV at 300 K). Occupied levels lying near the CB can be easily ionized, promoting electrons into the CB and are called donors; similarly unoccupied levels near the VB can easily accept electrons from the VB, leaving behind a hole, and are called acceptors. The (small) ionization energy of the donors and acceptor makes the electrical conduction an activated process.

Deep electronic levels have energies more near the center of the bandgap, with a stronger localization of the electrons (down to the size of an interatomic bond length). The corresponding defects are called deep defects. Some of them are effective electron-traps and/or hole-traps, in the sense that they can capture electrons from the CB or donate electrons to the VB, with the following effects: reduction of carrier density, reduction of carrier mobility via ionized defect formation and increase of non-radiative carrier recombination.

Foreign atoms creating donor or acceptor levels are called dopants; the range of *intentional* doping concentration spans several orders of magnitude, approximately from 0.1 to 1000 ppm. Dopant atoms for III-V semiconductors come usually from the neighboring groups II, IV and VI. They are incorporated in the crystal in the sites normally occupied by a group III or a group V atom, so that the number of the valence electrons of the impurity is higher or lower than that normally contributed by a constituent atom in that lattice position: p doping is obtained decreasing the number of electrons with II→III or IV→V substitutions, n doping increasing it with IV→III or VI→V substitutions. Concerning the specific selection of dopants species:

- VI→V substitution: S, Se and Te all introduce shallow donor levels; O introduces deep levels and cannot be used as a dopant.
- II→III substitutions: Be, Mg (group IIa) Zn and Cd (group IIb), introduce shallow acceptor levels; the other group II elements have not been used successfully for p-doping.
- IV→III and IV→V substitutions: group IV elements are electrically amphoteric, since they can be simultaneously incorporated in both III and V sublattices, leading to opposite kind of doping. The actual behavior depends on the specific material and from the doping process conditions. In GaAs, C is prevalently incorporated in the group V sites and is a shallow acceptor, Si is incorporated

prevalently in the group III sites where behaves as a shallow donor, Ge is strongly amphoteric, Sn is prevalently a shallow donor.

Silicon, carbon and zinc have been used as dopants in the experimental part of this work, while oxygen has been used to deliberately introduce deep levels.

2.4 III-V semiconductors and optoelectronics

III-V semiconductors as GaAs and InP are used in high-speed electronics applications, where their characteristics – in particular higher electron mobility and peak saturation velocity - make them superior to the otherwise more utilized silicon. In the framework of *opto*electronics, there are two fundamental reasons for their success: first, many of them are efficient light emitters, and second, it is technologically possible to monolithically integrate different material compositions having distinct electrical and optical properties.

The electrons can be excited, in particular by means of current injection or photon absorption, from the – largely occupied – valence band to the – largely empty – conduction band; the reverse disexcitation process can occur through different mechanisms, radiative (photon emission) or non-radiative. Efficient photon emitters are those semiconductors where the photon emission corresponding to the electronic transition from the lowest-energy states of the CB to the highest-energy states of the VB is possible without requiring a change in the momentum (or more properly in the *crystal* momentum) of the electrons. These transitions are called direct, and the semiconductors whose electronic band structure allows such transitions – as is the case for GaAs, InP and many other III-Vs, but not for Si - are called direct semiconductors. Under high-excitation conditions, which are relevant for laser devices, the total recombination rate of a bulk semiconductor (not including stimulated photon emission) is often expressed with the approximate "ABC equation":

$$R_{tot} = An + Bn^2 + Cn^3, \qquad \text{E1.1}$$

where n is the CB electron density; the first term represents the non-radiative recombination caused by the presence of deep levels, the second the (spontaneous) radiative recombination and the third the non-radiative recombination related to carrier-carrier scattering processes (Auger). In a direct semiconductor, the B coefficient is orders of magnitudes larger than in an indirect one. Of the three coefficients in E1.1, B and C stem from the fundamental properties of the material, while A depends on the kind and density of the defects, and is consequently the only one impacted by the material *quality*.

The monolithic integration of different III-V material compositions in a multilayered, epitaxial crystal structure is fundamentally limited by the necessity to guarantee similar average interatomic distances, or equivalently a similar lattice parameter, in order to avoid the formation of extended defects. Since for technological reasons the available bulk-grown substrates correspond to the binary compounds, the constraint is essentially that of combining the different atomic species, each of them characterized by a different size, in such a proportion that the resulting lattice parameter does not differ too much from that of the substrate, leading to "families" of devices indicated for example as "GaAs-based" or "InP-based". This leaves anyway a considerable room for the engineering of the material properties.

In the epitaxial growth of a multilayer stack, the transition between different materials can be a technologically critical point, because formation of defects or interlayers with undesirable composition might occur.

Once a laser is realized and tested, the task of determining to which extent the quality of the material and interfaces do actually condition its performance and reliability is – in many cases – not trivial; this will be an important topic in the following parts of the thesis.

3 MOVPE growth of III-V compounds

3.1 Introductory remarks on the MOVPE technique [6]

Epitaxy is the process of growing a crystalline material layer on a crystalline substrate, with the layer conforming to the lattice structure of the substrate. It is called homoepitaxy when substrate and layer materials are identical[1] and heteroepitaxy otherwise. Epitaxy is called pseudomorphic when any mismatch between the lattice parameters of the layer and of the substrate is elastically accommodated.

Metalorganic vapor phase epitaxy (MOVPE, also known as metalorganic chemical vapor deposition MOCVD) is the most widespread production method for the realization of light-emitting compound semiconductor devices.

In the MOVPE technique the reagents (or "precursors") are typically volatile metalorganic compounds (MO) and hydrides, which are transported in the deposition chamber highly diluted in a carrier gas, in most cases hydrogen. The chamber pressure can vary approximately in the range 10 to 1000 mbar, with values between 50 and 150 mbar more commonly used; carrier gas and precursors are kept under laminar flow conditions. The substrates are positioned on a rotating graphite susceptor, whose temperature depends on the material to be grown, and is in the range 550°C-850°C for III-V arsenides and phosphides, while higher temperatures are used for nitrides and lower temperatures for antimonides. Group III precursors are always introduced in lower amount than those of group V, and the growth rate is normally controlled by the diffusion of group III precursors from the gas phase to the substrate; the rate can reach typically some µm/h.

The technique is extremely flexible in terms of achievable material compositions, and allows good control over layer thickness, typically in the order of 2% for thick layers (i.e. approximately above 50 nm). Abrupt interfaces between layers of different composition can be obtained: with properly optimized switching sequences, the transition regions can be reduced to the monolayer range. This makes possible the realization of well-defined nanometer size structures, as needed in modern quantum-well lasers.

The experiments described in the present work have been conducted using MOVPE reactors produced by Aixtron SE, of a particular type called *planetary*; two different models were used, AIX2400G3 and AIX2800G4 (abbreviated in the following to **G3** and **G4**). Reactors and precursors are briefly described in the next sections. A more general description of MOVPE can be found in appendix 2.

---

[1] Alternatively, depending on the context, the terms homoepitaxy/heteroepitaxy are used to indicate that substrate and grown layer have/don´t have the same crystal structure.

3.2 Planetary reactors AIX2400G3 and AIX2800G4 [7-12]

3.2.1 Reactor chamber

Planetary reactors look very much alike vertical reactors, but are better described - in terms of flow configuration - as radially-isotropic horizontal reactors. A simplified scheme of the growth chamber is shown in Fig. 3.1; in the drawing, the height of the chamber has been stretched vertically for ease of representation, the actual chamber height/susceptor radius values are 2.6/16.5 cm for G3 and 3.35/31.5 for G4.

The susceptor has a comparatively slow rotation speed, normally in the range 5-25 rpm. The carrier gas with diluted reagents enters the chamber horizontally, through a central injector, and spreads radially; hydrides and MO flows are kept separated in the piping and mix only after injection. The vertical flow velocity - away from the central injector and the edges - is zero, while the radial flow velocity decreases with increasing distance x from the center. Fluid dynamics modeling [7, 8] shows that in this case the flux of group III to the susceptor - and consequently the growth rate – increases rapidly near the edge of the central injector, and then decreases almost linearly along the radius; more precisely, the decreasing trend can be described with an empirical relation of the form: $G(x) = a - bx + c/x$. The group III isobar lines shown in Fig. 3.1 correspond to concentration boundary layer profiles (not to the 99% boundary)[2].

This radial disuniformity is averaged out by an additional rotation of each wafer around its own center – whence the name planetary.

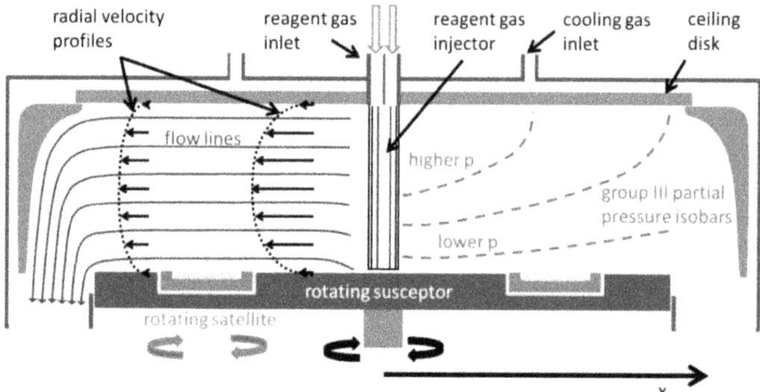

Figure 3.1: simplified scheme of a planetary reactor growth chamber, with qualitative representation of flow lines and group III isobar lines; the height of the chamber has been stretched with respect to real proportion for ease of representation.

The wafers are positioned over small graphite disks, called satellites, which are inserted in recesses within the susceptor. The surface of these recesses is not flat, but contains shallow spiral grooves, each with a small gas outlet, as can be seen in Figure 3.2a which shows the planetary susceptor with satellites removed.

The satellites are free to rotate around a metal pin protruding from the center of the recess, while a gas flow, normally of H_2, coming from inside the susceptor lifts them (by some tens μm) eliminating the friction and providing a viscous shear force for their rotation (gas-foil rotation). The gas is then

[2] For a discussion of boundary layer approximation, see appendix 2.

(mostly) collected in a circular groove surrounding the spirals and extracted through a larger extraction hole in the groove. Figure 3.2b shows the radial growth rate profile with and without satellite rotation. Under operative conditions the satellite rotation speed is in the range of 150-300 rpm, so that at typical deposition rates a full rotation takes place during the growth of a monolayer or less.

Depending on the shape of the growth profile over the susceptor radius, the shape of the growth profile over the wafer can be slightly concave or convex, and has to be optimized varying the operating parameters: in particular adjusting the total gas flow rate is an effective way to obtain an almost perfectly flat profile.

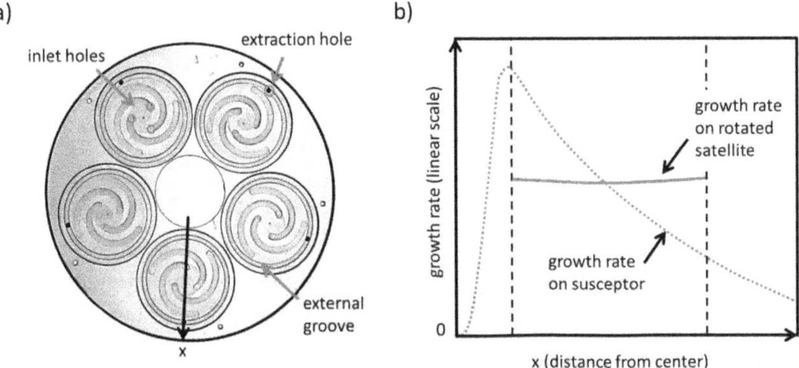

Figure 3.2 **a)** Top-down image of a graphite planetary susceptor with satellites removed; **b)** qualitative growth rate profiles over a wafer integral with the susceptor (dotted line) or with extra satellite rotation (continuous line).

The susceptor is rotated mechanically by a shaft with a magnetic liquid rotary sealing, and is heated with a water-cooled inductor coil connected to an RF generator (Fig. 3.3), its temperature is monitored from the lower side by an optical pyrometer. The actual temperature on the wafer surface can significantly differ from this value, and can be monitored by means of emissivity-corrected pyrometry.

Figure 3.3 Partially dismounted G3 reactor, the susceptor and the underlying induction coil are clearly visible.

The stainless-steel walls of the reactor (lateral, top and bottom) are water-cooled; the chamber ceiling is shielded by a disk of quartz (AIX2400G3) or graphite (AIX2800G4), whose temperature is monitored by a second pyrometer, and is gas-cooled by a mixture of H_2 and N_2 injected into a thin slit between ceiling and disk: by changing the H_2/N_2 ratio, it is possible to regulate the ceiling disk temperature during the growth (approximately in the range 100°C-350°C) thanks to the different thermal conductivity of the two gases. Accumulation of deposits over the ceiling disk leads to detachment of particles or to delamination and fall of large flakes, in both cases compromising the quality of the material. A low temperature reduces the deposition of polycrystalline semiconductor over the surface, but an excessively low temperature can lead to condensation of group V elements, so an optimal compromise temperature has to be found[3].

The chamber ceiling can be opened for manual loading/unloading of wafers, as shown in Fig. 3.4.

To protect the reactor chamber from contamination (and the operators from poisonous gases and - especially - particulates) the ceiling lid opens within a glove-box kept under nitrogen atmosphere; the wafers are taken in and out of the glove-box via a load-lock chamber. Configurations allowing automated satellite load/unload via a lateral opening in the chamber wall are available, but not installed in FBH.

All around the susceptor an annular exhaust collector deflects the flow downwards; its shape is different in the two reactors, being a large molybdenum duct in reactor G3 and a graphite ring in reactor G4.

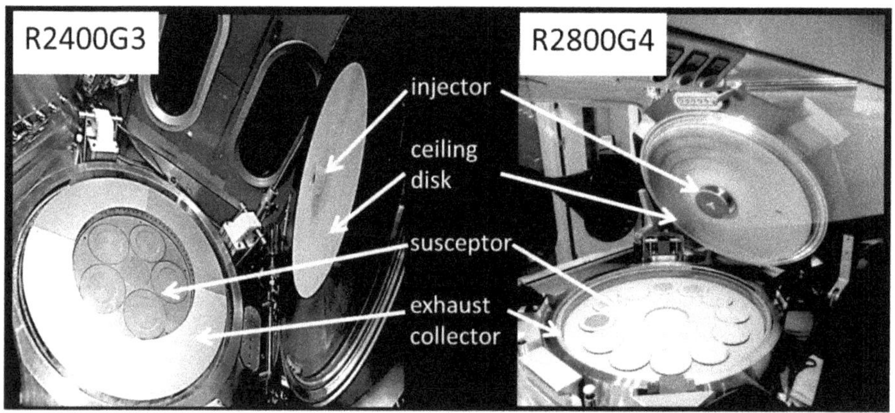

Figure 3.4 Images of the open growth chambers of FBH planetary reactors R2400G3 and R2800G4.

[3] An upside-down configuration would solve the problem of falling particles, and is actually produced by Taiyo Nippon Sanso Corporation (TNSC); some TNSC models include a "rotation and revolution" system similar to the planetary concept, but without use of gas foil rotation.

3.2.2 MOVPE gas mixing system and exhaust [6]

Figure 3.5 is a simplified scheme of gas mixing system and exhaust, providing only the basic elements (a detailed scheme would actually include hundreds of valves, gas lines and other components); it reflects the specific configuration of FBH reactors.

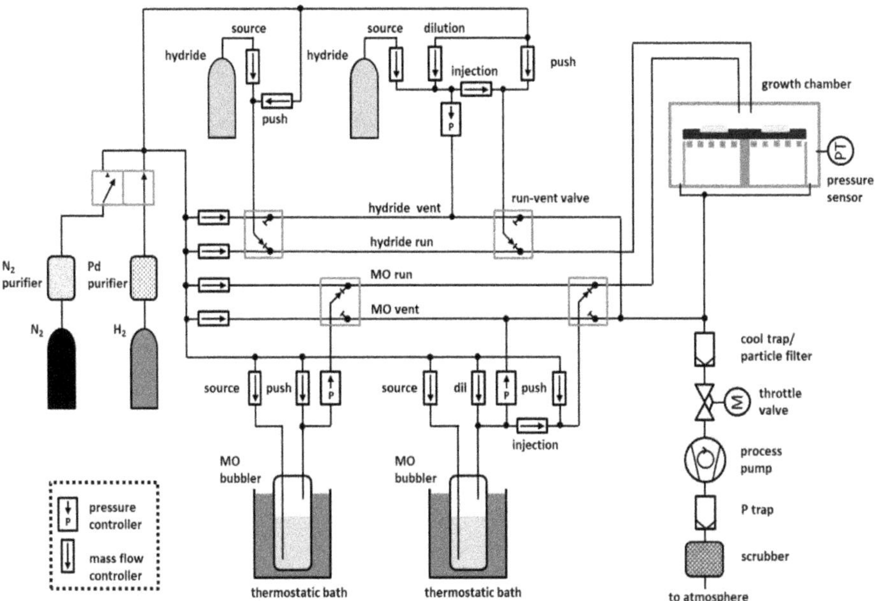

Figure 3.5: simplified representation of gas mixing system, reactor chamber and exhaust in a MOVPE reactor. The scheme reflects the configuration of FBH reactors.

Hydrogen is normally used as carrier gas, and a palladium purifier is inserted in the gas line as shown in the scheme. The purifier consists of a palladium alloy membrane, heated at about 400°C: the hydrogen molecules dissociate over the membrane surface on the low-purity side and recombine on the high-purity side, giving the desired 9N purity (99.9999999%); these purifiers can have a very high flow capacity, do not need regeneration and are ideally suited for the task. Historically, the early availability (since the 1980') of Pd purification systems, combined with the comparatively low cost of hydrogen gas with respect to possible alternatives like He or Ar, have probably been decisive in the adoption of H_2 as preferred carrier gas; nonetheless, nitrogen can currently be purified as much effectively as hydrogen, and represents a viable alternative. Other factors that can impact the choice between H_2 and N_2 are the different fluid-dynamics related properties (H_2 has smaller mass, higher thermal conductivity, higher diffusion coefficient, lower viscosity) and differences in chemical reactivity. The last point applies in particular to the case of III-V nitrides growth, where hydrogen behaves at high temperature (>1000°C) as a strong etchant of gallium nitride due to the formation of volatile NH_3 molecules, while in the case of arsenides and phosphides the chemical effect of hydrogen is mainly that of marginally promoting MO precursors pyrolysis.

The gas entering the growth chamber is mostly carrier gas, with a low concentration of group V precursors, in the range 3-0.3% in FBH reactors; the concentration of group III precursors is lower, approximately in the range 0.1-0.01%, the concentration of dopant precursors is even lower, from 0.005% down to the ppb range. Pressure, total flow rate, and chamber volume, determine the residence time of the reagents in the chamber (t_r) which is approximately 0.7 s in reactor G3 and 1.1 s in reactor G4 under the operating conditions used.

The precursor are divided in gases, which are hydrides as AsH_3, contained in pressurized cylinders, and volatile liquids or solids, which are mostly metalorganics as TMGa (trimethylgallium), contained in bubblers through which a bubbling gas (normally the same as the carrier gas) flows. The bubbling gas saturates in the bubbler with the precursor vapor, and the exiting gas contains a certain concentration of precursor molecules, as better detailed in appendix 2.

The gas flows along the lines are regulated by means of mass-flow controllers (MFCs) and the pressure by pressure controllers (PCs); in both cases, the regulation is obtained by means of a variable orifice along the line.

The precursor are delivered either to a gas line called *run*, that leads to the growth chamber, or to a line called *vent*, that bypasses the chamber and goes directly to the exhaust. There are two run lines, one for the hydrides and one for the MOs, and similarly two vent lines. To obtain an abrupt change in the precursor supply to the growth chamber – and consequently an abrupt interface in the growing material - a precursor can be switched from run to vent or vice-versa by operating the corresponding run-vent valve. Run-vent lines and their valves are collectively called *manifold*.

The gas flow exiting the growth chamber is joined with that of the vent line, and passes through a combined filter, which consists in an upstream cooling trap and a downstream particle filter. The filter is positioned immediately after the chamber to avoid that the highly reactive decomposed precursors, which are still abundant in the exhaust flow, might form deposits in the subsequent gas-line elements. After the filter there are in sequence: throttle valve, process pump, phosphorus trap and scrubber. The throttle valve regulates the pressure in the chamber and is driven by a stepper motor commanded by a PID controller unit, which is connected to a pressure sensor.

The process pump is a multistage Roots dry pump, which allows avoiding the oil back-streaming associated with conventional rotary vane pumps.

The additional cooled trap installed after the pump captures the elemental phosphorus, which condensates at atmospheric pressure forming a fine, pyrophoric particulate; after that, the scrubber captures the unreacted precursors, mostly hydrides, so that only clean gas is emitted in the atmosphere.

3.3 Precursors selected for the experimental work

In this work, a well-established selection of precursors for the growth of phosphides and arsenides has been used: arsine and phosphine (group V), trimethylaluminum, trimethylgallium and trimethylindium (group III), disilane (n-doping), dimethylzinc and carbon tetrabromide (p-doping). Hydrogen has been used as carrier. The precursor compounds are listed Table 3.1, along with their chemical formula, standard state, vapor pressure at 20°C, molecular weight, homolytic dissociation enthalpies and main role in the growth.

name (abbreviation)	formula	state at standard T, P	P_{vap} at 20°C (mbar)	molecular weight (g/mol)	homolytic dissociation enthalpy (kJ/mol)		used as precursor for
hydrogen	H_2	gas	≈10^4	2.02	H-H	436	used as carrier
arsine	AsH_3	gas	14.7	77.9	H_2As-H	319	As
phosphine	PH_3	gas	35.2	34.0	H_2P-H	351	P
disilane	Si_2H_6	gas	3.45	62.2	H_5Si_2-H H_3Si-SiH_3	373 321	Si (n-dopant)
trimethylaluminum (TMAl)	$Al(CH_3)_3$	liquid	12.5	72.1	$(CH_3)_2Al$-CH_3	272	Al
trimethylgallium (TMGa)	$Ga(CH_3)_3$	liquid	245	114.8	$(CH_3)_2Ga$-CH_3	260	Ga
trimethylindium (TMIn)	$In(CH_3)_3$	solid	1.55	159.9	$(CH_3)_2In$-CH_3	233	In
dimethylzinc (DMZn)	$Zn(CH_3)_2$	liquid	400	98.5	CH_3Zn-CH_3	236	Zn (p-dopant)
carbon tetrabromide	CBr_4	solid	0.80	331.6	Br_3C-Br	242	C (p-dopant)

Table 3.1 Selected properties of the precursors; CBr_4 can be used as solid source or in solution with $CH_3(CH_2)_{14}CH_3$. Dissociation enthalpy data are averaged values from [11, 12]; the associated uncertainty is rather high - about 10%. The C-H bond dissociation energies of methyl groups are not well known, but can be expected to be ≈ 430 kJ/mol.

The precursor molecules have only a limited time to react heterogeneously with the wafer surface, release their constituents into the gas phase and incorporate their group III or V atoms into the solid phase: the chemical reactions do not to take place in general under equilibrium conditions. Considering for example the case of GaAs growth from TMGa and AsH_3, the overall growth reaction - assuming the formation of the most stable volatile byproduct methane - can be written:

$$Ga(CH_3)_{3(g)} + AsH_{3(g)} \rightarrow GaAs_{(s)} + 3CH_{4(g)} \quad \text{R3.1}$$

This reaction has a favorable standard Gibbs free energy $\Delta G° \approx -340$ kJ/mol and from a thermodynamic point of view it could be expected to reach completeness already at room temperature, but its actual degree of completeness in the reactor is limited by the chemical kinetic mechanism with which it takes place and by mass and energy transport conditions.

3.4 Dopants and impurities incorporation

3.4.1 Intrinsic/unintentional

Carbon and hydrogen are contained in the precursors' molecules of Tab. 3.1, and are always incorporated to some extent in the grown layers; in the case of carbon, this effect is called *intrinsic* doping[4].

Intrinsic carbon doping can be controlled varying the group V partial pressure, because it competes for the same sites and because it reacts with hydrogen, which is released from the hydrides, forming methane and other volatile CH_x molecules; when p-doping is unwanted, it can be minimized using a high V/III ratio, but in compounds with high Al content it cannot be completely avoided due to the strong Al-C bond.

Oxygen is present in small amounts as contaminant in the precursors and on the wafer surface, and can enter the reactor when the chamber lid is open – since the nitrogen atmosphere in the glove-box still contains traces of O_2 and H_2O. Similar to the case of carbon, its incorporation is difficult to avoid in aluminum-rich materials. It never behaves as an n-dopant, and rather introduces deep levels in the bandgap as mentioned in Chapter 2.

As in the case of intrinsic carbon, oxygen incorporation can be minimized using high V/III ratios, probably because of the same reasons: competition for the same sites and formation of hydrogenated (H_2O and OH) volatile molecules. In AlGaAs, the oxygen content is higher at low growth temperature, but it does not decrease monotonically with increasing T. The minimum value of oxygen concentration achieved in FBH, according to SIMS (secondary ion mass spectrometry) measures, is 10^{16} cm^{-3} in $Al_{.85}Ga_{.15}As$, but values as high as 1×10^{18} cm^{-3} have been obtained even in nominally optimized conditions, when a residual contamination was present in the reactor chamber, due for example to water absorbed in recently cleaned quartz and graphite elements. In GaAs the oxygen concentration is normally below the detection limit of the SIMS (about 10^{15} cm^{-3} when special precautions are taken to eliminate background contamination).

3.4.2 Dopants from precursors

Carbon, zinc and silicon can be introduced in FBH reactors using the precursors CBr_4, DMZn and Si_2H_6. In all the three cases, the resulting doping increases linearly with the precursor flow, up to some saturation value which is dependent on the material and the growth conditions. The doping saturation can be due to several phenomena: saturation of the substitutional incorporation, incorporation in both crystal sublattices (especially in the case of C and Si) and formation of other electrically-compensating defects for example due to limitation of solubility as substitutional dopant or formation of complexes with e.g. vacancies.

Using CBr_4, the achievable p-doping in GaAs, in the appropriate growth conditions, can exceed 10^{20} cm^{-3} [13]. At high flows, CBr_4 strongly reduces the material growth rate: this aspect will be treated extensively in Chap. 4.

---

[4] It has nothing to do with the intrinsic carrier concentration of semiconductors.

4 In-situ etching with CBr_4

4.1 Motivation for in-situ etching

To fabricate III-V compound semiconductor devices, intermediate processing of wafers before or in between epitaxial growth steps is often necessary. The processing might consist for example in patterning of the semiconductor surface by means of lithography followed by wet or dry etching, deposition and etching of dielectric masks (as in selective area growth) and ion implantation. During these preparations, oxidation of the surface and adsorption of impurities inevitably occur, which may cause the degradation of device performance and reliability. To a certain extent, oxides and contaminants can be eliminated during the pre-growth heating in the MOVPE reactor under reducing atmosphere (hydrogen + hydrides) depending on the nature of the oxides and contaminants, and on the temperature that can be reached before the growth starts; this temperature can be limited by the need of avoiding damage to the structures present on the surface.

The possibility to perform an etching inside the MOVPE reactor chamber can be of advantage, as it could be used either for pre-growth cleaning to remove residual contamination and oxide layers, or for pattern transfer inside the reactor avoiding contamination of the regrowth interface. While for cleaning a shallow etch of a few nanometers might be sufficient, for pattern transfer deeper etching would be needed. Moreover, the etching should preferably be conducted at low temperature, in order to minimize diffusion of impurities within the semiconductor and to preserve patterns on the surface.

Thermal de-oxidation of GaAs wafers has been extensively studied especially for MBE and MOVPE applications: in general, the stability of the mixed Ga and As oxygen compounds depends on their oxidation states, composition and structure, which in turn depend on their formation conditions and ageing. The following points summarize the main results of the investigations [14-21].

- The oxides to be found on GaAs wafers surface can be approximately understood as a mixed-phase of amorphous oxygen-containing compounds: Ga oxides, As oxides and Ga arsenates; actually, the different phases are normally not separated and the local composition is non-stoichiometric, typically with a vertical gradient from the underlying GaAs to the surface.
- Thermal oxidation of GaAs produces a Ga-rich oxide, while UV (ozone) oxidation at low temperature produces an As-rich oxide, which is more easily removed (a technique used to prepare epi-ready wafers); acid etch of GaAs has been reported to leave a Ga-rich oxide on the surface, and alkaline etch an As-rich oxide.
- Ga^{III} and As^{III} are the most stable oxidation states; under ageing or low-temperature thermal treatments (<300°C), the other oxygen compounds tend to decompose forming Ga_2O_3 and As_2O_3.
- At temperatures approximately >300°C, the underlying GaAs starts to react with the arsenic oxides, forming the more stable gallium oxides, arsenic molecules (As_2, As_4) are released.
- At temperatures approximately >500°C, GaAs starts to react with gallium oxides, forming the volatile Ga_2O and arsenic molecules.
- A complete, purely thermal de-oxidation requires temperatures around 580°C, provided that the oxygen compounds are As-rich and not too stable: for the most stable, Ga-rich oxides, temperatures above 700°C can be required.

- In presence of atomic hydrogen, de-oxidation becomes easier, because of the formation of stable and/or volatile compounds as H_2O and GaOH; this reduces the temperature needed by roughly 100-150°C with respect to a purely thermal process.

The case of other arsenides and phosphides that do not contain aluminum is similar to that of GaAs, with the In-containing compounds more easily de-oxidized because of the lower In-O bond strength (approximately 100°C lower temperature required). In presence of Al, the high strength of Al-O bonds makes the thermal de-oxidation more difficult: for a stability comparison, the cohesive energies of the (bulk, crystalline) group III oxides M_2O_3 are: 513 kJ/mol for In_2O_3, 566kJ/mol for Ga_2O_3 and 734 kJ/mol for Al_2O_3.

Even using very high temperature, and in presence of atomic hydrogen, a complete oxygen decontamination of Al-containing compounds is probably impossible, because of the simultaneous process of in-depth diffusion of the oxygen, which leads to the formation of a heavily oxygen-doped semiconductor layer in proximity the surface, instead of a separate oxide phase.

4.2 Pre-existing research on in-situ etching

4.2.1 Chlorine compounds

In-situ (MOVPE) etching of III-V phosphides and arsenides using chlorine-containing species has been investigated by several research groups [22-33] on InP-based and GaAs-based epitaxial structures; a large number of molecules has been tested, including HCl, C_2H_5Cl, C_3H_7Cl, CH_2Cl_2, $CHCl_3$, CCl_4, TBC (tertiary-butyl chloride), $AsCl_3$ and PCl_3. In all these molecules, Cl is in the –I oxidation state. The main volatile reaction products are expected to be, according to thermodynamic calculations, the group III chlorides and phosphorus or arsenic molecules; in other words, the chlorine atom is expected to bond to - and remove - the group III, while the group V leaves the surface by a thermal process, and/or because of the removal of group III atoms. Experimentally, when the etching is performed on partially masked surfaces, it delineates crystal planes, mostly {100} {110} and {111}, indicating that surface kinetics plays an important role in limiting the rate (a purely diffusion-controlled etching would lead to round profiles near the masked areas). As could be expected, the etch rate increases with the partial pressure of the etchant, indicating that the reaction with group III or desorption of the resulting products limit the etching rate. At the same time, depending on the specific material and etching conditions, the $\{111\}_B$ planes can become the slowest-etching planes [23], which probably indicates that the group V removal becomes in these cases rate-limiting. In ultra-high vacuum experiments [34] of GaAs etching with HCl, in the temperature range 400-600°C, the main products were found to be GaCl, H_2 and As_2, and the rate-limiting step was identified in the desorption of As_2.

The etch rate on InP, GaAs and other Al-free compounds has been consistently found to strongly increase with increasing temperature [23-25, 27, 32]. The temperature dependence in the case of Al-containing compounds is less clear-cut: for AlAs the etch rate has been found to weakly increase with T when etched with CCl_4, and to first weakly decrease and then weakly increase when etched with HCl [24, 32].

In general, the etch rate of Al-containing compounds has been found to be much lower than the etch rate of Ga and In compounds; this effect has been variously explained in relation to the stronger Al-As bonds, to the lower volatility of Al chlorides (which should make $AlCl_x$ desorption the rate-limiting step and the $(111)_A$ planes the slowest etch planes – but there is no experimental confirmation of these effects) and to the formation of a surface layer of highly stable oxide that stops completely the etch. In-situ use of HCl (with or without AsH_3) to de-oxidize an air-exposed layer of $Al_{0.5}Ga_{0.5}As$ surface has proved ineffective [33].

The de-oxidation effect of Cl-based treatments is expected to be related to reactions of O with atomic hydrogen as in the thermal de-oxidation, with chlorine playing an ancillary role: for example Ref. [33] reports that oxygen removal from GaAs surface, in presence of HCl and H_2 but in absence of AsH_3, was negligible.

4.2.2 CBr_4

MOVPE in-situ etching based on bromine has been studied by some research groups [35-40], in all cases using CBr_4, a compound commonly employed for carbon doping. While the motivation of the first studies was only the carbon incorporation, an interest in possible applications of the etching properties has later developed.

The species responsible for group III removal have been indicated either as Br radicals or HBr molecules, both produced by CBr_4 pyrolysis, the main volatile reaction products are generally expected to be phosphorus or arsenic molecules and the group III bromides, although even group V bromides could form, and the tribromomethyl group CBr_3 could form volatile species with group III atoms. McEllistrem and White [41] conducted a temperature-programmed desorption experiment in UHV from a GaAs (100) surface covered with CBr_4, and concluded that CBr_4 did fully decompose on the surface already at temperatures around 500 K, and detected GaBr as the only gallium-containing desorbed species. The etching of Al-containing compounds has been found, as in the case of chlorine-based etch, particularly difficult.

The in-situ etch rate has been indirectly evaluated by Tateno et al. [35] as *reduction*, in presence of CBr_4, of the growth rate of GaAs, AlAs and AlGaAs grown on GaAs substrates. The precursors were TMGa, TMAl, AsH_3 and the carrier gas H_2. In all cases, the growth rate reduction has been found to be proportional to the CBr_4 flow and proportional to an inverse power of the AsH_3 concentration, to $[AsH_3]^{-0.5}$ for GaAs and to $[AsH_3]^{-1.1}$ for AlAs, both at 650°C. The growth rate reduction of GaAs did increase (approximately) exponentially with increasing T, while the opposite occurred for AlAs; GaAs growth rate reduction was higher than that of AlAs at 750°C, but the situation reversed at 600°C. Moreover, the growth rate reduction of AlGaAs did not vary linearly with the Al fraction, but did show a positive deviation (increment) with respect to a linear interpolation for intermediate Al content.

A first *direct* evaluation of the etch rate on the AlGaAs system for compositions ranging from GaAs to AlAs has been provided by Maaßdorf and Weyers [36], using CBr_4 in H_2-AsH_3 atmosphere at a fixed partial pressure, varying AsH_3 flows and temperature. The results only partially align with those of the previous study: the etch rate did decrease with increasing AsH_3 partial pressure, similarly to the growth rate reduction previously mentioned, and the etch rate decreased with increasing Al content. The etch rate of AlAs and AlGaAs did decrease with increasing T, but GaAs etch rate was almost temperature-independent. The etch rate of AlGaAs did vary approximately linearly with the Al fraction at 600°C, and did show a *negative* deviation (decrement) with respect to a linear interpolation for intermediate Al content at 650°C and 700°C

CBr_4 etch on bulk InP in H_2-PH_3 atmosphere has been tested by Arakawa et al. [37]; etch of InGaAsP and AlGaInAs quantum wells has been additionally tested under unspecified H_2-V atmosphere (presumably PH_3 and AsH_3 in different proportions). On InP, a smooth surface was obtained even after 500 nm of etching depth, the rate was proportional to the CBr_4 flow, independent of PH_3 flow and independent of temperature (tested range 540°C-660°C). InGaAsP quantum wells were slowly etched but the resulting morphology was very poor, which is attributed by the authors to a higher difficulty in desorbing Ga and As reaction products; AlGaInAs was not etched at all, which is similarly explained; based on these results, the following qualitative sequence of etching "easiness" results: In>Ga>Al and P>As. The "easiness" of InP etching is confirmed by Ebert et al. [38], who report to have obtained a smooth InP surface after 200 nm of etching with CBr_4 under PH_3 protection (exact conditions not specified).

In Ref. [39] Décobert et al. describe the results of growing lattice-matched AlInAs on InP at T=540°C, in H_2-AsH_3 atmosphere from TMAl and TMIn, using CBr_4 as dopant: not only the growth rate was reduced roughly in proportion to the CBr_4 flow, but even the Al content was reduced (as if the Al fraction was preferentially etched over the In fraction). They then separately compared the in-situ etch *and* the growth rate reduction of the binaries AlAs (grown on GaAs substrate) and InAs (grown on InAs substrate). InAs was etched at a rate equal to its growth reduction rate at the same CBr_4 flow, and the growth reduction was independent from the growth rate (TMIn flow). AlAs was not etched, but it did show a growth reduction proportional to the growth rate (TMAl flow). The authors tentatively conclude that CBr_4 does not have any etching effect on AlAs but reacts with TMAl in the gas phase, reducing the Al supply to the surface.

The crystallographic dependence of CBr_4 etching of InP has been investigated by Kuznetsova et al. [40], in presence of PH_3 and at 610°C; the (100) wafer surface was partially masked with resist stripes aligned along $[0\bar{1}\bar{1}]$ and $[0\bar{1}1]$ directions, the in-situ etch did delineate the crystal planes (100) and $\{111\}_B$ in the first case, (100) and $\{111\}_A$ in the second. The etch rate on these planes followed the order $\{111\}_B < (100) < \{111\}_A$ so the $\{111\}_B$ planes were identified as the slowest-etch planes.

The successful use of in-situ etch with CBr_4 to realize a pattern transfer has been reported by Maaßdorf et al. in Ref. [42], where the Bragg grating of a DFB laser, originally defined ex-situ into an upper GaAs layer, has been transferred in-situ into an underlying InGaP layer and subsequently buried with AlGaAs. The etching has been carried out at 600°C in presence of AsH_3 and PH_3, sufficiently preserving the grating shape, and has given a low oxygen contamination at the regrowth interface.

4.3 Investigation of CBr_4 etching of GaAs

4.3.1 General experimental details

The in-situ etching has been first investigated[1] on the planetary reactors G3 and at a later time (the reactor G3 being no more available) on reactor G4. The substrates used were all epi-ready 2" (only on G3) and 3" (100) GaAs wafers having different specifications in terms of quality (laser-grade, LED-grade and test-grade) corresponding to different maximum etch-pit density (EPD) - as declared by the supplier.

The etch rates have been determined in the following way: first, a sacrificial layer (500-1000 nm) of the material to be etched (mostly GaAs, but some tests were done on AlGaAs and InGaP) was grown on GaAs epi-ready (100) substrates, and then the layer was etched with CBr_4, measuring the etch rate by in-situ reflectance at λ=948 nm, using two Epicurve® TT systems from LayTec; the one installed on reactor G4 allowed to monitor simultaneously the center and the edge of the wafers. Shorter and longer wavelengths have been simultaneously monitored (489, 633, 1080 nm).

In the case of GaAs etching, an optical marker layer of $Al_{0.3}Ga_{0.7}As$ was introduced under the sacrificial layer – and above a thin GaAs buffer - in order to ensure the necessary refractive index contrast, while this was not necessary for the other materials.

The epitaxial material to be etched was in most cases grown and etched within the same run avoiding exposition to air, but some tests were done in two parts, growing the epitaxial layers in a first run, extracting them from the reactor and exposing them to air for different times and then proceeding with the etching run.

The same LayTec tools were used to measure the temperature T_w at the wafer surface with the method of emissivity corrected pyrometry. Because of the extreme sensitivity of the etching to the temperature, it has been deemed more physically significant to consistently use the measured wafer-temperature, instead of the more usually found set-point (or susceptor) temperature T_{sp}, through this chapter.

---

[1] Part of the results have been published by Della Casa et al. in Ref. [43]

Set-point temperatures are anyway also specified, and a general calibration of the relation between T_{sp} and T_w is shown in Fig. 4.1a. It can be seen that $T_w < T_{sp}$, and that the difference is approximately 25°C at T_{sp}=500°C, increasing to about 75°C at T_{sp}=800°C. It must be specified that the values measured in each case can deviate from this general calibration by a few °C, due to several factors: small differences in the used satellites, fine adjustments of the gas-foil rotation of the satellites, different characteristics of the substrates (thickness, and doping).

The wafer-temperature range tested on reactor G3 for the in-situ etching did span from ≈500°C to ≈700°C; the first tests were done at 500°C, but at this temperature it was not possible to obtain any measurable etching: the resulting sample surface appeared brownish and slightly rough after unloading, probably indicating the deposition of a thin layer, which was no further investigated. Significant etching rates (above 100 nm/h) were obtained at $T_w \geq 545$°C, the highest etch rates (up to 1750 nm/h) were obtained at the highest temperatures. Temperatures above 700°C were not tested.

The wafer-temperature range used on reactor G4 did span from ≈565°C to ≈710°C, and etching rates comparable to those of reactor G3 were obtained, as better detailed in section 4.3.3. Lower and higher temperatures were not tested.

Calibration of CBr_4 flows. On reactor G3 a constant hydrogen flow in the CBr_4 bubbler was used, and 100% efficiency was initially assumed in order to calculate the CBr_4 flow and its partial pressure p_{CBr4}. Subsequent analysis led to the conclusion that this assumption was not correct. Since no ultrasonic cells directly measuring the concentration in the gas phase [44] were installed, the efficiency of the CBr_4 bubbler could be checked only indirectly, measuring with electrochemical capacitance-voltage (ECV) profiling the p-doping of GaAs as a function of the hydrogen flown through the bubbler, checking its linearity and comparing the results with those obtained in the same conditions with a different bubbler. One year after the first group of in-situ etch measurements had been done, the (very) old CBr_4 bubbler – which contained solid CBr_4 - was exchanged with a new one containing a solution of CBr_4 in hexadecane, which does not significantly influence the vapor pressure but addresses the ageing problems due to the formation of "channels" in the solid CBr_4. The same kind of bubbler was installed on reactor G4. Checking the p-doping and comparing it with the values obtained with the old bubbler, an increment of a factor 1.5 was observed at all tested H_2 flows. This proves a sub-ideal efficiency of the old bubbler of (at least) a factor 1.5. Moreover, the dopant incorporation shows a sublinear behavior at high flows, even with the new bubblers (Fig. 4.1b).

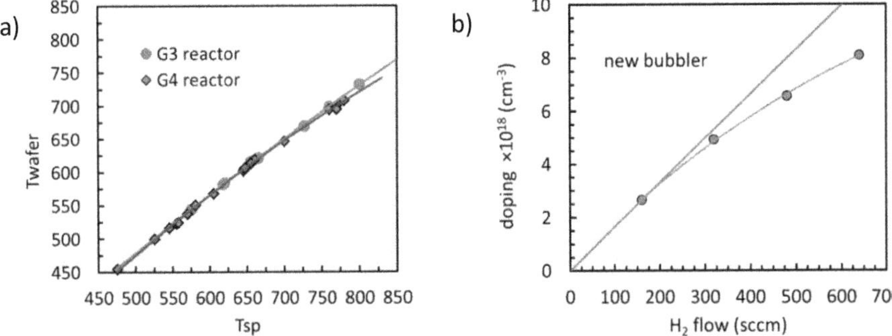

Figure 4.1 **a)** Relation between set-point temperature and wafer-temperature measured with emission-corrected pyrometry; **b)** doping concentration in GaAs grown on reactor G4 vs. H_2 flow through the bubbler; the straight line passing the origin and the first point indicates the ideal linear behavior assumed in dopant incorporation, the separation of the experimental points from this line is used to calculate the bubbler efficiency.

Since this behavior might be due to carbon solubility saturation, the growth conditions for the test have been chosen in order to (presumably) avoid this effect, staying below a dopant concentration of 1×10^{19} cm^{-3} at T_{sp}=620°C (T_w=589°C). Although this is still no *definitive* proof that the sublinearity is due to the bubbler efficiency, it is highly plausible due the very low vapor pressure of CBr_4 and the high hydrogen flows used. Based on this assumption, and on the further assumption that the new bubbler's behavior *is* ideal at low flows, an efficiency function can be obtained from the doping curves, and this correction has been introduced in the calculation of CBr_4 flows and partial pressures.

In-situ monitor of surface morphology. The surface morphology did always show at least some degree of degradation after the etching; while in certain cases this was limited to the development of a few isolated defects, in others the wafer surface did sensibly degrade, developing a high defect density or a more or less uniform roughness: the progress of this degradation was – in the worse cases - already observable from the intensity of the reflectivity, as depicted in Fig. 4.2, which illustrates the oscillations at different wavelengths during the growth and the subsequent etching of a layer.

During the growth, the short-wavelength oscillation amplitude rapidly decreases because the material is absorbing, and the reflectivity finally stabilizes at a constant value, while at long wavelength (photon energy less than the bandgap) the amplitude remains constant. During the etching, the pattern should reverse, stretched along the time axis because of the different growth and etching rates, and with a change in phase and amplitude when the etch temperature is different from the growth temperature. In presence of increasing surface roughness, the oscillation amplitude is reduced and the average reflectivity drops due to light scattering; the short wavelengths are more sensitive to the presence of a fine roughness and are the first to clearly signal the beginning of important surface degradation.

Differential interference contrast optical microscopy (DIC), scanning electron microscopy (SEM), energy dispersive X-ray analysis (EDX), cathodoluminescence (CL), and atomic force microscopy (AFM) have been employed to characterize the surface of the samples after the etching.

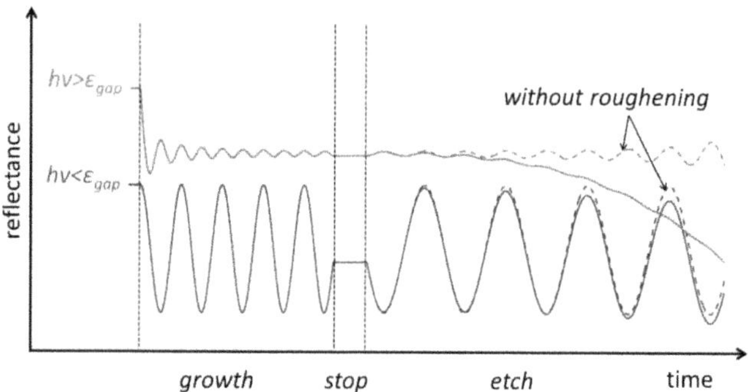

Figure 4.2 Scheme of reflectivity signal during growth and etch, showing the ideal behavior and the intensity drop in presence of surface roughening.

Differences between the two reactors. In comparing the results from G3 and G4, the following points can be preliminary considered:

- The reactor chambers are of different sizes, the total pressures and the carrier flows used for growth and etch were 100 mbar and 20 l/min on G3 vs. 50 mbar and 28 l/min on G4; this results in G3 having a flow speed ≈12% smaller and a residence time ≈40% smaller than G4. An estimate based on the boundary layer approximation gives an expected *growth* rate ratio near to 1 in the diffusion-limited region: this is mainly due to the fact that the effect of the larger diameter of G4 should be compensated by the effect of the lower pressure. Actually, the experimental growth rates of GaAs at the same group III partial pressures are almost the same - within 5% - at low temperature (T_W ≈550°C), and become moderately higher in G3 as the temperature is increased (up to 35% higher at T_W ≈700°C).
- The extent of gas-phase radical reactions could be influenced by the said differences in the total pressure *P* (being the precursor-carrier collision frequency proportional to *P*), by the different residence times and by differences in the thermal profiles above the susceptor. Moreover, being the diffusion coefficients proportional to P^{-1}, the concentration of the volatile products of the etching above the surface should be lower at lower total pressure (G4), possibly leading to a higher etch rate in case of reversible desorption of the etching products.
- There can be "hidden" accidental factors, as the efficiency of the installed CBr_4 bubblers previously discussed, or differences in mass-flow controllers and pressure-controllers calibrations.

In conclusion, the behavior of the in-situ etching could be reasonably expected to be qualitatively and quantitatively similar in the two reactors, but not necessarily identical.

4.3.2 GaAs etching: surface morphology

The morphology of the surface after a "deep" etching (i.e. in excess of 150 nm and up to about 800 nm) did vary according to the etched depth, the etching conditions and – especially - the substrate quality, as explained in the following. A limitation of this part of the study is that many of the tests were conducted varying the etching parameters in steps during the run, for example changing the CBr_4 flow or the AsH_3 flow, in order to measure the corresponding etch rate variations while keeping all the other conditions identical: consequently in these cases the final morphology could not be associated to a specific set of parameters values, but rather to a range. The kind of surface degradation found on the etched surfaces can be divided in two categories: roughness and isolated defects.

Roughness development and catastrophic degradation

Except for some roughness occasionally found at the wafer's edge, all the samples deep-etched in the temperature range T_w=550-630°C had only isolated defects; the next higher temperatures tested were in the range 695-712°C: five of the nine wafers etched in this range had similarly only isolated defects, while the other four had very rough surfaces. All these latter four wafers were etched on reactor G4, and the reflectivity signal slowly worsened during the first 100-400 nm of etching, dropping then more rapidly, first in the external part of the wafer successively in the center. Inspection of the etched surface shows a catastrophic degradation, consisting in innumerable localized islands of different sizes at the wafer's center, and progressively evolving into a continuous roughness towards the edge; the surface among the islands appears smooth and the islands are loosely aligned, approximately along the [010] direction (Fig. 4.3). The islands are "hills", up to 250 nm in height according to AFM measurements.

The rapidity of the degradation – in terms of the thickness that could be etched before the short-wavelengths reflectivity signal dropped below 50% of the initial value – did worsen at higher etch rates, which were obtained using higher CBr_4 flows and lower AsH_3 flows. Based on their irregular shape, it might be speculated that the hills are the consequence of tri-dimensional material re-deposition, starting from defects (acting as nucleation centers) diffusely generated on the surface.

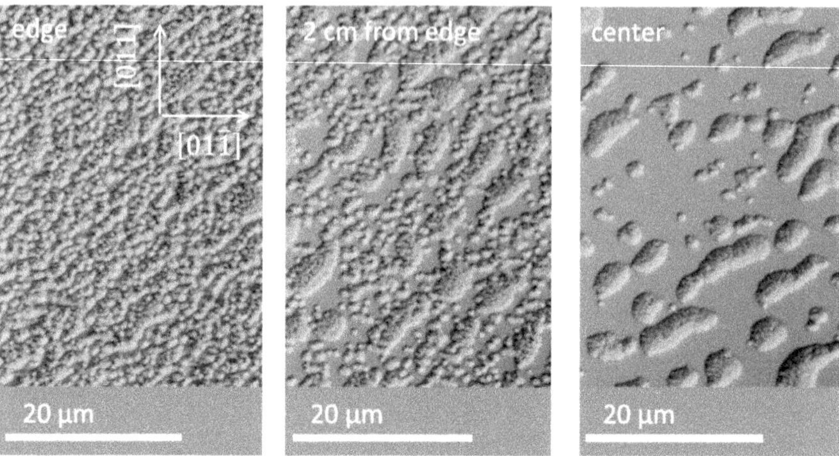

Figure 4.3 Optical microscope DIC images of the roughening of the surface, taken at three different positions after ≈0.4 µm etch at T_w=712°, the etching was done in two steps changing the CBr_4 flow, with rates ≈750 and 1450 nm/h, on reactor G4; the etching was stopped in the middle of the epitaxial GaAs layer.

An investigation of possible correlations that could shed some light on this issue did lead to the conclusion that the onset of catastrophic degradation is triggered by the characteristics of the used substrates, possibly due to a failed "EPI-ready" preparation of their surface and/or high density of extended defects. This conclusion is based on the following observations and tests:

- all the four wafers did belong to the same batch of test-grade substrates;
- when the etching corresponding to the sample shown in Fig. 4.3 was repeated simultaneously on another substrate from the same batch and on a laser-grade substrate, the former did still develop the same roughness while the latter showed no signs of roughening;
- GaAs layers grown on substrates of this batch did look normal (no morphological defects) but growing 100 nm of GaAs buffer followed by a 1000 nm of $Al_{0.5}Ga_{0.5}As$ did lead to a rough surface – although not catastrophic as that obtained in the etching experiments - contrary to previous experience with test substrates from the same supplier; introducing an extra 7 min long waiting step under arsine at T_w=700°C before starting the growth did partially reduce the roughness;
- an epitaxial structure containing an InGaAs quantum well embedded between $Al_{0.25}Ga_{0.75}As$ layers was grown simultaneously on one test substrate and on one laser-grade substrate, and the electroluminescence of the QW was measured: while the intensities were similar, the width of the peak was 60% larger on the test substrate.

Given the above evidence of pathology in the said batch of wafers, this kind of catastrophic defect formation was no further investigated. It is nonetheless interesting to notice that the in-situ etch appears to be strongly sensitive to the surface quality, a characteristic that has been repeatedly confirmed when it was used to remove sacrificial layers from epitaxial structures that had been previously processed ex-situ, as will be described in the next chapters in relation to the fabrication of laser structures. Another point is that some test wafers from the same batch had been previously used for the etching tests in the lower temperature range, and at lower etch rates, without obtaining a similar catastrophic degradation, but rather a moderate density of isolated defects of various shapes, mostly concentrated near the wafer's edge; as a tentative conclusion, it might be suggested that high temperature and high rate etching conditions are more prone to lead to morphology issues in presence of a non-ideal starting substrate's surface.

Isolated defects

Several kinds of isolated defects have been found on the surface after a deep etch, and they are here subdivided into three categories: those developed around particles fallen onto the wafer surface before or during the process, those not related to particles but present only on LED-grade and test-grade wafers, and finally those not related to particles and present *even* on laser-grade substrates. Examples of defects falling into the first two categories are shown in Fig. 4.4.

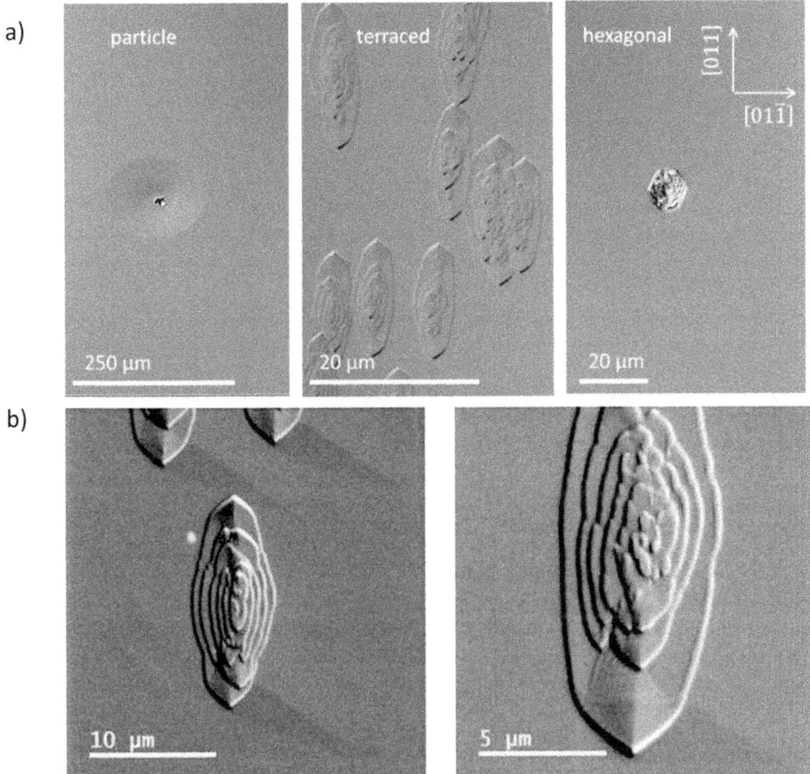

Figure 4.4 **a)** Optical microscope DIC images of three different defects found after a deep etch: round hill around a particle, a group of terraced protrusions and a hexagonal defect; **b)** SEM images of the terraced defects: a triangular pattern can be seen neat its lower extremity. All pictures are aligned in the same way with respect to the crystal orientation.

The first is a particle, observed after about 150 nm of etching; the particle might have fallen onto the sample already before or during the growth preceding the in-situ etch. The important thing about particles is that they have been found only in limited numbers even after a long etch, indicating that the detachment of material from the chamber ceiling during the etching is not a major problem.

The second is a group of terraced protrusions, ≈100 nm in height and with terraces spaced vertically by ≈20 nm according to AFM measurements; these defects, are approximately aligned along the [011] direction, with slight clockwise or counterclockwise tilts, and were observed in large groups, near the edge of several test-grade wafers, etched at different temperatures.

The third is a hexagonal defect, with sharp edges well aligned with the crystal axis, observed only on a test-grade wafer etched at T_w=589°C. In both the last two cases, a relation with pre-existing defects of the underlying substrate can be suspected. The terraced defects have been further investigated with SEM, CL and EDX; no compositional differences were noticed between the defects and the surrounding flat surface, and no systematic correlation with dislocations detectable in cathodoluminescence as recombination centers; triangular patterns were noticed near one extremity of the defects (Fig. 4.4b) which are suggestive of some kind of underlying extended defect with a screw component in its translation vector.

The third category - defects not related to particles and present even on laser-grade substrates - is represented by a single type: it will be referred-to as "reverse pyramid", and is illustrated in Fig. 4.5. The reverse pyramid defects are pits, their sizes vary from a few microns to 20-40 microns from sample to sample, increasing with the etched depth; their distribution has been found to be approximately uniform on each sample, with a tendency to group in small clusters.

They have a slightly rounded rectangular shape with the longest side aligned along the [011] direction, and sloped sides, with the slope usually becoming less steep moving from the defect's edge towards the center and then increasing again creating a deeper central depression, as shown by the AFM profiles in Fig. 4.5b; on samples with shallower etching they are smaller and appear more like simple pyramids with sides having a constant slope, and on samples with deeper etching they are larger and the external edge becomes less symmetric and more rounded.

This kind of defects was found on all the samples etched on both reactors, but their density did vary by orders of magnitude in correlation with the etch-pit density declared by the substrate supplier, from less than one per square centimeter on laser-grade to thousand per square centimeter on test-grade substrates. To compare the effect of the substrate in exactly the same conditions, 500 nm of GaAs were simultaneously grown on two substrates in reactor G3, and then 150 nm were immediately etched at T_w= 545°C.

One of the substrates was a semi-insulating wafer having max EPD < 25000 cm^{-2} and the other a laser-grade wafer having max EPD < 100 cm^{-2}: the resulting defect density was ≈10000 cm^{-2} and ≈1 cm^{-2} respectively. Except for the very rare reverse-pyramid defects, no other defects were noted on the laser-grade wafer after the etching.

Energy-dispersive X-ray analysis has been used to probe the center and the area immediately outside the defects (Fig. 4.5c): the spectra were superimposable, showing the peaks corresponding to Ga and As. This result contributes to exclude that the origin of the defect is due to some kind of contamination, as for example small particles detached from the ceiling of the reactor chamber (that would presumably contain some traces of other elements as In, Al or P).

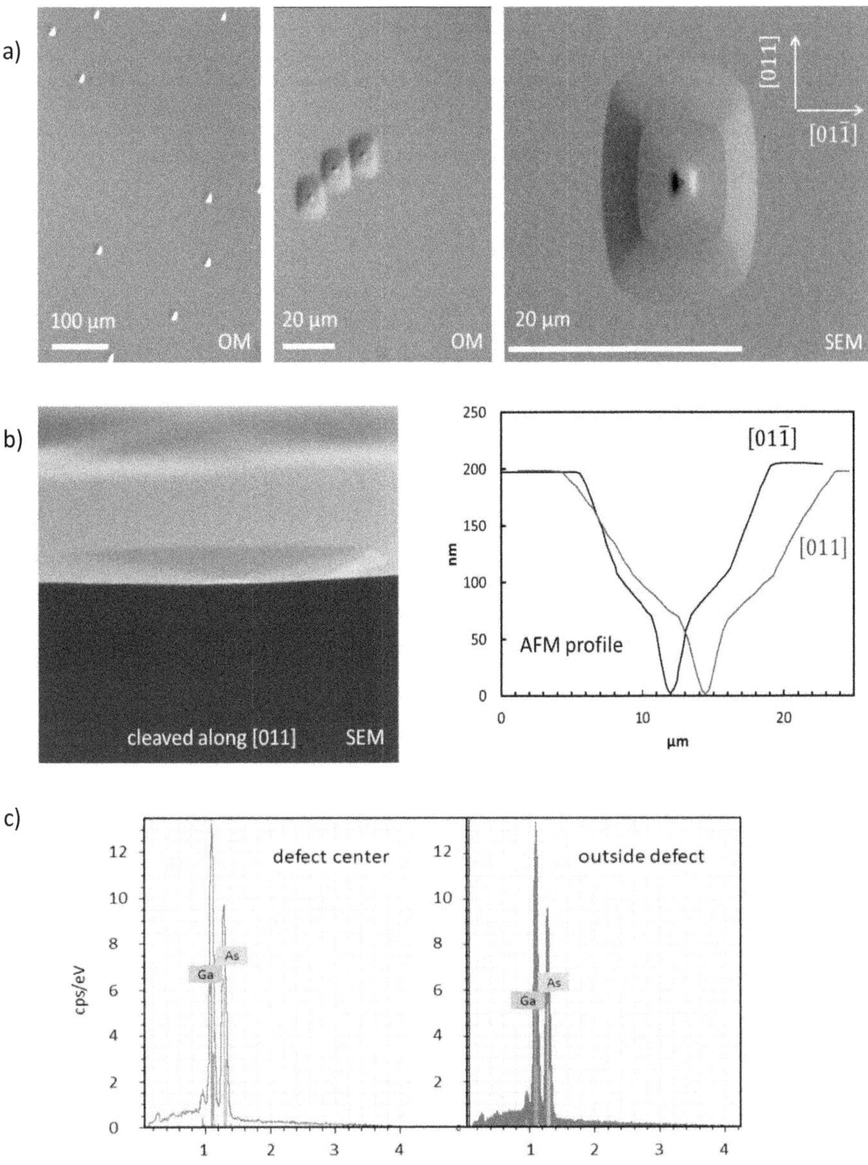

Figure 4.5 **a)** optical microscope DIC images and – to the right – SEM image of reverse pyramid defects; **b)** bird's eye SEM image of a defect (cut by cleavage) and AFM profiles; **c)** EDX analysis, values acquired at the center of the defect represented by SEM image in a), and 10 µm outside the defect's edge.

Cathodoluminescence analysis (Fig. 4.6) shows a dark spot at the center of each pit. This, in conjunction with the statistics previously outlined, is a strong indication that the origin of these defects are dislocations, which act as preferential etching sites, due to the associated increase of the local chemical potential caused by the strain field around the dislocations. They appear to be revealed by the in-situ etching as in the so-called orthodox wet etch for defect delineation, where the reaction is kinetically controlled by bond dissociation and formation on the surface, and no generation or transport of carriers within the semiconductor is involved [45]. The shape of the pits does actually strongly resemble that of the pits caused in correspondence to InP dislocations by wet etching with HBr.

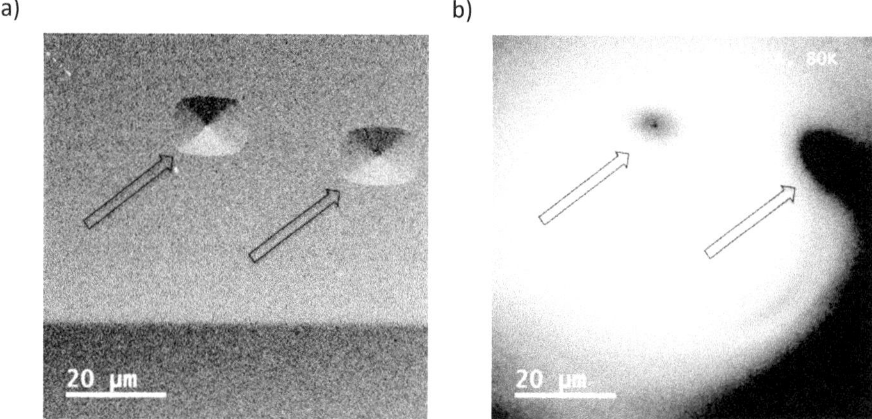

Figure 4.6 **a)** secondary electrons SEM image and **b)** panchromatic CL image of reverse-pyramid defects, recorded at 80 K simultaneously. The position of the defects corresponds to that of the CL dark spots, which are interpreted as dislocations.

4.3.3 GaAs etching: kinetics

Reactor G3

The kinetics of GaAs etch has been experimentally investigated on reactor G3 keeping a fixed CBr_4 flow of 400 sccm; the corresponding partial pressure, evaluated using the bubbler efficiency factor previously described, is 313 mPa for the old bubbler and 473 mPa for the new.

Figure 4.7a shows the etch rate values as function of arsine partial pressure p_{AsH3}, each of the two series of data corresponds to constant values of T_w and p_{CBr4}. It can be seen that the etch rate decreases with increasing arsine partial pressure, in qualitative accord to previously published results [35-36]. For comparison with the findings of Tateno et al. [35], the experimental points have been empirically interpolated with a power law of the form $y = c \cdot x^{-m}$ where y is etch rate and x is p_{AsH3}: the fitting gives a value of $m \approx 0.25$, to be compared with $m \approx 0.5$ found by Tateno et al. It can be noted that the power law would predict a very strong increase of the etch rate when the arsine partial pressure approaches zero.

Figure 4.7b shows the Arrhenius plot of data collected at p_{AsH3}=25 Pa, p_{CBr4}=313 mPa. The obtained apparent activation energy is E_a=119 kJ/mol. This result contrast with those of Ref. [36] where a temperature independent etch rate of GaAs was found, but is in good agreement with those of Ref. [35], which provides a similar activation energy of 116 kJ/mol, calculated from the Arrhenius plot of the GaAs growth rate *reduction* in presence of CBr_4.

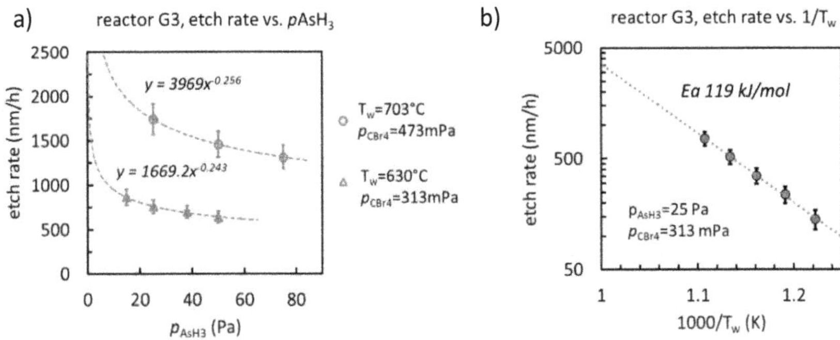

Figure 4.7 **a)** etch rate of GaAs vs. arsine partial pressure for two different combinations of T_w and p_{CBr4}, the dotted lines are empirical interpolations; **b)** Arrhenius plots of the etch rate at p_{AsH3}=25 Pa, p_{CBr4}=313 mPa.

The vertical bars in the above plots indicate the uncertainty on the measured etch rates; the different sources of uncertainty (on both reactors) are discussed more broadly in the following.

- Etch rates. Based on cross-checking of the thickness of layers as evaluated with in-situ reflectivity during the growth, and a-posteriori with SEM and XRD, the uncertainty on the *growth* rate can be expected to be ± 5% in ideal conditions (several full oscillations observed); in the case of measurements done during the etching, this value is generally worse because in many cases – due to the relatively slow etch rates – only a single full oscillation was observed. Moreover, in case of surface degradation, noise and distortions were introduced in the oscillation pattern, which have worsened the evaluation. It is deemed realistic to consider ±10% uncertainty on the etch rates values, up to ±15% in a few cases.

- Temperature. The precision of the kind of emissivity-corrected pyrometry system used has been evaluated to reach in ideal conditions ±1°C [46]. The accuracy (closeness to the "real" value) should depend mostly on the device calibration, done using a primary standard provided by Laytec (AbsoluT) which declares ±1°C for the latest version. Emission from the susceptor and different characteristics of the substrates might to some extent impact the accuracy. The variations of the measured values of T_w, observed during a run while keeping the same conditions are normally within ±2°C, provided that no surface roughening occurs. The repeatability from run to run can be significantly poorer (5-6°C), but this is presumably explained by real small drifts in the growth or etch conditions rather than by instability of the measurement. In conclusion, the possible systematic error is estimated to ±5°C; the uncertainty in terms of precision is dictated by the (visible) temperature drifts and the (possible) substrate effects, and is cautiously estimated to be ±5°C.
- The uncertainty on the flows of the active species is related only to the calibration of the MFCs in the case of AsH_3, and it is probably safe to consider ±2%. For CBr_4 it depends additionally on the efficiency of the bubbler, which has been estimated from the doping calibrations as already explained; the accuracy of the obtained values is unfortunately difficult to estimate. Further source of instability are small variations of the thermostatic bath temperature, which might cause ±1% run to run variations in CBr_4 flow.

Reactor G4

A more extended set of etch rate data was collected using reactor G4, and a semi-empirical model was developed to interpret the results. The experimental etch rates at different temperatures are plotted in Fig. 4.8a as function of AsH_3 partial pressure, and in Fig. 4.8b as function of CBr_4 partial pressure; all the indicated values of p_{CBr4} have been corrected based on the doping calibration. The continuous lines are not fits of the individual series, but "global fits" of all the data obtained at the same temperature, according to the model as will be explained. The temperature dependence of the etch rate is represented in the logarithmic plots of Fig. 4.9 as a function of $1/T_w$, the points corresponding to the same values of both p_{AsH3} and p_{CBr4} have been fitted with the Arrhenius equation.

The etch rate at constant p_{AsH3} (Fig. 4.8b) increases linearly with p_{CBr4}; this would not be the case if the partial pressure values had not been corrected for the cell efficiency, and a presumably spurious saturation effect would appear at high p_{CBr4}.

At constant p_{CBr4} (Fig. 4.8a) the rate decreases with increasing p_{AsH3}. Using again a power law to interpolate the points (not shown in the figure) values of the exponential factor m in the range 0.2-0.5 are obtained, indicating that a simple proportionality of the rate to ${p_{AsH3}}^{-m}$ does not hold.

All the values in the plots refer to the wafer center; comparison of the reflectivity signal from the center and the periphery did show differences within 5% of the etch rate: the etch rate was *often* found to be faster at the center. This could be related to the slightly lower temperature near the wafer's edge, due to the upwards bowing of the substrates, so that the edge is slightly lifted from the underlying satellite. The bowing is caused by a vertical thermal gradient which induces a gradient in the lateral thermal expansion. This result was not well reproducible, probably depending on the specific substrate and satellite.

Observing the Arrhenius plots in Fig. 4.9, a certain scattering of the slopes is evident. Unfortunately in most cases there are only two points for each combination of arsine and tetrabromide partial pressures, so it is difficult to decide whether the scattering is significant or not. There is no clear evidence of trends in the distribution of the slope values with respect to the partial pressures of either carbon tetrabromide or arsine. The apparent activation energy is spread in the range 106-151 kJ/mol, the median is 119 kJ/mol – which is the same value obtained on reactor G3. The etch rates on the two reactors are reasonably similar at similar values of T_w, p_{CBr4} and p_{AsH3} (Fig. 4.9c) in spite of the above mentioned differences, including the total pressure.

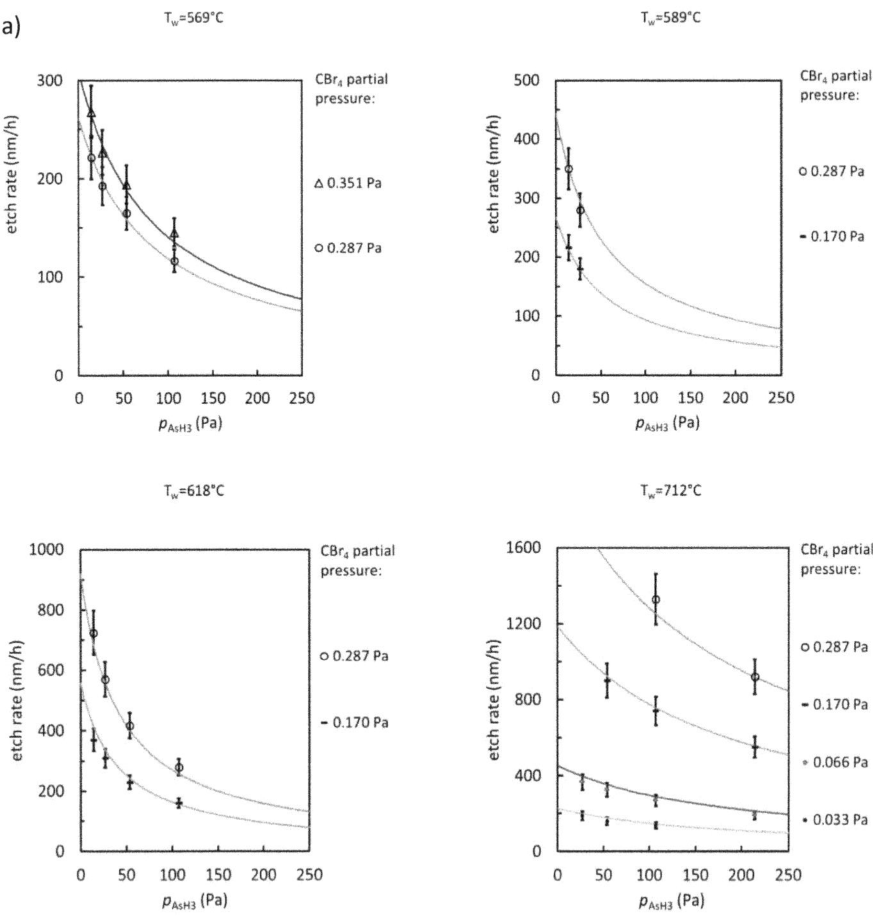

Figure 4.8 **a)** etch rates plotted vs p_{AsH3} at four different wafer-temperatures, grouped according to p_{CBr4}. The symbols represent experimental values, the continuous lines are fits obtained using the model described in the text. All data are obtained on reactor G4.

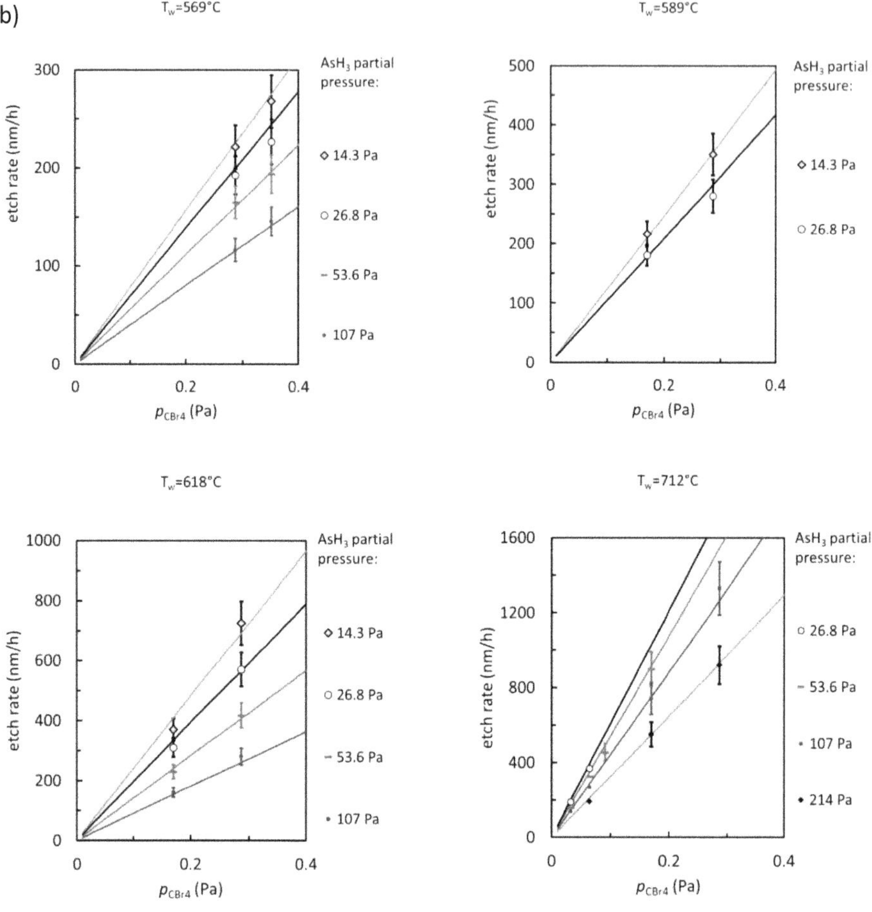

Figure 4.8 **b)** the same etch rate values are grouped according to p_{AsH3} and plotted vs. p_{CBr4}: the error bars are omitted in a few cases for graphical reasons.

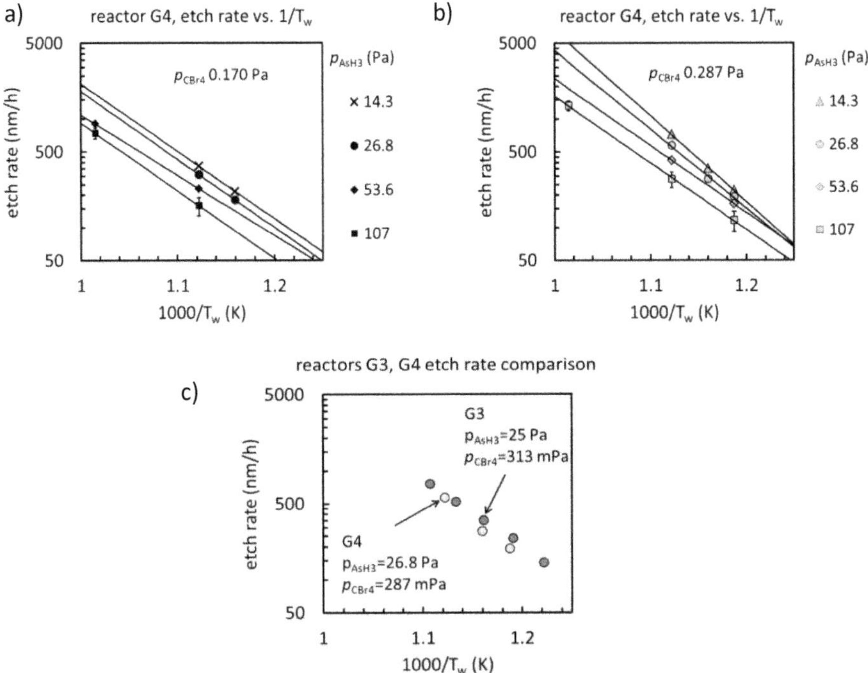

Figure 4.9 **a)** and **b)**: logarithmic plots of etch rate vs. inverse temperature ($T=T_w$) at two different CBr_4 partial pressures: **a)** 0.170 Pa and **b)** 0.287 Pa. Data are further grouped in series according to AsH_3 partial pressures. The symbols are experimental values, the lines are Arrhenius fits; the uncertainty on the rates is indicated only in one series for each diagram. **c)** comparison of the etch rate on the two reactors at similar conditions.

Based on the kinetic results, the etching process can expected to be limited by an activated step. Since there is no evidence of rate saturation with increasing temperature, the process appears to be in the kinetic-limited region in the whole temperature range under study. This is a bit surprising, since the *specific* etch rate η_{CBr4}, defined here as the ratio of the etch rate E to CBr_4 partial pressure ($\eta_{CBr4} = E/p_{CBr4}$), is very high at the highest temperature, up to 3800 $nm{\cdot}h^{-1}Pa^{-1}$ (at p_{AsH3}=26.8 Pa); this value can be compared with the analogously defined TMGa specific growth rate, which is about 1100 $nm{\cdot}h^{-1}Pa^{-1}$ in the whole temperature range tested. Even dividing η_{CBr4} by 4 on the assumption that *each* bromine atom can *alone* remove a gallium atom, the specific etch rate remains comparable to the specific growth rate. If GaAs growth is limited by transport, even the high temperature etch should become - at least partially - limited by transport, especially when p_{AsH3} is low.

The dependence of the etch rate on p_{AsH3} might suggest that the activated step could involve arsenic, and Tateno et al. [45] proposed that the etching is limited by desorption of arsenic molecules As_2 in analogy with the findings of Su et al. [44] concerning the etching of GaAs by HCl in UHV. Desorption should occur according to the reaction:

$$GaAs(a) \rightleftarrows Ga(a) + \frac{1}{2}As_2(g) \qquad \text{R4.1}$$

where (a) indicates adsorbed (surface) atoms. It is further suggested that the proportionality of the rate to arsine partial pressure (to $1/\sqrt{p_{AsH3}}$ according to their experiments) might be related to by the factor ½ in the above equation (but is not specified how). It is further suggested that CBr_4 decomposes in hydrogen atmosphere into HBr and CBr_3, and that the $Ga(a)$ atoms, rapidly react with HBr (unspecified whether adsorbed or gas) forming volatile GaBr and H_2:

$$Ga(a) + HBr \rightleftarrows GaBr(g) + \frac{1}{2}H_2(g) \qquad \text{R4.2}$$

It can be observed that, for the etching to be strictly limited by arsenic desorption, the rate should be independent of CBr_4 partial pressure: this is not consistent with the experimental data from reactor G4, and since a saturation of the etch rate with increasing p_{CBr4} has not been reached, it can be safely asserted that in the tested conditions arsenic desorption cannot be defined as "the" limiting step, although it probably *contributes* to reduce the rate.

A semi-empirical model is proposed to interpret the data, based on strongly simplifying assumptions.

It is assumed that the removal of Ga and As atoms from the surface occurs independently, the former as a consequence of a reaction with bromine species and the latter as desorption of arsenic species.

The Ga atoms are assumed to be removed by *one* reactive bromine species, indicated as Br^*, originated by CBr_4 pyrolysis. The term "pyrolysis" is used here (not entirely properly) to indicate any kind of fast pre-reaction of CBr_4 not involving Ga abstraction *and* producing Br^*. No a-priori assumption is made on the identity of Br^*; still, based on the high specific etch rate η_{CBr4} at high temperature, the preferred candidates are species containing a single bromine atom as HBr or Br rather than CBr_x or Br_2.

The partial pressure of Br^* near the surface is related to the input CBr_4 partial pressure p_{CBr4} through the number n_B of Br^* molecules that form from each CBr_4 molecule when the pyrolysis is complete (maximum yield), and a further proportionality factor c_B which must take into account the degree of completeness of the pyrolysis and the effect of mass transport (the mass transport might become slow compared to the rate of Br^* consumption at the surface when the specific rate is very high). To further simplify the treatment, it is assumed that c_B is a function of the temperature T_w, but not of p_{CBr4} or p_{AsH3}. This implies that the pyrolysis either occurs as a real or pseudo monomolecular process, or that it is fast enough to reach anyway completeness in the whole range of temperature and partial pressures used. The latter condition is probably satisfied, since there is wide evidence in literature that CBr_4 pyrolize easily, see the previously mentioned results from McEllistrem and White [41] and appendix 2.

The Br^* molecules attack the Ga atoms that have available dangling bonds, i.e. those corresponding to free group V sites, *and* remove them forming a volatile species indicated with $GaBr^*$; again for the sake of simplicity, it is assumed that $GaBr^*$ does not remain adsorbed on the surface, a hypothesis that is consistent with the fact that no rate saturation effect has been noticed with increasing p_{CBr4} in the tested range of T_w and p_{CBr4} (although it might appear at lower T_w or higher p_{CBr4}).

The etching reaction is considered irreversible, a hypothesis that might be justified assuming fast removal of the volatile product from the surface. It is treated as a simple "bimolecular" reaction:

$$Ga^{Ga} + Br^*(g) \rightarrow GaBr^*(g) \qquad \text{R4.3}$$

In R4.3 Ga^{Ga} should be intended as a *reactive* group V site: it is not known whether the etching takes place uniformly over the surface or preferentially at surface steps, proceeding layer-by-layer; the second possibility seems more in line with the fact that extremely flat surfaces can be obtained after deep etchings (at least on laser-grade substrates) and in this case Ga^{Ga} would represent only the group V sites on the steps. Another simplification embedded in using the bimolecular reaction R4.3 is related to the possibility that the species Br^* is dissociatively chemisorbed, transferring a leaving group to a neighboring surface site: in this case, the reaction would become properly termolecular, but this aspect is not included in the model.

The effect of AsH_3 is described in terms of etching inhibition through competition with Br^* for the reactive group V sites. A *reversible* chemisorption of *one* reactive arsenic species As^* originated by AsH_3 pyrolysis is assumed:

$$Ga^{Ga} + As^*(g) \rightleftarrows Ga^{Ga}As^{*As} \qquad \text{R4.4}$$

It is further assumed that a Langmuir isotherm can be used to describe As^* adsorption on the reactive sites, treating R4.4 as an *equilibrium* reaction.

Possible candidates for the role of As^* are the AsH_x and As_x species. The near-surface partial pressure of As^* is taken to be proportional to the input p_{AsH3}through the number n_A of As^* molecules that form from each AsH_3 molecule when the pyrolysis is complete, and a temperature-dependent proportionality factor c_A, defined as a function that takes into account the effect of pyrolysis completeness; c_A is assumed to be only a function of the temperature T_w. It can be noted that As^* cannot be "consumed" by the equilibrium R4.4, and its near-surface concentration is not limited by mass transport.

To approximate the dynamics of arsenic species on the surface with an equilibrium involving a single reactive species previously formed by pyrolysis, is of course more a heuristic tool than a realistic description, and As^* is best seen as an "effective" molecule than a real one. A related difficulty in using the equilibrium R4.4 to represent the occupation of the reactive sites is that the removal of Ga atoms – which consumes the reactive sites uncovering the underlying As atoms - must be slow enough to allow the equilibrium to take place and restore the supposedly rate-independent density of reactive sites. If this is not the case, the rate will be limited by the *kinetics* of arsenic desorption, making the equilibrium approximation inappropriate.

In the above hypothesis, the etching process is formulized as follows.

The fraction of (reactive) group V sites occupied by As^* is indicated with θ, and is given by the Langmuir equation:

$$\theta = \frac{n_A \cdot c_A \cdot K_A \cdot p_{AsH3}}{1 + n_A \cdot c_A \cdot K_A \cdot p_{AsH3}} \qquad \text{E4.1}$$

The etch rate E is proportional to the free sites fraction $1 - \theta$ and to the near-surface Br^* partial pressure $n_B \cdot c_B \cdot p_{CBr4}$ through a kinetic constant k_{Br}:

$$E = k_{Br} \cdot (1 - \theta) \cdot n_B \cdot c_B \cdot p_{CBr4} \qquad \text{E4.2a}$$

It can be noted that $k_B \cdot (1 - \theta) \cdot c_B$ is the specific etch rate η_{CBr4} defined above. Combining E4.1 and E4.2a the etch rate becomes:

$$E = n_B \cdot c_B \cdot k_{Br} \cdot p_{CBr4} \cdot \left(1 - \frac{n_A \cdot c_A \cdot K_A \cdot p_{AsH3}}{1 + n_A \cdot c_A \cdot K_A \cdot p_{AsH3}}\right) \qquad \text{E4.2b}$$

The last equation can be used to interpolate the etch rate for any combination of p_{CBr4} and p_{AsH3} at a fixed temperature, using the products $K_A' = n_A \cdot c_A \cdot K_A$ and $k_{Br}' = n_B \cdot c_B \cdot k_{Br}$ as fitting parameters.

To each temperature correspond a value of K_A' and one of k_{Br}': the first parameter, K_A', determines the separation between the straight lines E vs. p_{CBr4} in Fig. 4.8b, and the curvature of the lines E vs. p_{AsH3} in Fig. 4.8a. The second parameter, k_{Br}', determines the maximum slope of E vs. p_{CBr4} and the maximum separation of the curved lines E vs. p_{AsH3}, both conditions found in the limit $p_{AsH3} \rightarrow 0$. These relations make the iterative optimization of K_A' and k_{Br}' comparatively straightforward. The optimized values obtained at the four different temperatures investigated are shown in the logarithmic plots of Fig. 4.10 (the black dots).

The experimental points of Fig. 4.8 are reasonably interpolated (within the experimental errors) by the curves obtained with the model: it can then be concluded that the mathematical form of the relations among E, p_{CBr4} and p_{AsH3} is compatible with the experiments, and this supports the idea that the proposed mechanism might actually approximate the real one. It remains to discuss whether the temperature dependence of the fitting parameters can be physically meaningful. Let's consider first the case of proportionality factors c_A and c_B that are simply constants (i.e. independent of temperature).

Within the logic of the model, this would happen if the following conditions are satisfied:

- the pyrolysis of the precursor (pre-reactions leading to the formation of Br^* or As^*) is complete at all temperatures;
- in the case of Br^*, the mass transport is fast with respect to surface consumption at all temperatures.

With constant proportionality factors, $K_A^{'}$ should behave as an equilibrium constant, and $k_{Br}^{'}$ as kinetic constant for a single-step reaction: in both cases their logarithm should be approximately proportional to $1/T_w$, respectively through the reaction energy ΔU_A of R4.4 and the activation energy E_a of R4.3: this is evidently not the case, at least not on the entire interval of T_w = 569-712°C.

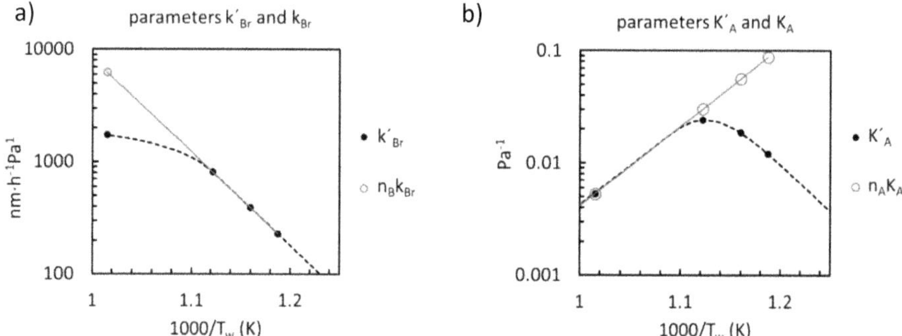

Figure 4.10 Logarithmic plots vs $1/T_w$ of the optimized values of the adjustable parameters $K_A^{'}$ and $k_{Br}^{'}$ obtained from the fitting of the experimental data at the four investigated temperatures (black dots); the dashed lines represent the *continuous* behavior of $K_A^{'}$ and $k_{Br}^{'}$ deduced as described in the text. The straight lines represent the behavior of the products $n_A \cdot K_A$ and $n_B \cdot k_{Br}$, the empty circles mark the positions corresponding to the investigated temperatures, and coincide with the black dots when c_A or c_B are =1.

As can be seen from the red straight line in Fig. 4.10a, an Arrhenius interpolation of $k_{Br}^{'}$ passes perfectly through the 3 lower-temperature points, indicating that c_B is constant in this range and suggesting complete pyrolysis and fast transport (c_B=1). The high-temperature value lies well below the line: a possible explanation consistent with the model assumptions is that the etching is entering the diffusion-controlled regime, and that the near-surface concentration of Br^* drops at high T_w; as previously noted, this hypothesis is compatible with the very high efficiency of the etching at high temperature. It is then assumed that c_B=1 in the lower temperature range, which allows to use the low-temperature Arrhenius interpolation to extract the activation energy of R4.3. The value E_a=159 kJ/mol is obtained, which is slightly above the highest apparent activation energy obtained from the Arrhenius fittings of the etch rates. The red straight line represents the product $n_B \cdot k_{Br}$ at all temperatures, and coincide with $k_{Br}^{'}$ in the low-T_w range. The black dashed line represents $k_{Br}^{'}$, the bended part of this line is only drawn *empirically* to connect the points with no discontinuity in the first derivative, but is not based on a model of the transport effects: although the values cannot be much in error, the first derivative might decrease more or less rapidly with increasing temperature than what indicated, and no extrapolation at higher temperatures is possible.

Tentatively assuming that $GaBr^*$ in R4.3 is actually the radical GaBr, the activation energy should correspond to its desorption barrier from the surface; in Ref. [47], Jenichen and Engler estimate the activation energy for GaBr desorption from (100) GaAs in the range 160-190 kJ/mol based on density functional theory simulations. This range is compatible with the value obtained from the fitting of k'_{Br}.

It can be seen in Fig. 4.10b that K'_A first increases and then decreases with increasing temperature. The increase is tentatively attributed to the increase of the factor c_A due to arsine pyrolysis, while the decrease is explained assuming an exothermic character of reaction R4.4 (in the direction of chemisorption).

To test these hypotheses, the degree of pyrolysis $c_A(T_w)$ has been expressed using a first order kinetic - with a (still unknown) kinetic constant of the form $k_p = A_p \cdot \exp(-E_p/RT_w)$ - and the residence time t_r, giving:

$$c_A(T_w) = 1 - exp\left[-A_p \cdot t_r \cdot exp\left(-E_p/RT_w\right)\right] \qquad \text{E4.3}$$

The product $n_A \cdot K_A$ can then obtained in correspondence to the four values of K'_A derived from the fits dividing K'_A by c_A. If the assumptions are correct, the four values of $n_A \cdot K_A$ *must* align along a straight line in an Arrhenius plot. Assuming – as it is reasonable – that at T_w=712°C the pyrolysis is complete, it is set c_A(712°C)=1. It is than possible to determine the values of A_p and E_p - and consequently the pyrolysis kinetic constant k_p and the function $c_A(T_w)$ - which optimize the linearity condition. It turns out that the condition *can* be very well satisfied, obtaining four aligned values of $n_A \cdot K_A$ (the red circles in Fig. 4.10b) which supports the correctness of the hypotheses.

This Arrhenius interpolation of the four $n_A \cdot K_A$ points (the red straight line passing through the circles) is the function $n_A \cdot K_A(T_w)$; the energy ΔU_A of the reaction R4.4 can be calculated from its slope. Multiplying $n_A \cdot K_A(T_w)$ by $c_A(T_w)$ one obtains $K'_A(T_w)$, which is the dashed line of Fig. 4.10b.

The optimized pyrolysis kinetic constant k_p has pre-exponential factor A_p=5×10^{17} s^{-1}, and apparent activation energy E_p=300 kJ/mol, which is not unreasonable because it is near to the first bond dissociation energy of AsH_3 = 319 kJ/mol [12]. The calculated degree of pyrolysis (more properly: of As^* formation) is shown in Fig. 4.11a, along with the similarly calculated degree of pyrolysis of CBr_4, the latter based on the apparent kinetic constant provided in Ref. [48].

The fraction θ of reactive sites occupied by As^* estimated in this way is shown in Fig. 4.11b, compared with the values that would be obtained with complete pyrolysis: below ≈600°C, θ strongly decreases. It must be said that this evaluation cannot be extended to very low temperatures because the arsenic would not desorb anymore from the surface, making the equilibrium assumption unsustainable.

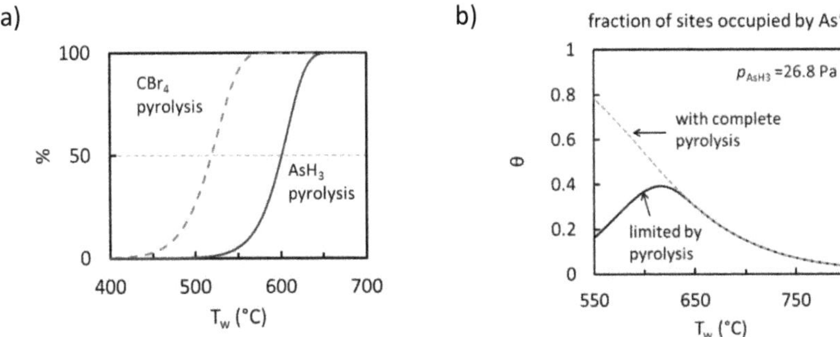

Figure 4.11 **a)** Calculated temperature profile of AsH_3 pyrolysis, the evaluation is based on its impact on GaAs etch rate by CBr_4; for comparison, the profile of CBr_4 pyrolysis calculated with the kinetic data of Ref. [48] is also show. **b)** calculated fraction of reactive sites occupied by arsenic taking into account an incomplete AsH_3 pyrolysis (continuous line) and with 100% pyrolysis (dashed line).

The reaction energy of R4.4 is ΔU_A=-135 kJ/mol (exothermic). Tentatively assuming that the effective As^* species appearing in the equilibrium equation can be identified with AsH, the reaction energy would correspond to the binding energy of AsH to the surface. In Refs. [49-50] an experimental activation energy for As_xH_y species desorption from (001) GaAs of 144 kJ/mol is reported; moreover, according to an evaluation made with DFT computations by Fu et al. [51], the binding energy of AsH to a (001) GaAs surface should be -121 kJ/mol. It can be concluded that the calculated value of ΔU_A has a reasonable order of magnitude.

The logarithmic plots of the etch rate E vs $1/T_w$ in Fig. 4.11b illustrate how the model predicts a partially non-Arrhenius behavior. The upper, straight continuous line indicates the etch rate in the limit of p_{AsH3}=0 and with no saturation effects due to mass transport. With increasing p_{AsH3} there is an increasing downwards bending due to the inhibitor effect of arsenic, which is strongest near the 50% AsH_3 pyrolysis temperature; beyond this temperature, the arsenic coverage starts to decrease and the etch rate would ultimately return to the p_{AsH3}=0 values if the mass transport would not intervene limiting the maximum achievable rate. The apparent activation energy is expected to depend on p_{AsH3} and on the selected temperature interval, and to be in general lower than the "true" value 159 kJ/mol.

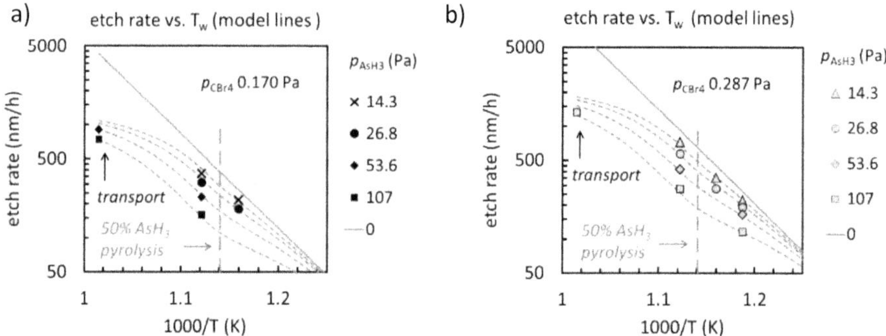

Figure 4.12 Experimental etch rates (symbols) and model simulations (lines): the straight line represents the rate in the limit for 0 arsine partial pressure.

To summarize, the in-situ etching of GaAs with CBr_4 in AsH_3 atmosphere is an activated process, the etch rate increases exponentially with temperature and decreases with increasing AsH_3 partial pressure. A semi-empirical model has been used to interpret the experimental data: it is proposed that arsenic species inhibit the reaction competing with bromine species for the same surface sites, and that the extent of arsine pyrolysis influences this competition. It is further proposed that at high temperatures the etch rate is limited by mass transport. An activation energy of 159 kJ/mol is estimated for the activated step, and GaBr desorption is suggested as the reaction product.

4.4 Investigation of CBr_4 etching of GaAs assisted with TMGa and TMAl

In this set of experiments, the in-situ etching of GaAs with CBr_4 has been investigated adding moderate quantities of either trimethylaluminum or trimethylgallium during etching, to study the effect of additional reaction pathways [43]. One of the motivations was the speculation that combining the etching with a small "growth component" *might* have a positive impact on the final morphology: both the experiments and literature information indicated that the etching is strongly sensitive to the local state of the surface (defects, different crystal planes) and the growth component might have introduced a compensating effect, with preferential growth occurring on the sites of preferential etching. Another motivation was the hope that the tests might contribute to shed some light on the partially contradictory results found in literature concerning the etching of Al-containing compounds and the parallel phenomenon of growth reduction in presence of CBr_4.

4.4.1 Experimental details

The in-situ etching experiments have been carried out in the planetary reactor G3, using H_2 as carrier gas and the reagents AsH_3, CBr_4, TMGa and TMAl. The substrates were epi-ready 2" and 3" (100) GaAs wafers.

For calibrating the etch rate with in-situ reflectometry, the following structure has been first grown on several GaAs test wafers: 200 nm GaAs buffer layer, 50 nm $Al_{.25}Ga_{.75}As$ optical marker layer and a 900 nm GaAs sacrificial layer. The etch rate of CBr_4 has then be determined in-situ from the oscillations of the reflectivity signal as previously described.

During all the etching runs the following parameters have been kept constant: overall pressure p=100 mbar and p_{CBr4}=0.31 Pa; arsine partial pressure has been kept fixed at p_{AsH3}=25 Pa. Two set-point temperatures T_{sp} have been used, 575°C and 675°C, corresponding to the wafer temperatures T_w 545°C and 630°C. Part of the etching tests did include the simultaneous use of CBr_4 and either TMAl or TMGa: p_{TMAl} and p_{TMGa} have been varied from 0 to 0.7 Pa.

In a second set of experiments, making use of the previously calibrated etching rates, the surface morphologies resulting after a 150 nm single-step etch, have been compared. The comparison has been done using both bare substrates and wafers with a 500 nm thick GaAs epitaxial layer grown in a separate run (without AlGaAs marker layer). Furthermore, substrates of different quality in terms of the maximum EPD have been compared.

4.4.2 Assisted etching: kinetics

In this section the following conventions will be used:

- $E^{GaAs}(0)$ represents the etch rate of GaAs when only CBr_4 is present (no metalorganics added during the etch).
- $E^{GaAs}(p_{Ga})$ represents the etch rate of GaAs when TMGa is present along with CBr_4, at the partial pressure p_{TMGa}. A *negative* value of $E^{GaAs}(p_{Ga})$ corresponds to GaAs growth. This situation occurs if the TMGa partial pressure is high enough to overcome the etching effect of CBr_4.
- $G^{GaAs}(p_{Ga})$ represents the growth rate of GaAs at the partial pressure p_{TMGa} when no CBr_4 is present during growth.
- A correspondent nomenclature is used in the case of TMAl.

The case CBr_4 + TMGa

Fig. 4.13a shows the effect of TMGa: at both temperatures the dependence of the etch rate $E^{GaAs}(p_{Ga})$ on TMGa partial pressure p_{TMGa} is approximately linear, decreasing with increasing p_{TMGa}, up to the point when growth dominates over etching. The horizontal line corresponding to zero etch rate is a reminder that negative values of the etch rate correspond to growth.

The growth rate of GaAs has been independently calibrated with the same reactor conditions but without adding CBr_4 (Fig. 4.13b). It is moderately sub-linear with respect to p_{TMGa} (probably due to a non-linearity in the bubbler efficiency or in the MFCs) and almost independent from the growth temperature. Assuming that the etch rate can be simply interpreted as the difference between the etch rate due to CBr_4 (in absence of TMGa) and the growth rate due to TMGa (in absence of CBr_4) we have:

$$E^{GaAs}(p_{Ga}) \approx E^{GaAs}(0) - G^{GaAs}(p_{Ga}) \qquad \text{E4.4}$$

Equation E4.4 is plotted in Fig. 4.13a at the two investigated temperatures, and matches reasonably well the experimental data, both in the region where etching prevails and in the region where growth prevails. This suggests that this simple approach is valid for the description of the etching in presence of TMGa.

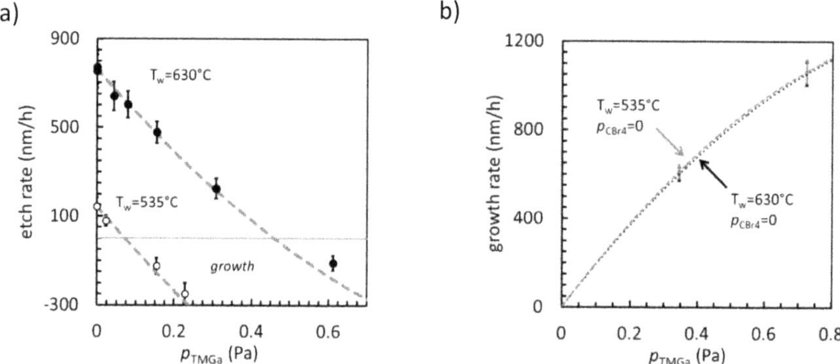

Figure 4.13 **a)** experimental values for GaAs etch rate with addition of TMGa, $E^{GaAs}(p_{Ga})$, at T_w=630°C (full circles) and at T_w=545°C (empty circles); dashed lines represent the etch rate as calculated from Eq. E4.4; the full line at $E^{GaAs}(p_{Ga}) = 0$ represents the transition from etching to growth; **b)** calibration of GaAs growth rate without CBr_4, $G^{GaAs}(p_{Ga})$; these data are used in Eq. E4.4 to calculate the etch rate plotted in a). The vertical bars indicate the uncertainty on the rates.

The case CBr_4 + TMAl

Let us now consider what happens when adding TMAl instead of TMGa. Fig. 4.14 shows the experimental etch rate $E^{Ga(Al)As}(p_{Al})$ at T_w=630°C and 545°C: the superscript Ga(Al)As indicates that the Al atoms possibly deposited on the surface must be etched along with the Ga atoms, and that in the region where growth prevails it is AlAs and not GaAs that is actually grown.

It is immediately clear that there are fundamental differences with respect to the previous case at both temperatures, but especially at 545°C, where initially the etch rate *increases* with the addition of TMAl.

In order to model the kinetic behavior in a way similar to the case of the addition of TMGa, the growth rate $G^{AlAs}(p_{Al})$ of AlAs has been independently calibrated with the same reactor conditions but without adding CBr_4 (Fig. 4.15a). The etch rate $E^{AlAs}(0)$ of AlAs with CBr_4 and no metalorganics has been also calibrated (Fig. 4.15b) and deserves to be commented before proceeding with the modeling of GaAs etch.

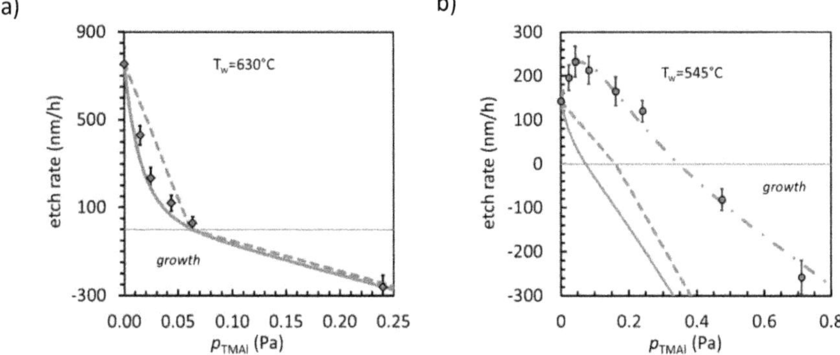

Figure 4.14 Experimental values for etch rate, $E^{Ga(Al)As}(p_{Al})$ at **a)** T_w=630°C and **b)** T_w=545°C. Lines represent the etch rate as calculated with different models: dashed lines represent Eq. E4.5 and Eq. E4.6 (linear model), solid lines the Langmuir model of Eq. E4.10, the dashed-dotted line in **b)** represents the modified Langmuir model of Eq. E4.11.

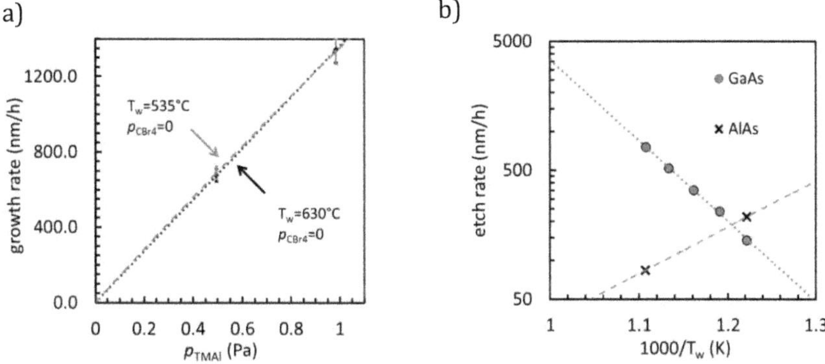

Figure 4.15 **a)** calibration of AlAs growth rate without CBr_4, $G^{AlAs}(p_{Al})$; these data are used in in Eq. E4.10 and E4.11 to calculate the etch rate, and the results are plotted in Fig. 4.14a and 4.14b; **b)** Arrhenius plots of AlAs etch rate with only CBr_4, $E^{AlAs}(0)$, compared with that of GaAs, $E^{GaAs}(0)$, at the same CBr_4 and AsH_3 partial pressures.

AlAs growth and etch. The growth rate of AlAs is linear with respect to p_{TMAl} and independent from the growth temperature in the tested range. The etch rate of AlAs is comparatively low and *decreases* with increasing temperature, in agreement with previous findings [35, 36]. Moreover, no isolated defects – as those seen in the etching GaAs – were found on the AlAs after etching, but only a moderate degree of roughness. This indicates an etching mechanism fundamentally different from that of GaAs. The As-Al bond energy is stronger than the As-Ga bond, as can be seen from the cohesion energies of the two compounds (appendix 1); another important factor to evaluate is the stability of the Al-Br bond in the AlBr(g) radical with respect to that in the adsorbed bromine group $Al^{Al}Br^{As}$. According to a rough estimate, the activation energy for AlBr desorption should be approximately 300 kJ/mol, which would make the etching via AlBr desorption negligibly small in the considered temperature range (Br should rather passivate the Al than removing it). While these considerations can explain why the etching mechanism is different from the case of GaAs, they do not give an explanation of the "negative activation energy" observed. In Ref. [36] it is proposed that the radical CBr_3 could be the responsible for AlAs etch, and for the etch of the AlAs fraction in the AlGaAs ternary, with formation of $Al(CBr_3)_x$ molecules: this could explain the temperature dependence, because the CBr_3 radical concentration can be expected to rapidly decrease with increasing temperature, due to the pyrolysis leading to the formation of CH_x, HBr and possibly Br. The weakness of Ga-CBr_3 bonds would make this mechanism unfeasible in the case of GaAs, explaining why at low temperature the etching of AlAs becomes *faster* than that of GaAs. On the other hand, it is not completely evident that $Al(CBr_3)_x$ molecules would be much better leaving groups than $AlBr_x$ molecules: bromine ion-beam-assisted etching experiments reported by Goodhue et al. [52] indicate that AlAs is effectively etched by Br_2 with no limitations related to the volatility of aluminum bromides ($GaBr_3$ and $AlBr_3$ vapor pressures are almost identical).

CBr_4+TMAl etch: linear model

Coming back to the GaAs etching in presence of TMAl, the assumption is tentatively made that the rate, $E^{Ga(Al)As}(p_{Al})$, is reduced due to simultaneous deposition of AlAs, which is then etched at a rate different from that of GaAs, corresponding to that calibrated on the AlAs layer. This idea leads to the following approximate expression:

$$E^{Ga(Al)As}(p_{Al}) \approx E^{GaAs}(0) \cdot \left(1 - G^{AlAs}(p_{Al})/E^{AlAs}(0)\right) \qquad \text{E4.5}$$

The derivation of E4.5 is detailed in appendix 3.

When the partial pressure of TMAl is high enough to overcome the etching, the growth rate of AlAs in analogy with Eq. E4.4 is simply given by:

$$E^{Ga(Al)As}(p_{Al}) \approx E^{AlAs}(0) - G^{AlAs}(p_{Al}) \qquad \text{E4.6}$$

Etch and growth rates, respectively calculated from E4.5 and E4.6, are plotted in Fig. 4.14a and 4.14b (dashed straight lines): at high temperature (T_w=630°C) the model seems roughly adequate, while it obviously fails at low temperature (T_w=545°C).

CBr_4+TMAl etch: Langmuir model

As a possible refinement of the previous "linear model", the fraction of surface which in the steady-state (during etching) is covered by AlAs has been estimated using a pseudo-Langmuir equation (the derivation of E4.7, E4.8, E4.10 is better detailed in appendix 3):

$$\theta = K \cdot p_{TMAl}/(1 + K \cdot p_{TMAl}) \qquad \text{E4.7}$$

where θ is the fraction of group III surface sites occupied by Al and K is given by:

$$K = G^{AlAs}(p_{Al})/\left(p_{TMAl} \cdot E^{AlAs}(0)\right) \qquad \text{E4.8}$$

It is then assumed that a Ga atom cannot be removed from the surface by CBr_4 if at least one of the four group III sub-lattice sites lying directly over it is occupied by an Al atom. The probability of a surface Ga atom of being entirely free from Al adatoms is $(1-\theta)^4$.

The etch rate of GaAs when the surface is partially occupied by Al atoms[2] will be indicated with $E^{GaAs}(p_{Al})$; in the interval of p_{TMAl} where etching prevails we have:

$$E^{GaAs}(p_{Al}) \approx E^{GaAs}(0) \cdot (1-\theta)^4 \qquad \text{E4.9}$$

[2] Not including AlAs etch and growth components

Combining Eq. E4.9 with Eq. E4.6 we can get an expression that approximates the etch rate (or negative of growth rate) $E^{Ga(Al)As}(p_{Al})$ in both regions:

$$E^{Ga(Al)As}(p_{Al}) \approx E^{GaAs}(0) \cdot (1-\theta)^4 + E^{AlAs}(0) \cdot \theta - G^{AlAs}(p_{Al}) \qquad \text{E4.10}$$

The rate calculated with Eq. E4.10 is plotted in Fig. 4.14a and b (solid lines). It is apparent that this "Langmuir model" interprets very well the experimental data at high temperature, but - like the previous "linear model" – fails completely at low temperature.

Considering the behavior at T_w=545°C, comparison of the experimental points with equations E4.5, E4.6 and E4.10 leads to two remarks:

- In the region where etching prevails, its rate is much higher than what is predicted either from E4.5 or E4.10, and some sort of homogeneous or heterogeneous catalysis of GaAs etching due to the presence TMAl must occur.
- In the region where the growth of AlAs prevails, the growth rate is much lower than what is predicted from either E4.6 or E4.10, which could be interpreted equivalently as due to higher than expected AlAs etch rate or to lower than expected AlAs growth rate.

CBr_4+TMAl etch: catalysis model

Tentatively, Eq. E4.10 has been modified in order to introduce an empirical *heterogeneous* catalysis term for GaAs etching, $f_1 \cdot \theta \cdot E^{GaAs}(0)$ (with $f_1 \geq 0$), and an empirical correction factor f_2 for AlAs growth rate (with $0 < f_2 \leq 1$) in presence of CBr_4:

$$E^{Ga(Al)As}(p_{Al}) \approx E^{GaAs}(0) \cdot (1 + f_1 \cdot \theta) \cdot (1-\theta)^4 + E^{AlAs}(0) \cdot \theta - f_2 \cdot G^{AlAs}(p_{Al}) \qquad \text{E4.11}$$

In Eq. E4.11 the fraction of occupied sites θ is calculated from E4.7 using a consistently corrected modification of Eq. E4.8, where the AlAs growth rate is corrected by the factor f_2 as well:

$$K = f_2\, G^{AlAs}(p_{Al}) / \left(p_{TMAl} \cdot E^{AlAs}(0)\right) \qquad \text{E4.12}$$

Note that Eq. E4.11 and Eq. E4.12 reduce to Eq. E4.10 and Eq. E4.7 when f_1=0 and f_2=1.

As can be seen in Fig. 4.14b (dash-dotted line), Eq. E4.11 provides a reasonable fit to the experimental points when f_1 is chosen as 14.4 and f_2 as 0.41, but their empirical nature is emphasized.

The coefficient f_2 in E4.11 represents a reduction in AlAs growth rate in presence of CBr_4. This reduction is supposed here to be unrelated to the etching of AlAs, which is represented by the independent term $E^{AlAs}(0) \cdot \theta$. The reduction in the growth rate of AlAs is reminiscent of the similar reduction reported by Décobert et al. [39] when growing AlInAs at 540°C in presence of CBr_4 (section 4.2.2) and was attributed by the authors to gas-phase reactions between TMAl and CBr_4. This interpretation is not entirely satisfactory: a consistent mechanism should take into account that experimentally such a growth rate reduction is not present – or at least not relevant – at high temperature, and that it is not present when adding TMGa instead of TMAl. It might be better interpreted as due to *surface* reactions of adsorbed $Al(CH_3)_x$ and adsorbed CBr_x species or HBr, with formation of volatile $AlBr_3$; at higher temperatures the faster desorption of bromine species and increasing homogeneous pyrolysis of TMAl would reduce this effect.

The coefficient f_1 in Eq. E4.11 represents an enhancement in GaAs etch rate in presence of TMAl. Experimentally, such an etch rate enhancement is not present at the higher temperature, and a similar effect is not to be seen when adding TMGa instead of TMAl. The enhancement is supposed here to be related to a heterogeneous mechanism, and is assumed to be simply proportional to Al surface coverage θ, at least at low temperature and in the tested interval of p_{TMAl}/p_{CBr4} from 0 to 2.

The assumption that the low-temperature GaAs etch rate increases with the fraction of Al present on the surface is partially supported by an observation in the previously mentioned work of Tateno et al.

[35]: the growth rate reduction of $Al_xGa_{1-x}As$ at 600°C in presence of CBr_4 did show a non-linear dependence on x, and reached its maximum value for x=0.75. On the other hand, in their case the effect was seen similarly even at 750°C.

It can be speculated that the presence of Al on the surface might affect the activation energy for the etching of GaAs; if the mechanism of GaAs etching is controlled by GaBr desorption, as suggested in the previous section, a possibility could be that the activation energy for this desorption is lowered in presence of a vicinal Al atom - i.e. when the Ga to be extracted and the Al atom are bonded to the same As atom - which might weaken the Ga-As bonds.

4.4.3 Assisted etching: morphology

As anticipated in Section 4.4.1, after the calibration of etch rates, the surface morphology after a 150 nm deep etch, using wafers with no epitaxial layers deposited and wafers which had been previously covered with 500 nm GaAs buffer have been compared.

The etching conditions tested were: CBr_4 with no metalorganics, CBr_4 + TMAl (p_{TMAl}= 0.04 Pa), CBr_4 + TMGa (p_{TMGa} = 0.04 Pa); the partial pressure of CBr_4 was again p_{CBr4}= 0.31 Pa and the temperatures T_{sp}=575°C and 675°C.

The substrates used were: Si-doped laser grade (EPD < 100 cm^{-2}), Si-doped LED grade (EPD < 5000 cm^{-2}) and semi-insulating wafers (EPD < 25000 cm^{-2}).

After etching with CBr_4 only, the reverse pyramid defects (pits) previously described were clearly visible on the surface, independently of whether the epitaxial GaAs was present or not. Changing the temperature did not have a significant impact on their shape, size and number. The density of the defects did correlate with the maximum EPD declared by the supplier, being of the same order of magnitude. While the addition of TMGa during etching did not influence the development of pits, the addition of TMAl had a dramatic impact, as shown in Fig. 4.16 for the case of semi-insulating wafers (which suffer the highest defect density).

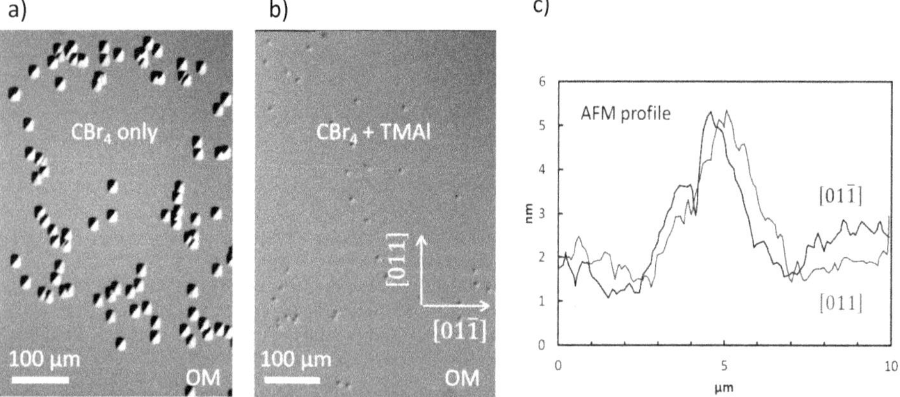

Figure 4.16 **a)** and **b)** optical microscope DIC images of the defects on the SI wafer surface after deposition of 500 nm GaAs, and subsequent 150 nm in-situ etching at T_{sp}=575°C: **a)** etching with CBr_4 only, **b)** etching with CBr4 and TMAl, both images are oriented as shown. **c)** AFM profiles of the defects on the sample etched in presence of TMAl: the defects are shallow hillocks, with height 5-10 nm.

At both the tested etching temperatures, no etch pits are visible in the TMAL-assisted etch. It is nonetheless still possible to locate the dislocations as very shallow, circular hillocks of a few nm in height. This kind of morphology is characteristic of the so-called electroless wet etching for defect delineation, which typically involves a redox reaction with carrier transport within the semiconductor; the resulting morphology is determined by the formation of a passivating layer, whose thickness increases near the exit point of the dislocation on the surface, due to to the associated increase of the local chemical potential caused by the strain field around the dislocations. It is the "symmetric" case with respect to the orthodox etching that causes the formation of pits at dislocations outcrops. The similarity is supportive of the hypothesis that a passivating layer forms on the surface during TMAl assisted etch with CBr_4, slowing down the etching especially at dislocations.

To summarize the results obtained with metalorganic-assisted etching: adding TMGa during the etching of GaAs with CBr_4 reduces the etch rate due to competition with growth. The two phenomena are apparently independent. The development of etch pits at the outcrops of dislocations is unaffected by the presence of TMGa.

Adding TMAl during etching has a much more complex effect: at high temperature the etch rate is strongly reduced, possibly due to a "masking" by AlAs forming on the surface, while at low temperature the etch rate is enhanced; the origin of this effect is not completely understood, but it is suggested that it might stem from an effect over the heterogeneous reactions involving adsorbed Al atoms. When the TMAl flow is sufficiently high, growth prevails; in this case, AlAs growth and etching appear to be independent at high temperature, as in the case of CBr_4+TMGa, while at low temperature the growth rate is strongly reduced. It is suggested that this might be due to surface reaction between adsorbed $Al(CH_3)_x$ species and adsorbed bromine species as HBr or CBr_x.

Deep etching of GaAs with CBr_4 and TMAl suppresses the formation of the etch pits observed with CBr_4 alone, which are substituted by extremely shallow hillocks.

4.5 CBr_4 etching of AlGaAs and GaInP

4.5.1 AlGaAs

The etching of the ternary $Al_xGa_{1-x}As$ with x=0.3 and 0.5 has been subject to a limited investigation on reactor G4. The etch rate of the ternary was lower than that of GaAs, being actually too low to be measured by in-situ reflectometry in most of the tested conditions, and only using low temperature, high CBr_4 flows and low AsH_3 flows it was possible to obtain significant etch rates.

In the tested range, the rate increased linearly with CBr_4 partial pressure (Fig. 4.17a), decreased with increasing Al fraction and - contrary to GaAs and similarly to AlAs – decreased with increasing temperature (Fig. 4.17b). It can be noted that at T_w=557°C the effect of the composition is very weak, and - based on the temperature dependence of GaAs and AlAs etching - at even lower temperatures the rate should increase with increasing Al fraction. The rate did even decrease with increasing AsH_3 partial pressure: when p_{AsH3} was increased from 14.3 Pa to 26.8 Pa (same values used in the tests on GaAs etch) keeping constant the other values indicated in Fig. 4.17, the etch rate dropped to near-zero values (too low to be precisely measured by in-situ reflectometry).

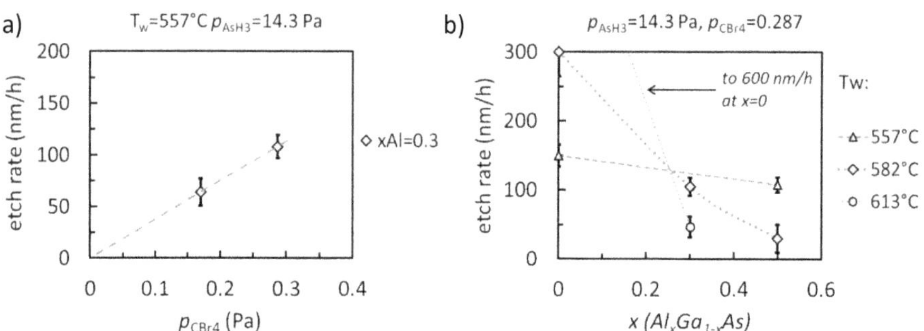

Figure 4.17 **a)** AlGaAs etch rate vs. CBr_4 partial pressure and a linear fitting; **b)** etch rate vs. Al fraction at two different temperatures, the lines connecting the points are *only* guides for the eye, the GaAs etch rate at 613°C is out of scale, at 600 nm/h.

As for the case of AlAs, no isolated defects – as those seen on GaAs - were noticed on the surface after the etching tests (the maximum etched depth was 260 nm) but the surface did become slightly rough. Even this behavior suggests a different etching mechanism with respect to GaAs.

As could be expected, once the AlGaAs surface was exposed to air, it was not possible anymore to etch the samples. A wafer terminating with a layer of $Al_{0.5}Ga_{0.5}As$ that had been exposed to air for one hour was first brought to T_w=710°C for half an hour – to test a possible deoxidizing effect - and then cooled down for the etching, but the only result was a roughening of the surface.

These (few) results, together with those on AlAs obtained on reactor G3, are qualitatively in line with what is reported in literature. It could be supposed that, to a first approximation, the etching of AlGaAs proceeds combining two different independent mechanisms, one for the etching of the AlAs fraction and one for the etching of the GaAs fraction, but both the literature data previously discussed, and the experiments on GaAs etching assisted by TMAl, indicate that this approximation needs probably to be corrected for some "alloy" effect. More extensive experimental information would be needed to verify this idea, and to develop a kinetic model.

4.5.2 GaInP

Even in the case of GaInP only a limited investigation has been done. $Ga_{0.52}In_{0.48}P$ (lattice-matched with GaAs) has been etched with CBr_4 under mixed PH_3/AsH_3 atmosphere, and the effect of using the same etching conditions has been tested on GaAs. The investigation was strictly focused on specific applications for the fabrication of laser devices (chapters 5-7), the goal was to find etching conditions that could be used on both materials obtaining a significant etch rate and good morphology. In general, it was difficult to avoid the roughening of a $Ga_{0.52}In_{0.48}P$ surface after a deep etch, which is possibly due to different etch rates of GaP and InP components, in line with the observations of Arakawa et al. [37] in relation to the etching of InGaAsP. The tests were done using etching temperatures T_w in the range 540-580°C, and AsH_3 to PH_3 ratios in the range 5-15. In this process window, the etch rates on GaAs and $Ga_{0.52}In_{0.48}P$ were similar. The etch rate on GaInP and GaAs did behave similarly, increasing with increasing temperature and CBr_4 partial pressure, and decreasing with increasing AsH_3 and PH_3 partial pressures.

4.6 Regrowth and interface contamination

At the beginning of this chapter, the study of in-situ etching has been motivated with its potential usefulness as a de-oxidizing/cleaning technique and/or as an in-situ pattern-transfer technique, in both cases before a regrowth. In this section, these aspects will be further elaborated.

A pattern created in a semiconductor layer 1 ("mask layer") by means of lithography and ex-situ etching can be transferred by in-situ etching to an underlying layer 2 which has not been directly exposed to air or chemical reagents. Ideally, it should be possible to completely remove the surface contamination associated with the ex-situ exposure to chemical reagents, water and air, and preserve pattern integrity. Two possibilities are shown in Fig. 4.18a and b.

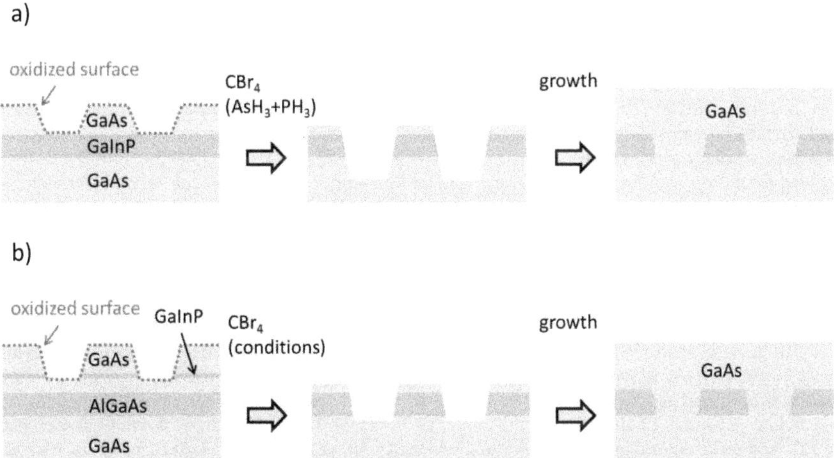

Figure 4.18 **a)** Scheme of in-situ transfer of a pattern from GaAs to GaInP and subsequent regrowth. **b)** Possible scheme of in-situ transfer of a similarly etched pattern into AlGaAs.

In both cases, a pattern – for example stripes - is created in the GaAs upper mask layer, exploiting the availability of wet etchants that selectively attack GaAs with respect to GaInP. The in-situ etch is then used in **a)** to transfer the pattern into the GaInP layer, and a GaAs regrowth is then used to bury the GaInP stripes. This strategy has been used by Maaßdorf et al. [42] to realize buried Bragg gratings over the active zone of laser devices, and variants including multiple GaAs/GaInP layers have been tested (one example is provided in chapter 7). In **b)**, the in-situ etch is used to transfer the pattern in into an AlGaAs layer: this second strategy has not yet been fully tested (to the writer's knowledge); if feasible, it would allow the regrowth over patterned AlGaAs avoiding surface oxidation.

In these examples, the binary GaAs is the regrown material, but this must not necessarily be the case, and for example AlGaAs could take its place. It must nonetheless be said that the regrowth over a patterned surface can introduce new issues of its own, in particular when a multinary alloy is used, because the material composition and the growth rates depend on the shape of the underlying geometries; these aspects will emerge in the following chapters.

SIMS tests

The effectiveness of the in-situ etching in removing the surface contamination will now be discussed, with the aid of the SIMS profiles reported in Figs. 4.19-4.22. The measurements have been done on several samples grown in two epitaxial steps, detecting the Al signal as a layer marker, and the signal of the species possibly present as contaminants after an intermediate ex-situ processing between the two epitaxial steps. The plots are taken in a 0.5 μm interval, approximately centered on the position of the regrowth interface; the upside end of the interval is always indicated as 0 on the horizontal scale, but corresponds to a depth with respect to the surface variable from sample to sample, in the range 0.5-2 μm. All the tests have been done using reactor G3.

In Fig. 4.19, the interface oxygen, carbon and silicon contamination of 3 similar structures is compared. The vertical structure is drawn (not in scale) in **a)** and the sample preparation is schematized: the first epitaxy step terminates with a GaAs layer, with thickness 20 nm, and an underlying GaInP layer, with thickness 40 nm; both are selectively etched ex-situ from *part* of the wafer surface (using a resist mask), the second epitaxy step starts with GaAs and then continues with AlGaAs. The resist used for the selective wet etch of the upper GaAs was removed *before* proceeding with the selective wet etch of GaInP, in order to optimize the surface cleanliness and to minimize the time during which the finally obtained surface is exposed to air before being loaded into the MOVPE reactor. The GaAs selective etchants was a solution containing tartaric acid and hydrogen peroxide, while the GaInP selective etchant was a solution containing hydrochloric acid and phosphoric acid. After the second wet etching, the wafers were rinsed in deionized water and dried according to the standards of semiconductor processing. The time elapsed between the ending of the drying procedure and the loading into the reactor's load-lock was in the range 15-30 min for these tests.

In Fig. 4.19b one sample regrown with no further treatments (reference) is compared with one on which a pre-growth shallow in-situ etch with CBr_4 was done at T_{sp}=575°C (T_w≈545°C), removing nominally 10 nm of GaAs based on calibrations. The temperature was then raised to T_{sp}=760°C before starting the subsequent growth. The oxygen profiles shown are taken on the wet-etched (1) and on the not wet-etched (2) areas of the reference sample, and only on the etched area of the in-situ etched sample (3). The plots are aligned based on the lower GaAs-AlGaAs interface. In all cases, oxygen peaks are visible, completely contained within the GaAs material. The concentration of oxygen in bulk AlGaAs is here 2-5×10^{16} cm^{-3}, while in GaAs it is below ×10^{16} cm^{-3}.

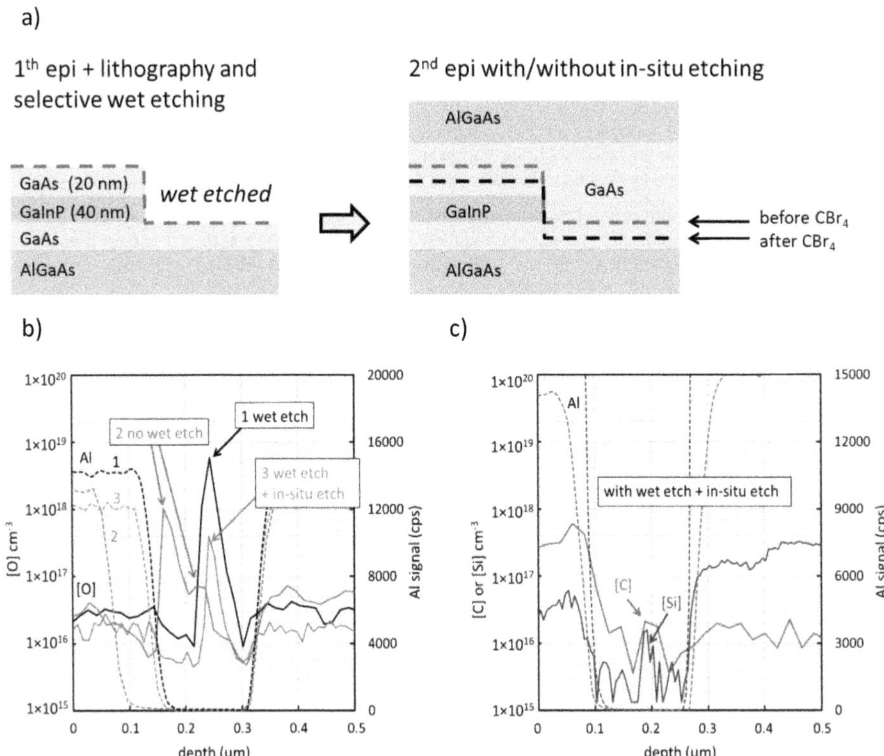

Figure 4.19 **a)** scheme of sample preparation; **b)**, **c)** SIMS profiles of oxygen, carbon and silicon concentration (continuous lines) and aluminum signal (dashed lines). The regrowth interface is in the middle of the GaAs layers, identified by Al profiles, and corresponds to the left side of the contaminants peaks.

Observing the reference peaks, it can be seen that the oxygen contamination is clearly higher in the areas that have seen both the GaAs and the GaInP selective etchings (1), with respect to the areas that were protected from the resist during the first etch (2); the reason for this difference is not clear. The oxygen peak is in the second case lower and broader, and is split in a primary peak on the left (above GaInP) and one on the right (under GaInP), possibly corresponding to a preferential accumulation of oxygen in the GaAs layers with respect to GaInP (but the SIMS concentrations in GaInP are calculated with a calibration for GaAs, so [O] in GaInP is not accurate). This result indicates that oxygen diffuses through the layers for tens of nanometers, most probably when the samples are brought to high temperature inside the reactor and during the growth.

The oxygen contamination is strongly reduced on the in-situ etched sample: the measurement was done in the wet-etched area, the peak value is at 4×10^{17} cm^{-3} and should be compared with the corresponding peak value of the reference at 6×10^{18} cm^{-3}.

Carbon and silicon concentration have been measured on a similar sample, not intentionally doped in the GaAs layers, after the same wet-etching process and with the same in-situ etching: the results are plotted in Fig. 4.19c. C and Si are present in the surrounding AlGaAs due to intentional doping; carbon concentration at the interface is about 2×10^{16} cm^{-3} and barely above the background signal (but the result was confirmed on a second sample). Silicon concentration is $1\text{-}2\times10^{16}$ cm^{-3}, significantly above the background. A comparison with a sample regrown without in-situ etching is unfortunately not available; it is possible that the in-situ etch introduces some C contamination (from the carbon in CBr_4) while the silicon contamination might stem from the wet process or from a "memory effect" of the reactor.

Bromine concentration was measured on the same sample: the measurement is not shown, the signal in GaAs did not rise above the background level at about 1×10^{15} cm^{-3}.

Since the main residual contamination, even after the in-situ etch, was oxygen, the efforts have initially concentrated on developing an ex-situ deoxidation and passivation procedure. After the second selective etch, a final treatment with either concentrated sulfuric acid (H_2SO_4) or ammonium sulfide $(NH_4)_2S$ (both followed by deionized water rinse) was tested, and the results are shown in Fig. 4.20, along with peak 3 of Fig. 4.19 for comparison. As in the previous cases, the samples were loaded in the reactor within 15-30 min after the drying procedure.

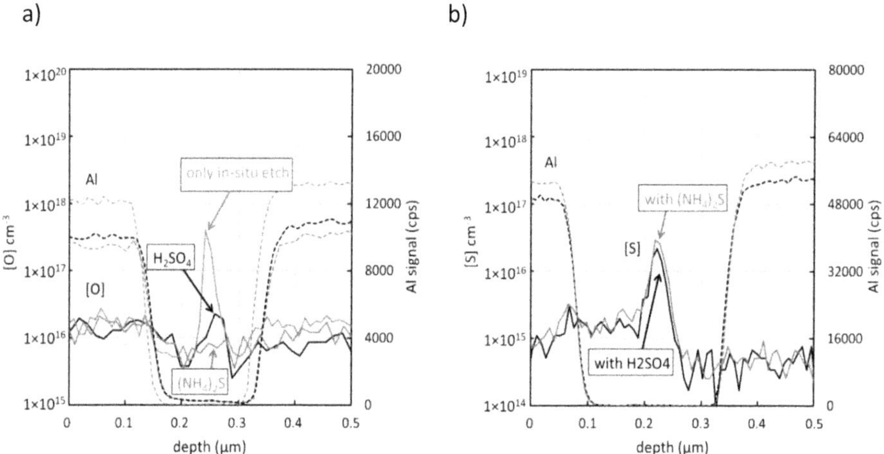

Figure 4.20 Effect of passivation with sulfuric acid and ammonium sulfide on: **a)** oxygen contamination and **b)** sulfur contamination.

It can be seen from Fig. 4.19 a) that both treatments reduce the interface oxygen contamination to values $\leq 2\times10^{16}$ cm^{-3}, with the $(NH_4)_2S$ performing slightly better than H_2SO_4. These results proved to be repeatable, provided that the samples did not stay exposed to air for a longer time. A longer waiting time (3.5 hours) was deliberately introduced, and in both cases the interface oxygen peak increased to levels comparable to those obtained in non-passivated samples, with the worse results obtained in the case of H_2SO_4. Residual sulfur contamination due to the passivation was detected and is shown in Fig. 4.20b.

Since the in-situ etching assisted with TMAl had given some interesting results from the point of view of the final morphology on non-patterned samples, its effect on interface contamination was tested. The sample was selectively wet etched as previously described, but no passivating treatment was done. The assisted etching was done at the same temperature of the previous cases. Fig. 4.21a shows the results of the measurements (again, peak 3 of Fig. 4.20a is present for comparison): the interface oxygen peak reaches a record value of 1×10^{19} cm^{-3}, which obviously makes this technique not useful in probably all foreseeable device applications. In the same figure, the Al signal scale has been increased to show the interface aluminum peak, which indicates that Al-O units have been "anchored" to the surface. The obtained oxygen concentration reflects the contamination that is present at the onset of the in-situ etching, and is even higher than the values obtained *without* any in-situ etching - peak 1 in Fig. 4.19a – probably because in that case more oxygen was desorbed from the surface during the further heating to the growth temperature.

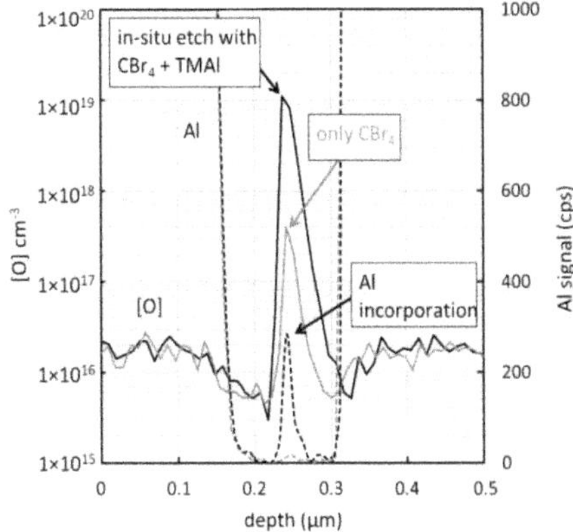

Figure 4.21 Effect of TMAl addition during the in-situ etch: oxygen and aluminum are both present at the regrowth interface.

From the above results, it is apparent, that the combination of an appropriate deoxidation-passivation procedure with the in-situ etch (without TMAl) can be used to obtain an almost oxygen-free regrowth interface in the case of a GaAs regrowth over a GaAs surface. The case is more complicated when an AlGaAs regrowth must be done over an AlGaAs surface. The fact that, once exposed to air, the AlGaAs cannot be etched by CBr_4 is a clear proof of the ineffectiveness of this in-situ etching in removing the oxide layer, a result predictable based on the oxide stability and the various literature presented at the beginning of this chapter. It is nonetheless possible to etch in-situ a protective cap of GaAs or GaInP, uncovering the underlying AlGaAs, and then proceed with the regrowth. The possible usefulness of this kind of procedure will be demonstrated in chapter 6, while here only the oxygen-contamination aspect is discussed.

Two variants have been compared, depicted in Fig. 4.22a: in the first case a GaAs/AlGaAs layer structure has been selectively etched on part of the surface, leaving only a thin GaInP etch-stop layer to protect the underlying GaAs. In the second, even the GaInP has been etched, but an additional thin GaAs protection layer has been inserted in the vertical structure. It can be noted that in both cases the upper AlGaAs remains unprotected on the side. After an ammonium sulfide passivation the samples have been rapidly loaded in the reactor. The in-situ etch with CBr_4 was done in this case at T_{sp}=650°C, corresponding to $T_w\approx$609°C; at this temperature, the etching is highly selective with respect to AlGaAs. The GaAs and GaInP protective layers have been completely removed using an overetch time with respect to the minimum time required according to calibrations. The temperature has then been raised to T_{sp}=760°C and AlGaAs has been grown; the Al fraction was 0.2 in the layers immediately below and above the regrowth interface.

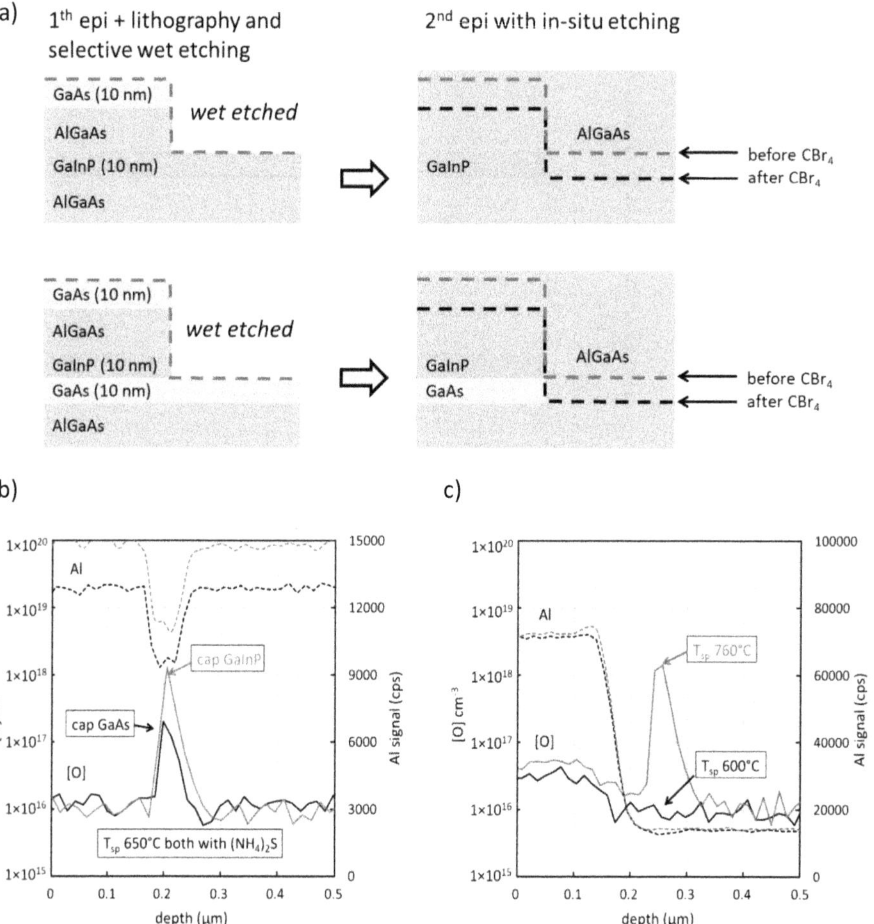

Figure 4.22 **a)** Scheme of the preparation and **b)** SIMS profiles of the "AlGaAs on AlGaAs" samples, showing the effect of the composition of the protective cap layer on interface oxygen concentration. **c)** Effect of the etching temperature.

The interface oxygen contamination of the two variants is shown in Fig. 4.22b. Using a GaAs cap, the peak value is 2×10^{17} cm^{-3}, an order of magnitude worse than in the case of GaAs/GaAs regrowth. Using a GaInP cap, the value increases above 1×10^{18} cm^{-3}.

A first conclusion that can be drawn is that GaInP much less effective than GaAs in protecting the AlGaAs layer from oxidation: this conclusion has been confirmed by independent tests done using different combinations of GaInP and GaAs protective caps [53].

The reasons for this difference are not clear: it might be speculated that it is related to the different chemical reactivity of GaInP and GaAs with oxygen, leading to the formation of phosphates in the first case and of oxides in the second, and/or to a different permeability of the materials to oxygen-containing species.

A second conclusion is that a 10 nm cap – even of GaAs - is not sufficient as a protective "screen" for AlGaAs in the applied conditions; the question arises then, whether the higher in-situ etching temperature used for this test might be part of the problem. A possibility is that, if the contaminated cap is not etched away already at low temperature, the oxygen will either diffuse and react with the aluminum before the start of the etching, or linger on the surface in the form of gallium oxide species until the CBr_4 etch uncovers the AlGaAs. In order to test these ideas, two samples having a 10 nm GaAs cap over an AlGaAs layer have been rinsed 15 min in deionized water, dried and then subjected to an in-situ etch followed by AlGaAs regrowth. In both cases the in-situ etch duration was long enough to remove 17 nm of GaAs (according to calibrations), but the etching temperatures were very different: T_{sp}= 600 and 760°C (T_w 566 and 699°C). The results are shown in Fig. 4.22b: after the low-temperature etching the interface oxygen concentration is $\approx 1\times10^{16}$ cm^{-3}, while with the high temperature a peak concentration above 1×10^{18} cm^{-3} is detected. It can be concluded that a low-temperature in-situ cleaning is actually a more efficient deoxidizing procedure.

To summarize: the in-situ etching of AlGaAs with CBr_4 is possible, provided that the surface is not oxidized; the etch rate decreases with increasing temperature, opposite to the case of GaAs and GaInP. This effect can be exploited to selectively etch GaAs (or GaInP) over AlGaAs at high temperature. The opposite case – selective etching of AlGaAs over GaAs – should be possible at low temperature for very high Al content (compare the etch rate of AlAs in Fig. 4.15b). With an appropriate selection of the etching conditions, in-situ pattern transfer between layers of these materials appears feasible.

The oxygen contamination resulting from wet-etch processing can be partially removed from GaAs by a shallow in-situ etching, and almost completely removed using a combination of a passivation pre-treatment with H_2SO_4 or $(NH_4)_2S$ and a shallow in-situ etch.

AlGaAs cannot be de-oxidized by CBr_4; nonetheless, a sacrificial protective cap of GaAs can be used to protect AlGaAs from oxidation during ex-situ processing, and be then removed in-situ. The oxygen contamination present on the GaAs surface is preferably removed at low temperature, to avoid diffusion during the heating in the MOVPE reactor; GaInP appears to be less suited as protective material than GaAs, possibly due to higher permeability to oxygen-containing species.

5 SG-DBR tunable lasers

5.1 Chapter introduction

In this chapter the realization of a particular kind of semiconductor wavelength-tunable lasers, the sampled-grating distributed Bragg reflector laser (SG-DBR) is described. Its general working principle is first illustrated, then the structure and technological steps actually used in this work for the fabrication of thermally tuned devices, and finally the main results obtained from these devices are discussed. A second approach that has been investigated, based on electronic tuning, has not been successful: the possible reasons behind these different outcomes are discussed. Part of the material presented in this chapter has been published in Refs. [54-57] (the writer is co-author) and parts of the work have been carried out under the Mid-TECH project funded by the European Union's Horizon 2020 research and innovation program under Grant Agreement No. 642661

5.2 SG-DBR lasers

Wavelength-tunable diode lasers are required in a variety of applications such as absorption spectroscopy, trace-gas detection and optical telecommunication systems. There are various types of tunable diode lasers with different wavelength-tuning mechanisms. Among the monolithic solutions, InP-based multi-quantum-well (MQW) SG-DBR lasers are well established devices, originally developed for optical telecommunication applications. They offer wide tuning ranges around 1.3 µm (O-band) and 1.55 µm (C-L-bands). While a conventional DBR laser can provide a tunability range up to approximately 10-15 nm, the integration with two sampled gratings can extend the tunability range above 50 nm, exploiting the Vernier effect [58-63]. The ultimate limit to the tunability range in these devices is given by the gain bandwidth of the MQW structure. The wavelength shift is obtained changing the effective refractive index of the tuning sections, either thermally (using micro-heaters) of electronically (by carrier injection). The main difference between these two approaches is the obtainable switching time, which is of the order of milliseconds for thermal tuning and nanoseconds for electronic tuning; another difference is that tuning by carrier injection introduces high optical absorption losses in the tuning sections due to free carrier absorption.

Widely-tunable lasers operating in the shorter wavelength range offered by GaAs-based devices are attractive components for a number of industrial and biomedical sensor applications [56]; the implementation of SG-DBR technology on GaAs is then potentially interesting, but its state of development is much less advanced with respect to InP.

A schematic of a typical SG-DBR is shown in Fig. 5.1a; it is composed of four sections, one active section which includes one or more quantum wells, and three passive sections, two DBR mirrors and one phase section. The refractive index of the three passive sections can be adjusted independently in each section. The reflectivity of the two mirrors is due to the presence of wavelength-selective elements (gratings), the end facets should not contribute to reflectivity and anti-reflective (AR) coatings, possibly combined with a tilt in the facets, are applied. The sampled gratings are obtained with the periodical blanking of Bragg gratings having the same period Λ_B in the two mirrors. The grating segments ("bursts") have lengths Z_{1f} and Z_{1b} in the front and back mirror respectively, the sampling periods (burst + blank) have lengths Z_{0f} and Z_{0b}.

Each mirror has a comb-shaped reflectivity, as qualitatively shown in Fig. 5.1b, with a main lobe composed of several tightly spaced peaks and very weak secondary lobes. It is centered on the wavelength λ_B, given approximately by the Bragg grating constructive interference condition:

$$m\lambda_B = 2n_e\Lambda_B \quad \text{E5.1}$$

where m is the grating order and n_e the effective refractive index (the equivalent refractive index for a propagating mode within the waveguide). The spacing $\Delta\lambda_s$ between the reflectivity peaks for each mirror is approximately given by:

$$\Delta\lambda_s = \frac{\lambda_B^2}{2n_g Z_0} \tag{E5.2}$$

where the effective group index n_g is defined as:

$$n_g = n_e - \lambda\frac{dn_e}{d\lambda} \tag{E5.3}$$

The larger the sampling period is, the smaller becomes the inter-peak spacing. The width of the comb envelope (main lobe) for each mirror is inversely proportional to the burst length of the mirror Z_1 and is approximately given by:

$$\Delta\lambda_{env} = \frac{\lambda_B^2}{2n_g Z_1} \tag{E5.4}$$

Shorter grating bursts lead to larger envelopes.

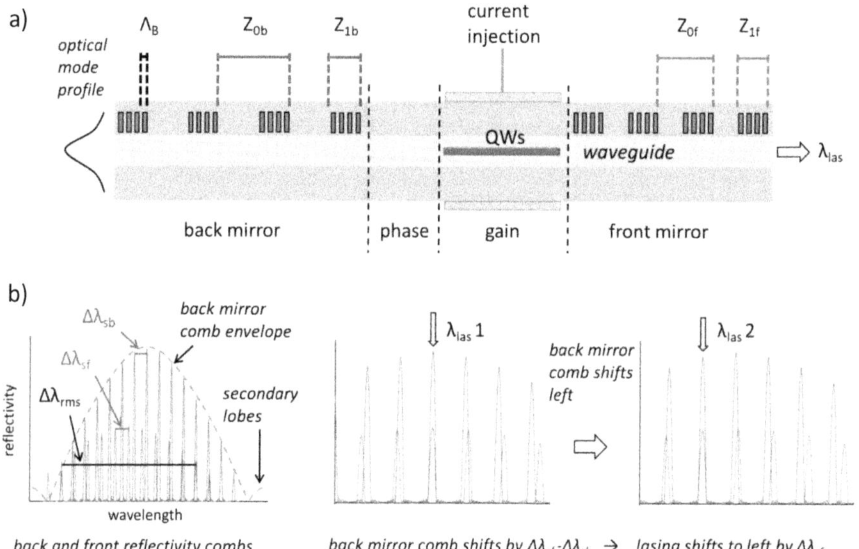

Figure 5.1 **a)** Schematic of a generic SG-DBR laser; **b)** mirror reflectivity combs (qualitative) and their relation with the lasing wavelength; $Z_{0b} > Z_{0f}$ has been assumed.

The lasing wavelength corresponds to the wavelength at which there is coincidence of a peak from the front mirror reflectivity and a peak from the back mirror. There might be more than one of such pairs at a time, in which case the relative intensity of the reflectivity and the wavelength dependence of material gain and internal optical losses will contribute to determine the lasing wavelength selection.

The tuning range cannot exceed the repeat-mode spacing $\Delta\lambda_{rms}$ (wavelength interval corresponding to the nearest 2 coincident peaks of the 2 mirrors). This is approximately given by:

$$\Delta\lambda_{rms} = \frac{\lambda_B^2}{2n_g(Z_{0b} - Z_{0f})} \qquad \text{E5.5}$$

Where it has been assumed $Z_{0b} > Z_{0f}$. The smaller the difference between the sampling periods, the larger is the repeat-mode spacing.

The tuning of the lasing wavelength can be done in principle by shifting the reflectivity comb of only one mirror, but in order to obtain a continuous wavelength change, and to optimize side mode suppression ratio (SMSR) and output power, a simultaneous control over the three passive sections is needed. This can be understood in the following way: first, the two combs must be aligned in order to superimpose two peaks, selecting a common wavelength reflected by both mirrors ("channel selection"), than this wavelength can be finely tuned – shifting it by few nanometers – moving simultaneously the two combs in the same direction; the resonance condition for the laser cavity can then be optimized adjusting its optical length through modification of the refractive index in the phase section.

The reflectivity of the front mirror is normally kept lower than that of the back mirror in order to obtain most of the optical output power from the front side of the device.

5.3 Thermally tuned SG-DBR lasers

5.3.1 Structure and process

The preliminary development of the devices did aim at an operative wavelength centered at about 1060 nm, and DBR lasers with passive section were realized and tested [54], but then the focus was shifted to a range centered around 975 nm to comply with the requirements of the European project Mid-TECH. Complete SG-DBR tunable lasers – described below - were fabricated and tested only at the shorter wavelength. The vertical structure and the process strategy remained essentially the same.

The developed SG-DBR laser is depicted in Fig. 5.2a. The chip has a footprint of 4.0 mm × 0.5 mm, and is a ridge-type laser. The device consists of six sections: a 900 µm long gain section in the device center and two SG-DBR sections with lengths of 1000 µm and 450 µm as rear and front mirror, respectively. Furthermore, there is a 100 µm long phase section, and two passive terminal sections with the ridge waveguide (RW) tilted by an angle of 3° with respect to the normal to the facets to help suppress disturbing reflections; the facets are cleaved and AR coated.

The RW introduces a lateral optical confinement because of the effective refractive index difference in the direction perpendicular to the propagation direction and parallel to the waveguide layers; its width is 2.2 µm: this ensures fundamental lateral mode operation. The vertical optical confinement is determined by the refractive index profile of the vertical (epitaxial) structure, which similarly ensures vertical monomodality. The longitudinal mode selection in a given channel is determined by the combined mirrors bandwidths and the resonance condition, which depends on the cavity optical length. In general, DBR-type lasers operated in continuous wave (CW) mode, are subject to longitudinal mode-hopping as the current injection increases, mainly because the temperature drifts in the active zone, detuning the lasing mode, until a neighboring mode is spontaneously selected [64, 65].

The current is injected from the top of the ridge, and since this is dry etched (with BCl_3-Cl_2 chemistry) almost down to the undoped waveguide (Fig. 5.2b), even the lateral current confinement is reasonably ensured, in spite of the lack of buried lateral current confinement structures. Metal stripes with a width of 8 µm are placed on top of grating and phase sections, close to the RW (Fig. 5.2c) and act as resistive heaters, their temperature is determined by the respective currents I_b, I_f, I_p. Based on an approximate mirror reflectivity temperature coefficient $d\lambda/dT$ ≈0.065 nm/K, and with a design spacing between the mirror peaks of about 2.3 nm, to exploit the full tuning potential of the laser a mirror temperature variation of ≈ 40°C is needed, corresponding to a variation of GaAs refractive index of ≈0.013.

Between the semiconductor and the p-metallization there is a SiN_x insulation layer, in which a contact window is opened only over the RW in the gain section, to allow electrical pumping.

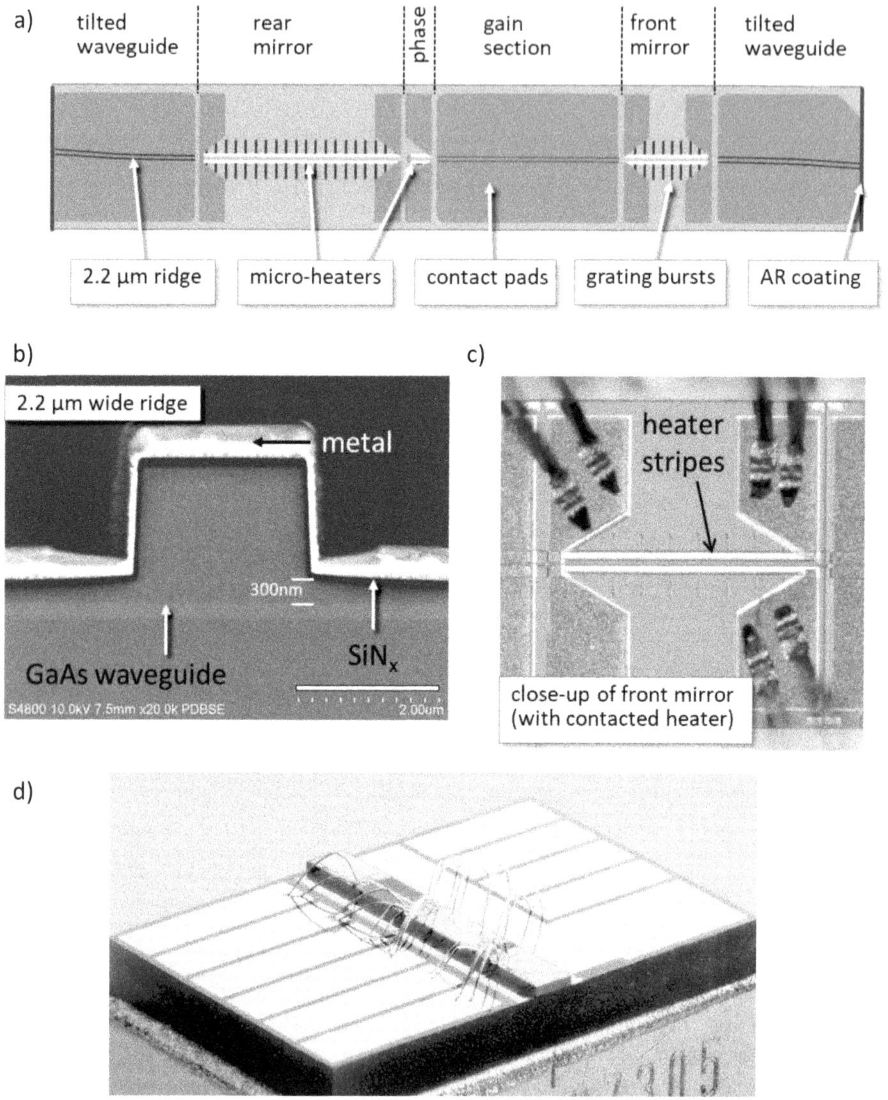

a) tilted waveguide | rear mirror | phase | gain section | front mirror | tilted waveguide

2.2 µm ridge | micro-heaters | contact pads | grating bursts | AR coating

b) 2.2 µm wide ridge | metal | 300nm | GaAs waveguide | SiN_x | S4800 10.0kV 7.5mm x20.0k PDBSE | 2.00um

c) heater stripes | close-up of front mirror (with contacted heater)

d)

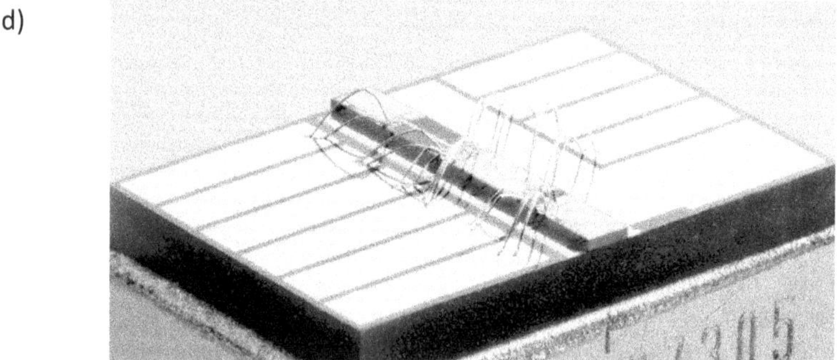

Figure 5.2 **a)** Drawing of SG-DBR (top view), showing the ridge shape, the position of the heaters, the galvanic pads (those on the tilted-waveguide sections are only for mechanical purposes) and the position of the gratings; **b)** SEM secondary electrons image of the uncoated facet; **c)** OM image of the front mirror, the interferometric contrast allows to see the lines corresponding to the buried grating bursts; **d)** bird's eye view of a chip soldered and bonded to a ceramic carrier platform, soldered in turn to a copper C-mount heat-sink.

The vertical structure, along with the main process steps for the definition of the different sections of the device, is shown in Fig. 5.3. During the first epitaxial growth step, the following layers are grown over the GaAs (100) substrate, in a down-up (growth-chronological) order: n-GaAs buffer, n-AlGaAs cladding (1), the undoped GaAs layers numbered 2, 4, 7 interleaved by GaInP grating layer (3), GaInP etch-stop layer (5) and active multi-layer comprising one or two QWs and GaAsP barriers (6). The composition of GaInP is near the lattice-match with GaAs ($Ga_{0.51}In_{0.49}P$), the QWs composition is approximately $Ga_{0.83}In_{0.17}As$ (≈12000 ppm compressive strain), and the barrier composition is approximately $GaAs_{0.9}P_{0.1}$ (≈3600 ppm tensile strain),

This base structure is then processed with E-beam lithography and two selective wet etching steps, to define the gain section; a second E-beam lithography and two further selective etchings define the mirrors and the other passive parts (phase and bended sections).

After the last etching, the surface is passivated with $(NH_4)_2S$ and the wafers are immediately loaded into the reactor. Before starting the second epitaxy, a shallow in-situ etch with CBr_4 is done to deoxidize/decontaminate the surface as described in the previous chapter.

The following layers are then grown: undoped GaAs (which completes the waveguide), p-$Al_{0.35}Ga_{0.65}As$ cladding and finally a highly p-doped GaAs contact layer. The resulting thickness of the GaAs waveguide is ≈260 nm in the passive sections, and ≈400 nm in the active section. Si and C have been used as n and p dopants throughout the structure.

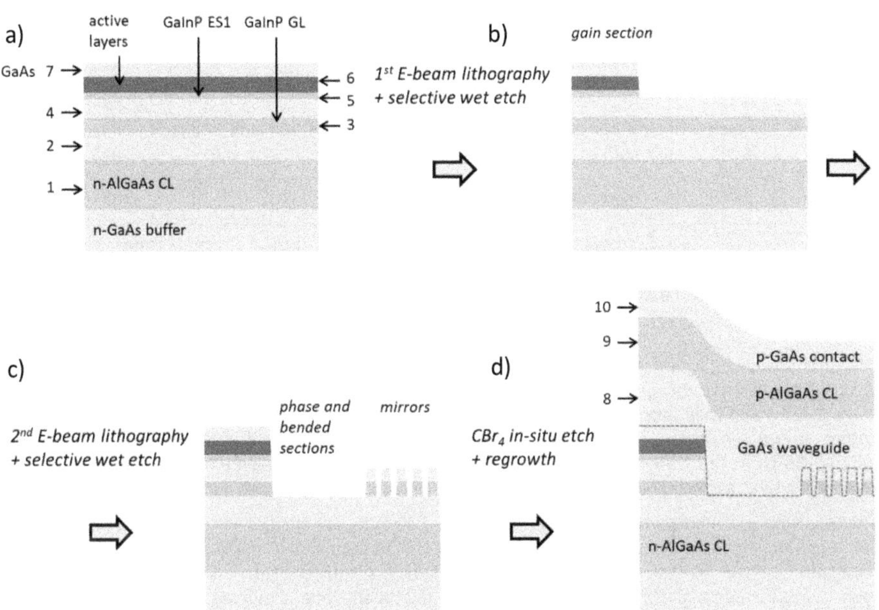

Figure 5.3 Vertical structure and process sequence of SG-DBRs: **a)** after 1st epitaxy, **b)** definition of the gain section, **c)** definition of the passive sections, **d)** 2nd epitaxy; the dashed line represents the regrowth interface.

The DQW structure has a lower aluminum fraction in the n-cladding layer with respect to the SQW structure (0.17 vs. 0.25) and correspondingly a weaker vertical optical confinement, which is associated with a narrower vertical far field (VFF), and with a smaller confinement factor in each QW (Γ_{QW}). The narrower far-field was desired to facilitate the efficient out-coupling of the emitted light into a semiconductor optical amplifier, specifically for the realization of a miniaturized master-oscillator power-amplifier (MOPA) as described in Ref. [57]. The thickness of the n-cladding in the DQW structure is larger than in the SQW structure (3.2 vs 2.6 µm) to avoid coupling of the guided mode into the substrate. The layer sequence is reported in Table 5.1. The GaAs/AlGaAs interfaces between layers 1-2 and 8-9 where graded across 30 nm.

epitaxy	#	layer	thickness (nm)	thickness remaining in the phase section (nm)
2nd	10	p-GaAs contact	100	100
	9	p-AlGaAs	1230	1230
	8	GaAs	130	130
1st	7	GaAs	20	0
	6	GaAsP/GaInAs	2×10/7 (SQW) 3×10/2×7 (DQW)	0
	5	GaInP etch stop	10	0
	4	GaAs	20	0
	3	GaInP grating layer	40	0
	2	GaAs	130	130
	1	n-AlGaAs cladding	2550 (SQW) 3150 (DQW)	2550 (SQW) 3150 (DQW)

Tab. 5.1 Epitaxial layers of the SQW and DQW structures.

For each structure, a first order and a second order Bragg grating (in the grating bursts) have been tested. The effective group index and the grating coupling coefficients κ_g have been first estimated theoretically, and then experimentally evaluated using standard DBR structures (gain + single mirror) with a *continuous* grating, realized with the same vertical structure and fabrication procedure of the tunable SG-DBR: the DBR sub-threshold spectra (amplified spontaneous emission - ASE) have been acquired, and κ_g has been obtained fitting the whole reflection pattern with a parametrized model [56].

The resulting coupling coefficient, in the range 150-500 cm^{-1}, are smaller for the second order grating, and – comparing the same grating order - for the DQW structure, but in all cases compatible with the realization of the SG-DBR according to the previously specified overall dimensions of the mirror sections. A further experimental evaluation of n_g has been done acquiring the ASE spectra of DBR structures with a *sampled* grating, and using E5.2.

The 6 grating parameters that had to be chosen for the SG-DBR tunable lasers are the duty cycle *dc* and the period Λ_B of the Bragg grating bursts, and the sampled grating parameters: Z_{0b}, Z_{0f}, Z_{1b}, Z_{1f}. They are listed in Table 5.2, along with the other calculated or measured parameters previously mentioned. The expected tunable range is about 24 nm.

<table>
<tr><td colspan="2" rowspan="2">parameter</td><td colspan="2">SQW</td><td colspan="2">DQW</td></tr>
<tr><td>1st order</td><td>2nd order</td><td>1st order</td><td>2nd order</td></tr>
<tr><td colspan="2">Γ_{QW} total (calc.)</td><td colspan="2">0.015</td><td colspan="2">0.022</td></tr>
<tr><td colspan="2">n_e mirrors (calc.)</td><td colspan="2">3.408</td><td colspan="2">3.424</td></tr>
<tr><td colspan="2">n_g mirrors (meas.)</td><td colspan="2">≈4</td><td colspan="2">≈4</td></tr>
<tr><td colspan="2">κ_g (cm^{-1}) (meas.)</td><td>497</td><td>292</td><td>251</td><td>151</td></tr>
<tr><td rowspan="6">set</td><td>Λ_B (nm)</td><td>143</td><td>286</td><td>143</td><td>286</td></tr>
<tr><td>dc</td><td>0.5</td><td>0.25</td><td>0.5</td><td>0.25</td></tr>
<tr><td>Z_{0f} (μm)</td><td colspan="4">45</td></tr>
<tr><td>Z_{0b} (μm)</td><td colspan="4">50</td></tr>
<tr><td>Z_{1f} (μm)</td><td colspan="4">5</td></tr>
<tr><td>Z_{1b} (μm)</td><td colspan="4">5</td></tr>
<tr><td rowspan="4">expected</td><td>VFF (FWHM)</td><td colspan="2">40°</td><td colspan="2">28°</td></tr>
<tr><td>$\Delta\lambda_{sf}$ (nm)</td><td colspan="4">2.6</td></tr>
<tr><td>$\Delta\lambda_{sb}$ (nm)</td><td colspan="4">2.4</td></tr>
<tr><td>$\Delta\lambda_{rms}$ (nm)</td><td colspan="4">24</td></tr>
</table>

Table 5.2 List of parameters related to the optical design of the SG-DBR tunable lasers; Γ_{QW} total is the confinement factor in the active (1 or 2 QWs). Note that $\Delta\lambda_{env}$=$\Delta\lambda_{rms}$ because of the particular values selected in the design. Values calculated for λ_B=975 nm.

5.3.2 MOVPE growth, and intermediate pattern-definition process

The epitaxial structures have been grown on the G3 reactor, using H_2 as carrier gas and the precursors TMGa, TMAl, TMIn, AsH_3, PH_3, SI_2H_6 and CBr_4, as described in chapter 3.

In the first epitaxial growth step, the realization of the base structure, the AlGaAs:Si lower cladding was grown at T_{sp}=760°C, then the temperature was lowered to T_{sp}=605°C for the growth of the following layers. The higher temperature selected for AlGaAs is required to obtain a high material quality, because of the low surface mobility of aluminum, and in order to reduce oxygen incorporation. The lower temperature used for the following layers was motivated by two different reasons. The first was that the GaInAs QW was initially designed for 1064 nm emission, and contained 25% of indium; at these values, the compressive strain is ≈18000 ppm, and a low growth temperature allows avoiding the risk of defect formation caused by InAs phase separation [66-68]. The second reason was to avoid the growth of ordered GaInP, which occurs at higher temperatures; since the GaInP layers are located under the QW, the worsening of the morphology usually associated with ordering might have negatively impacted the QW itself and consequently the device performance [69]. This second reason motivated to keep the same growth temperature even when the QW was redesigned to emit at 975 nm, lowering its indium content.

The second epitaxy was preceded by an in-situ shallow etching with CBr_4 at low temperature and under a mixed AsH_3/PH_3 atmosphere (the reason for selecting a mixed atmosphere will be explained in the following). This, in combination with the use of an ammonium sulfide pre-treatment of the wafers, allowed obtaining an almost oxygen-free surface, as is shown by the SIMS depth profile in Fig. 5.4, which has been taken over the grating section. This was considered important, because the active layers are only 20 nm below the regrowth interface, and oxygen contamination in this position would lead to significant non-radiative recombination.

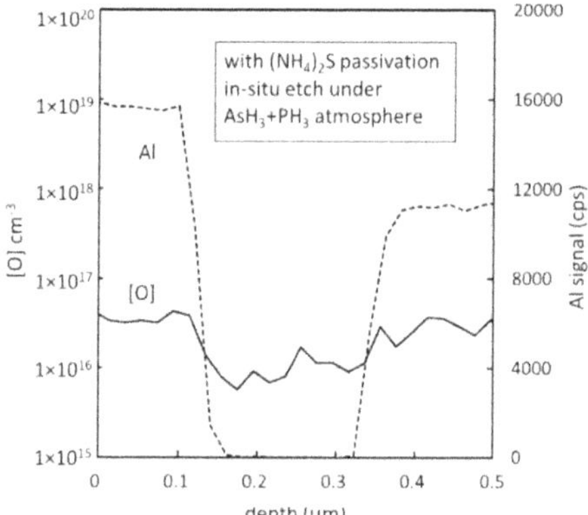

Figure 5.4 SIMS depth profile of oxygen concentration in the structure, taken in the mirror section; the plot shows part of the upper and lower AlGaAs claddings and the waveguide (Al signal=0), the regrowth interface is approximately in the center of the plot.

The introduction of CBr_4 treatment did cause – or more properly did put in evidence - an interface defectiveness that otherwise was not visible. Fig. 5.5 shows the surface of three different test structures after the second epitaxy, in a zone comprising a passive (wet-etched) area and an active (not wet-etched) area; when the regrowth was not preceded by the CBr_4 treatment, the resulting morphology was extremely smooth (a), while using CBr_4 several defects (shallow protrusions a few nm in height) where present, especially in the vicinity of the etched step (b). Introducing the ammonium sulfide pre-treatment, the defectiveness disappeared (c). The origin of these defects is not well understood, but since EDX analysis did not indicate the presence of surface contamination prior to the regrowth, it is possible that the defects are related to an inhomogeneous surface oxidation after the wet-etching steps, whose "pattern" might be transferred by the in-situ etching into the underlying material due to different etch rates of oxidized and non-oxidized areas; the subsequent growth could further amplify the resulting protruding shapes, due to a different growth rate on the different crystal planes of the protrusions. This explanation is consistent with the fact that the ammonium sulfide passivation treatment removed the morphology issue.

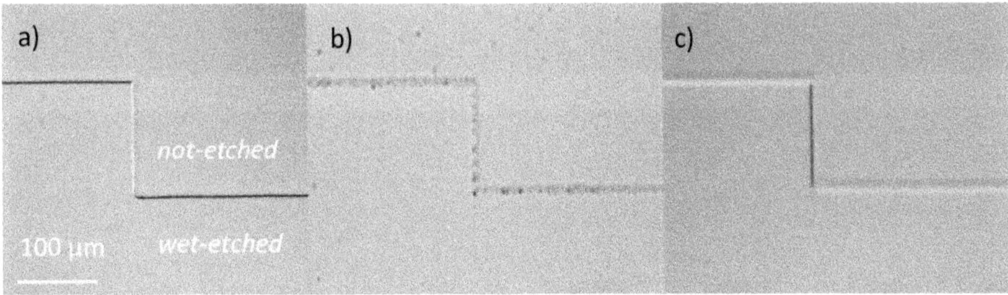

Figure 5.5 OM-DIC images showing the morphology of partially wet-etched surface after the second epitaxy: **a)** regrown directly, **b)** adding CBr_4 in-situ etch, **c)** adding $(NH_4)_2S$ ex-situ passivation before the in-situ etch.

The pattern definition was done with E-beam lithography using an SB251 tool from Vistec, ma-N 2401 negative resist and MF 26-A metal-ion-free developer. The first wet etching was done using an organic acid solution containing an oxidizing agent, which selectively etches GaAs, GaInAs and GaAsP over GaInP, and then the resist was removed with N-Methyl-2 Pyrrolidone (NMP) stripper. A second acid solution was then used to selectively etch GaInP. The etchings are surface-rate limited, and delineate crystal planes at the sides of the grating stripes, which run parallel to $[01\bar{1}]$ (while the device is aligned along [011]).

In the optimization of the conditions for the in-situ etching and the second epitaxy, three factors were taken into account: the necessity to reduce oxygen contamination, the preservation of the grating stripes, and the minimization of defect formation in the regrown material. During the heating phase preceding the growth, there can be outdiffusion of indium from GaInP, and arsenic-phosphorous exchange, with formation of quaternary GaInAsP phases adjacent to the grating stripes [70]. Moreover, when AlGaAs is grown over a patterned surface, its composition is not uniform, but is perturbed in correspondence to the underlying steps, an effect that has been interpreted as due to different growth rate combined with different Ga and Al surface mobility on different crystal planes. This effect is exacerbated at low growth temperatures, and can lead to the formation of stacking faults where two different growth fronts meet [70, 71].

Regrowth test were done on patterned base structures, having etched gratings with two different duty cycles. The temperature of the in-situ etch was kept fixed at T_{sp}=575°C (based on the results on oxygen contamination illustrated in Chapter 4) while for the GaAs growth two temperatures were tested, T_{sp}=605°C and 650°C. The subsequent growth of AlGaAs was done always at T_{sp}=760°C. During the initial heating to reach the in-situ etch temperature, the in-situ etching, and the second heating step to reach the GaAs growth temperature, the samples were kept either under AsH_3 or under a mixed AsH_3/PH_3 atmosphere, with 1/14 ratio. The etch time was regulated based on separate etch rate calibrations on (100) surface, in order to remove in each case 5 nm of GaAs from the areas that were left without patterns. The resulting grating shapes after the regrowth were inspected by SEM on cleaved sections, and are illustrated in Fig. 5.6.

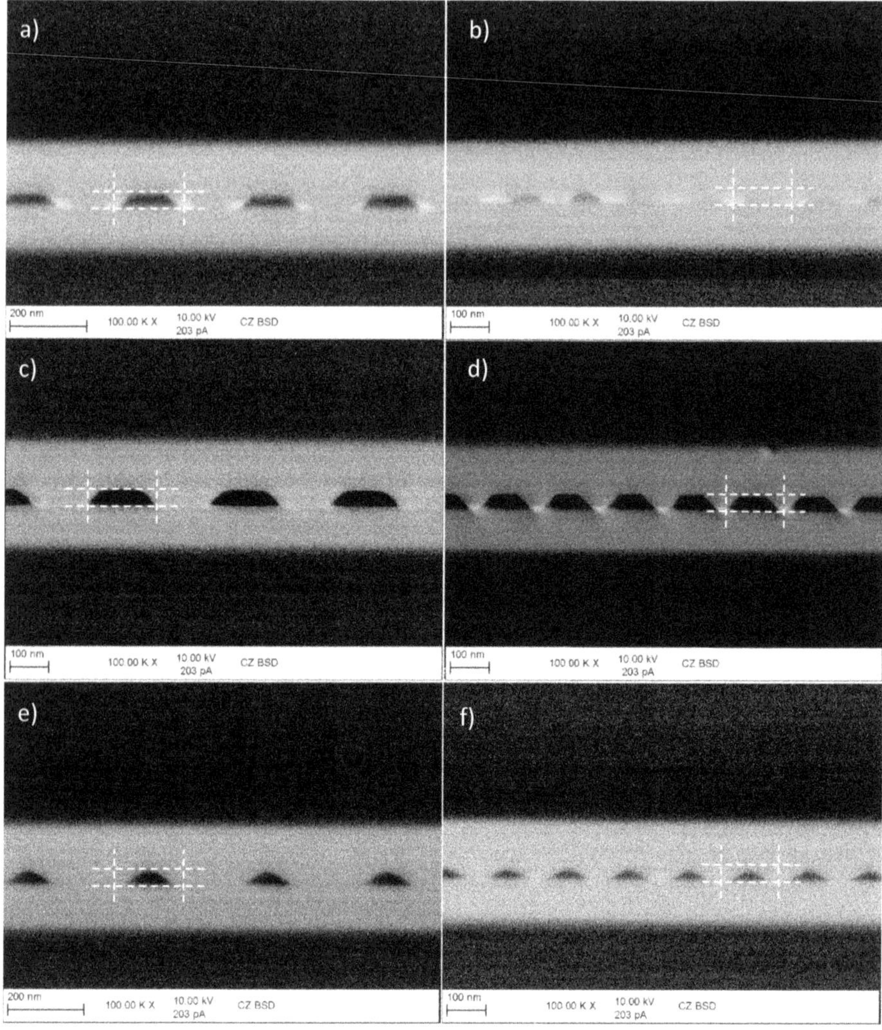

Figure 5.6 SEM backscattered-electrons images of the grating after the second epitaxy, with different duty cycles: **a)**, **b)** etch under AsH_3 atmosphere, regrowth at T_{sp}=650°C, **c)**, **d)** etch under AsH_3 atmosphere, regrowth at T_{sp}=605°C, **e)**, **f)** etch under AsH_3+PH_3 atmosphere, regrowth at 605°C. The heating before the regrowth was done under the same atmosphere as the corresponding etching step. The dashed lines indicate the original thickness of the GaInP grating layer (40 nm) and the approximate width of the stripes after the wet etch.

The in-situ etching under AsH_3 combined with the regrowth at 650°C (Fig. 5.6 a and b) did reasonably preserve the GaInP stripes in the low duty-cycle area (a) but definitely not in the high duty cycle area (b). Even in the first case, the height of the stripes is slightly reduced with respect to the original value, and in both cases a compositional perturbation is visible with white areas associated to high atomic number material (high content in indium and arsenic). When the regrowth temperature was lowered to 605°C, keeping all the other parameters constant, the grating shape preservation was almost perfect, but the compositional perturbation was still present, although less pronounced (Fig. 5.6 c and d). Using a mixed AsH_3/PH_3 atmosphere in all the steps preceding the growth, and using the lower growth temperature as in the previous case, there was a sensible reduction in the size of the GaInP stripes, both in the vertical and in the horizontal directions, leading to slightly rounded triangular shapes, but no signs of compositional perturbations were visible.

By comparison of a) b) with c) d), which had the same in-situ etch, it can be suggested that the main cause for shrinkage of the GaInP stripes is mass transport away from the stripes during the heating from 575°C to 650°C under arsine, which seems to proceed eroding mainly the sides of the stripes. It is possible that the upper GaAs (layer 3 of Fig. 5.5, 20 nm thickness) has been already completely removed by the nominally 5 nm in-situ etch on the narrower stripes, instead of only reducing its thickness to 15 nm; this would imply that the etch rate of GaAs is strongly enhanced when it proceeds laterally on narrow stripes compared to when it proceeds on a continuous (100) surface, possibly because of a layer-by-layer mechanism. This would explain the almost complete disappearance of the grating in b).

The differences between c), d) and e), f) can only be due to the different atmosphere used during the in-situ etching and heating. The reduced dimensions of the GaInP stripes can be explained with an increase in the in-situ lateral etch rate, while the absence of compositional perturbations seems to indicate the suppression of mass transport after the etching. These conditions were selected for the process, based on the fact that the grating was still sufficiently defined to guarantee a high coupling coefficient, avoiding at the same time the creation of phases that, assuming a composition near to InAs, would have been highly strained and introduce low-bandgap "quantum wires" in the mirror sections of the devices. The oxygen contamination obtained in these conditions at the regrowth interface was very low, as already shown in Fig. 5.4.

In Figure 5.6, the transition between the mirror and the gain section is shown.

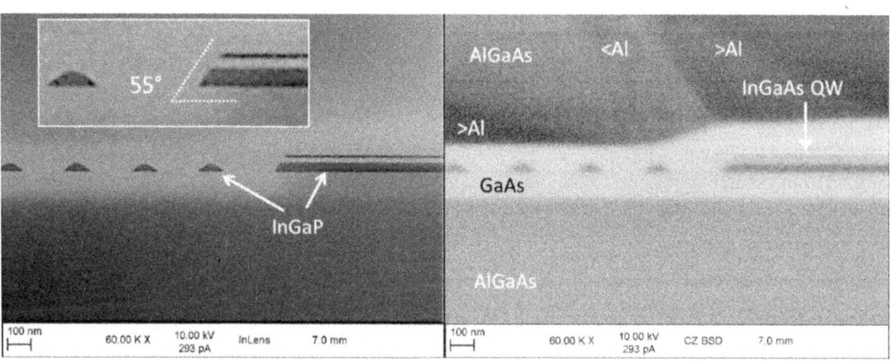

Figure 5.7 SEM images of the transition between the mirror section and the grating section of a device. A compositional perturbation in the regrown AlGaAs is visible in the backscattered electrons image, corresponding to the step in the underlying GaAs.

The inset in the secondary electron image allows appreciating the different lateral angles with which the grating layer terminates. This difference supports the idea that the GaAs layer 3 has been completely removed above the grating stripes during the in-situ etch, and that this has happened with a layer-by-layer mechanism, because the GaInP layers 2 and 4 in the gain section terminate with sharp angles and align well along the same line, corresponding approximately to a $(111)_A$ plane; if the GaAs layer 3 in between had been laterally etched any faster than GaInP, smooth angles and termination misalignment would be present. In the backscattered electrons image, the contrast between darker and lighter areas in the upper $Al_xGa_{1-x}As$ corresponds to a slightly different Al content (Δx evaluated to less than 0.01 based on CL analysis).

5.3.3 Device results

The SQW and the DQW structures had been preliminary tested, fabricating broad area laser (BAL) devices without passive or grating sections, but using a 2-step epitaxy process. The unmounted BALs have been characterized measuring in pulsed mode (1 µs, 5 kHz) their light-current-voltage (LIV) characteristics for different laser lengths. The SQW structure has intrinsically the advantage of a lower transparency current, about ½ that of the DQW, but the slope efficiency of the DQW structure (slope of the optical power vs. current above threshold) resulted to be 10% higher than that of the SQW structure, for the same device length (approximately 0.5 vs. 0.45 W/A). The optical losses were evaluated to be higher for the SQW structure, 1.9 cm^{-1} vs 1.5 cm^{-1}; this difference only partially explains the difference in the slope efficiency.

The single-mode lasers were measured bonded p-side up on AlN carriers, bonded in turn on C-mounts, whose temperature was kept constant at 20°C. When single-mode DBR lasers with uniform gratings (1st and 2nd order), comprising a front gain section, an intermediate mirror section and a backside bended section, were compared, the DQW devices resulted to be ≈35% more efficient than the SQW devices. The cause of the *increased* difference has been tentatively attributed to a higher leakage current in the SQW structure with respect to the DQW structure, when both are processed as single mode (narrow ridge) lasers [57] but the precise origin of the different leakage is not well understood. The tunable SG-DBR lasers did show a similar difference in the slope efficiency between SQW and DQW structures.

The slope efficiency of the DBR structures with 2nd order gratings was only slightly lower (7%) than that of the 1st order gratings, indicating that the different coupling coefficients κ_g (Tab. 5.1) have sufficiently high values to avoid that their differences impact strongly the grating reflectivity.

The vertical far-field determined at FWHM was reasonably in line with the expected values, resulting 41° for the SQW and 24° for the DQW.

All the tested devices proved to be capable of tuning, with mode spacing and tuning range of ≈2.3 nm and ≈21 nm respectively, close to the design values. Since the tuning range of each Vernier mode did exceed the inter-peak distance, a quasi-continuous tuning was achievable. The SMSR was in excess of 40 dB and a FWHM spectral linewidth below 100 pm has been obtained in the full tuning range on DQW devices. The SG-DBR spectra, obtained at different tuning currents on a DQW structure are shown in Fig. 5.8a; an example of LIV characteristics at constant heaters currents, together with the corresponding lasing wavelength variation are shown in Fig. 5.8b. The kinks due to longitudinal mode hopping are associated to small jumps in the wavelength. The series resistance is ≈1Ω and the slope efficiency ≈0.2 W/A.

To summarize: thermally-tuned SG-DBR lasers operating around 975 nm have been realized, using a process based on two-step epitaxy. The device performance, in terms of tuning characteristics, was satisfactorily in agreement with design parameters. The comparatively low efficiency suggests that the vertical structure, the process and the general design can still be improved, reducing the sources of current leakage, non-radiative recombination and optical losses.

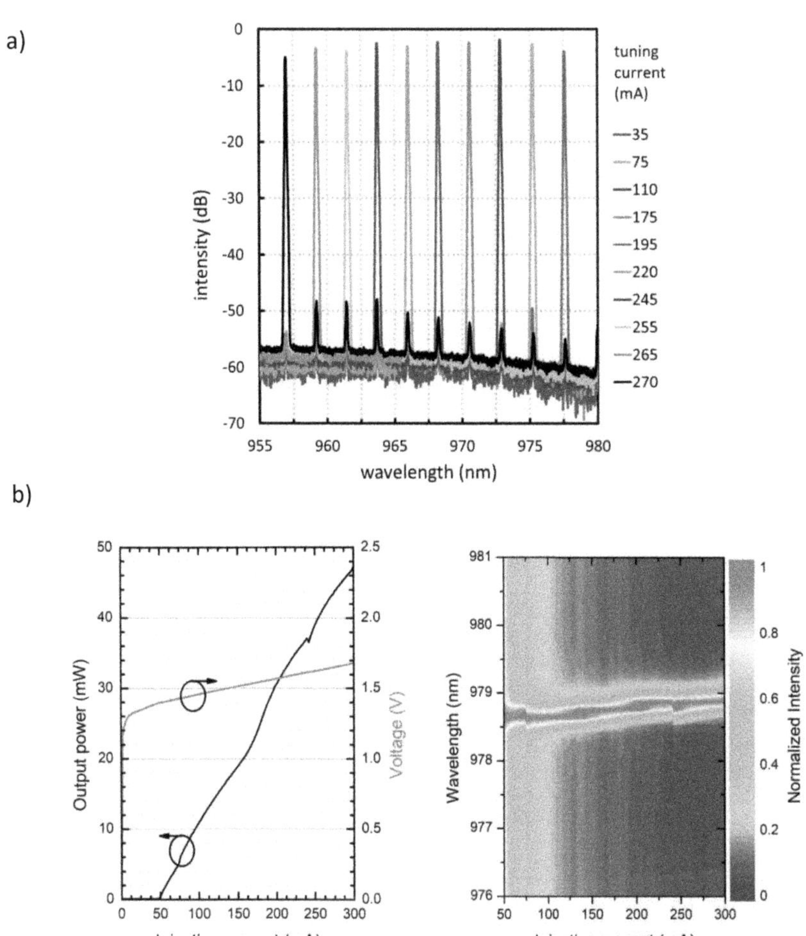

Figure 5.8 **a)** Spectra of SG-DBR lasers with DQW and 2nd order gratings, obtained varying the back-mirror heating current I_b keeping I_f and I_p fixed; CW measure at T(heat sink)=20°C, the obtained SMSR is >40 dB; **b)** example of power-current, voltage-current and wavelength-current characteristics at fixed I_b, I_f, I_p: the kinks indicate longitudinal mode jumps as the injection current I_G increases.

5.4 Investigation of electronic tuning

The refractive index of a semiconductor material is modified by the injection of carriers through several physical effects that impact the material absorption, which in turn is linked to the refractive index through the Kramers-Kronig relation. Three main electronic effects can be identified, which usually are treated as independent [72]: the band filling (or Moss-Burstein shift), the free-carrier absorption and the bandgap shrinkage. The first two lead to a reduction of the refractive index at energies below the bandgap, while the third causes an increase. The first two effects prevail, and the net result of carrier injection is a reduction of the refractive index. At the same time, in a real device the carrier injection causes inevitably some thermal dissipation, leading to a temperature increase, which represents an independent cause of bandgap shrinkage and refractive index increase: in order to use carrier injection to realize a SG-DBR laser as previously described, the thermal effect must be overcome by the electrical one, and a net negative variation of the effective refractive index in the tuning sections must be achieved. Since the Bragg wavelength shifts according to $\Delta\lambda_g/\lambda_g = \Delta n_e/n_e$, and a shift of at least 2.5 nm is desired to control the channel change, a negative variation Δn_e in excess of about 0.01 must be achieved in the mirror sections. Since the mirror sections can be expected to be subject to higher non-radiative recombination with respect to the phase section, because of the GaInP/GaAs interfaces and related surface recombination, higher values of Δn_e in the phase section should probably be obtained [73].

Preliminary simulations based on a theoretical model of the refractive index change with wavelength and carrier density [74] had indicated that the realization of such widely-tunable electronic-tuned SG-DBR lasers *might* have been possible. A series of explorative tests has been done using two-section lasers, based on a vertical structure similar to the one used for the SG-DBRs but with a SQW active emitting at ≈1064 nm and with 25% Al in both the upper and lower cladding. The lasers were fabricated with a 2-step epitaxy process similar to the one previously described: they comprised an active section, including the GaInP layers 2 and 4 of Fig. 5.3, and a passive, tuning section where the active and the two GaInP layers had been removed as in the phase section of the SG-DBRs. The gain section length was 200 μm and the tuning section length 400 μm, the ridge width was 2.2 μm; separate contact windows were opened over the ridge in the two sections.

The devices were mounted p-side up on C-mounts and measured in CW at 20°C (heat-sink temperature), keeping constant the gain current and varying the tuning current I_p. The shift of the cavity modes below threshold was used to evaluate the effect of the tuning current: based on the above considerations, a blue shift, of 4 nm, corresponding to $\Delta n_e \approx$-0.02, would have been considered fully satisfactory for the phase section.

The more significant results are shown in Fig. 5.9. The first attempt was done starting the regrowth on the wet-etched samples without any passivation or in-situ etching: the corresponding tuning did saturate at about -0.3 nm at I_p=50 mA (line 1). When the combination of ammonium sulfide passivation and in-situ etch was introduced, the tuning did reach -0.6 mA at the same current (line 2). This improvement corresponds to the suppression of SRH recombination due to oxygen defects.

A further, minor improvement was to reduce the depth of RIE etching of the ridge by 90 nm, in order to increase the residual thickness of the p-cladding layer from the original 220 nm to 310 nm at the sides of the ridge (line 3); the physical reason for the improvement is in this case not completely clear, but it might be related to the impact of non-radiative recombination at the p-AlGaAs-SiN_x interface.

Using abrupt interfaces between GaAs and AlGaAs (layers 1-2 and 8-9 in Tab. 5.1) did worsen the tunability (line 4), probably indicating an increased carrier trapping and recombination in correspondence to the band spikes forming at abrupt heterojunctions (compare appendix 1, A1.7.3).

The last incremental improvement was obtained increasing the Al content of the p-cladding to 35% (line 5); this is probably due to a reduction of the carrier leakage into the p-side. The tuning reached in this case was -0.8 nm at 50 mA, and saturated at about -0.9 nm, still well below the target -4 nm.

Other tests were done changing the V/III ratio during the epitaxial growth of GaAs, which should influence the formation of bulk stoichiometric defects, but the tests did not lead to any significant variation and are not shown in the figure.

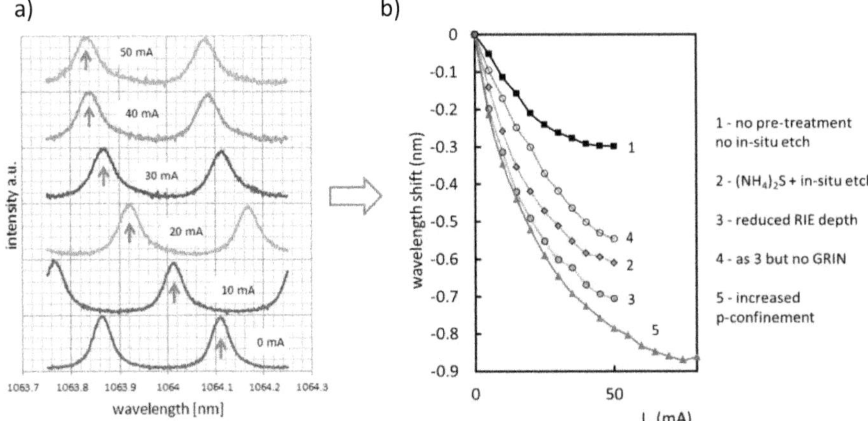

Fig. 5.9 **a)** Example of how the cavity modes shifts varying the tuning current I_p in the two-section lasers; **b)** tuning obtained for the five process/structure variants described in the text.

There are several concomitant effects that can limit the tunability.

- Since no buried electrical confinement structures are present, the current can spread laterally in the residual p-cladding above the GaAs waveguide, and there can be significant ambipolar diffusion inside the waveguide itself; this effect will dilute the carrier density reducing its impact on the effective index.
- There can be vertical leakage, i.e. electrons crossing the waveguide layer and reaching the p-doped cladding (or vice-versa, holes reaching the n-doped cladding) where they would essentially be lost with regards to the impact on the effective index.
- There can be SRH recombination in the bulk and at the interfaces, including the two GaAs/AlGaAs interfaces, the GaAs/GaAs regrowth interface and the AlGaAs/SiN_x interface.
- Auger recombination contribution to non-radiative recombination can be expected to be of minor importance (less than 10%) in the foreseeable operative conditions (carrier density possibly up to $\approx 4\times10^{18}$ cm^{-3}) because of the relative magnitude of the radiative bimolecular coefficient B and the trimolecular Auger coefficient C.
- Finally, the impact of joule heating depends on the device resistance and on the efficiency of thermal dissipation: in this sense, a device that uses a ternary compound as cladding material (as in the present case) will always be at disadvantage with respect to a device that uses a binary (as in the case of InP devices, where normally InP is the cladding material and GaInAsP the waveguide material); the reason is that the thermal conductivity of the ternary is lower, for example it is 0.44 $Wcm^{-1}K^{-1}$ for GaAs and 0.11 $Wcm^{-1}K^{-1}$ for $Al_{0.35}Ga_{0.65}As$ while it is 0.68 $Wcm^{-1}K^{-1}$ for InP [75].

In order to determine whether the tuning was limited by a lower than expected carrier lifetime, the passive sections of the lasers that had given the higher tunability, were separated by cleavage from the active sections, obtaining pin diodes (p-AlGaAs/GaAs/n-AlGaAs), which were mounted on RF transistor packages and electrically tested, evaluating the carrier lifetime with the reverse recovery-time technique [76]. This technique is based on the application of a negative bias pulse to a normally forward-polarized diode under high injection conditions, and on the measured time-decay of the reverse current as the carriers recombine and the junction becomes negatively polarized.

The lifetime values, obtained from 5 diodes with a forward bias current of 20 mA, were distributed in the rather large interval 3-7 ns, with average 5.5 ns. Using the bimolecular radiative recombination coefficient of GaAs B=2×10^{-10} cm^{3}s^{-1} (uncertainty probably ±1×10^{-10} cm^{3}s^{-1}), and considering an injected GaAs volume given by: diode length × GaAs thickness × ridge width × spread correction factor, where the last term takes empirically into account the carrier diffusion to a larger volume, a radiative lifetime of 2.7 ns is obtained assuming no carrier spread and 4.6 ns assuming a spread factor of 3, but these values would almost double using B=1.2×10^{-10} cm^{3}s^{-1} as suggested in [58]. What can be said based on these crude estimates is that the measured lifetimes are *compatible* with a predominant radiative recombination at the used excitation rate. It must be added that the spread in the results points either to a lack of reproducibility in the measurement or to a device-to-device disuniformity.

A critical point in the evaluation of the obtained tunability is that the quantitative relation between carrier density and refractive index change is not well known. A comparison of the predictions of the used model with an experimental evaluation of the injection effect in GaAs based on spontaneous emission measurements [77], showed that the model predicted about 60% more reduction with respect to the experimental values at λ≈1 μm [57].

It is interesting here to consider a fundamental difference that exists between the GaInAsP/InP system and the GaAs/AlGaAs system. In both cases, the negative shift of the refractive index of the waveguide material (GaAs or GaInAsP) is stronger when the wavelength approaches the bandgap wavelength; in the case of GaAs, the value is fixed at 871 nm (at room T) while in the case of GaInAsP it can be changed in a broad interval, nominally 920-1630 nm, maintaining the lattice-match with InP. This allows selecting the GaInAsP composition that represents the best compromise between tuning and optical absorption, but such flexibility is not given in the GaAs/AlGaAs system. Taking for comparison the work of T. G. B. Mason in Ref. [73], where tunable SG-DBR working around 1.56 μm were realized using a quaternary composition corresponding to a bandgap wavelength of 1.42 μm, and using the evaluations of the refractive index change with carrier density of Ref. [77] for GaAs and Ref. [78] for GaInAsP, it emerges that to obtain a comparable shift for the same carrier density of 10^{18} cm^{-3}, the GaAs operative wavelength should be reduced to approximately 900 nm. The situation is schematized in Fig. 5.10, where the variation of GaAs refractive index with photon energy is plotted for two values of the carrier density, and three possible operative wavelengths are indicated.

According to this data, the carrier density necessary to obtain a refractive index variation Δn= -0.01 in GaAs at 1064 nm or 975 nm should be about 2×10^{18} cm^{-3}, and probably above 5×10^{18} cm^{-3} for Δn= -0.02.

A further test was done using SG-DBR lasers based on the 975nm structures [57]; these lasers comprised a gain section (length 1000 nm), a phase section (500 nm), a grating section (500 nm) and a passive section with bended waveguide (1000 nm). The devices were fabricated together with the tunable SG-DBR previously described, as a layout variant on the same wafers. The electrical tunability was tested injecting current through a contact window opened on the ridge in the phase section.

The SQW structure did provide only a negligible (0.03 nm) electronic tuning (blue shift), reached at I_p=35 mA, then the wavelength started to increase, indicating a predominant thermal effect. The DQW structure did provide a modest electronic tuning of -0.3 nm at I_p=100mA. The different behavior of SQW and DQW structures parallels the slope efficiency differences previously noted, pointing to a higher carrier leakage in the first. Moreover, the limited tuning of both structures when compared to the results obtained on the 1064 nm structures suggests that more carriers are lost, which might be explained by an increased SRH recombination due to undetected process or material quality issues.

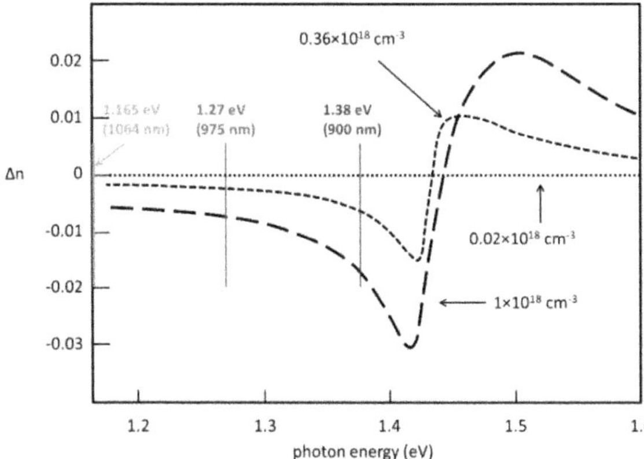

Fig. 5.10 Variation of the refractive index of GaAs with photon energy according to Ref. [77] plotted for 3 values of the carrier density (the values below 1.24 eV are extrapolated); the vertical lines indicate the operative points 975 nm and 900 nm.

To summarize: electrical tuning by means of current injection in the passive section of GaAs-based laser operating at 1064 nm and 975 nm has been demonstrated, but the obtained wavelength shift is too small for tunable-laser applications. The variability of the obtained results points to the necessity of further improvements in the growth and/or process conditions. It is foreseeable that minimization of carrier vertical and lateral leakage, minimization of the sources of non-radiative recombination, and optimization of the thermal dissipation, would allow to increase the tunability to the point of making the realization of SG-DBR widely-tunable lasers with electronic tuning possible, but their wavelength range might be limited to a narrow interval below the GaAs bandgap by fundamental material properties.

6 Buried-mesa broad-area lasers

6.1 Chapter introduction

Broad area lasers emitting near 940 nm are fabricated using a process based on 2-step epitaxy. The n-side of the layer structure and the active layer are grown during the first epitaxial growth step, the p-side during the second. Between the first and the second step an in-situ etching based on CBr_4 is used to remove the active layer from the two sides and at the two facets.

This approach allows the creation of buried-mesa lasers with non-absorbing mirrors and enhanced lateral current confinement; comparison with reference devices, fabricated with the same 2-step epitaxy process but without etching the active layer around the laser cavity indicates a reduced lateral current leakage, lower threshold current and higher efficiency, plus an increased robustness with respect to catastrophic optical damage. In comparison with standard lasers fabricated with single epitaxy, the process introduces a penalty on the efficiency which can be explained by increased non-radiative recombination.

The results of this investigation have been published by P. Della Casa et al. in Ref. [79].

6.2 High-power lasers

GaAs-based high power diode lasers are the most efficient technology for converting electrical power into optical power in the near-infrared range. Depending on the operative conditions – in particular on the power, on the heat-sink temperature and on whether CW or pulsed mode is used - record energy conversion efficiency from 65% to 75% has been reported in the 9xx wavelength range; broad area lasers[1] emitting in this range are especially required as optical pumps for fibre and solid state industrial lasers, which are typically used for cutting, welding and other material processing and whose power can reach tens of kilowatts. The emitted power of a single BAL diode does usually not exceed 10-20 W, but higher output levels are commonly obtained using one or more monolithic arrays of diodes (diode bars) and coupling their emission into a single passive optical fibre, which can in turn be coupled to the active medium [80-83]. The combination of the beams of a large number of single diodes or bars has allowed reaching the kW range in a passive medium (e.g. in air), enabling the realization of "direct diode" lasers for industrial applications, i.e. lasers where the output emission comes directly from the diodes and not from an optically pumped active medium [84].

The design of a laser is inevitably the result of a compromise among different and possibly conflicting requirements. Depending on the specific application, the focus for the realization of new and improved high power BAL is on one or more of the following aspects [81-83, 85-87]:

- energy conversion efficiency
- maximum output power
- reliability
- brightness and beam quality

In the rest of this section, selected basic aspects related to the above list are outlined.

---

[1] The term "broad" refers to the fact that the laser width is much larger than the emission wavelength and of the carrier diffusion length.

6.2.1 Brightness and beam quality

The term brightness is used to indicate the amount of optical power emitted per unit area and solid angle[2] (referred to the beam waist and far-field divergence respectively). The concept of beam quality is multi-fold and application-dependent, and is primarily related to the ease with which the laser emission can be coupled into a fibre or combined with that of other laser diodes. The far field of a high power BAL is typically mono-modal and Gaussian-shaped in the vertical direction (called "fast" axis because of the high far-field divergence), while it is multi-modal in the lateral direction (called "slow" axis because of the low far-field divergence); the profiles are qualitatively shown in Fig. 6.1. In the same figure, the relation between the far field angle and the beam waist of a Gaussian beam is also reported: note that the divergence increases as the beam waist decreases, a relationship that qualitatively holds even for non-Gaussian beam shapes.

An important parameter related to the laser beam shape, defined along one axis, is the beam parameter product $BPP = 0.25 \times W_{95\%} \times \Theta_{95\%}$, where $W_{95\%}$ is the measured near field width (beam waist) and $\Theta_{95\%}$ is the far field angle, both at 95% optical power content[3]; it is usually given in mm·mrad.

The minimum value of BPP for a diffraction-limited Gaussian beam is λ/π, and the ratio of the real BPP to this value $M^2=BPP/(\lambda/\pi)$, is called beam quality factor; the fundamental reason for this name is that Gaussian beam properties make them easy to collimate or focus, so a factor $M^2=1$ is considered to represent the "ideal" behaviour and values >1 indicate lower-quality beams. Since the limiting factor for coupling the beams from BALs is usually the lateral (=slow axis) profile which has typically $M^2>>1$, it is common practice in this context to define the beam quality in terms of the *lateral* beam parameter product BPP_{lat} [85].

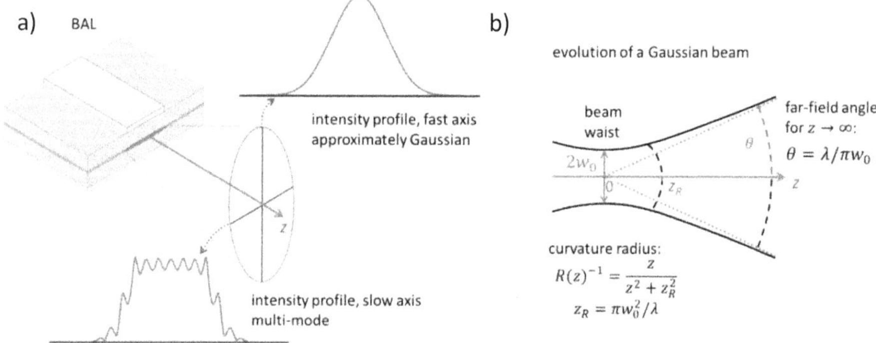

Figure 6.1 **a)** Schematic of BAL emission intensity profiles along the slow and fast axis; **b)** evolution of a Gaussian beam along the propagation direction z: the wavefront curvature is flat in the focus z=0, reaches a maximum at the Rayleigh range z_R and then asymptotically approximates a spherical surface; the far-field angle is inversely proportional to the waist radius w_0.

---

[2] This definition is equivalent to that of radiance, the corresponding units are $Wcm^{-2}sr^{-1}$.

[3] There are alternative conventions to identify W and Θ, including taking the values at $1/e^2$ of peak intensity.

6.2.2 Reliability and maximum output power

The maximum output power at which a laser diode can be operated is limited either by a reversible performance decrease, as a power roll-off, or by the onset of irreversible degradation. Which one of the two conditions is reached first depends on the specific design, since this can deliberately sacrifice the maximum performance in order to avoid the risk of degradation. In either case, reliability can be considered the ultimate limiting factor for high power.

Among the possible degradation mechanisms, catastrophic optical damage (COD) initiated at facets –or mirrors - (COMD) is probably the most critical one [88-90]. COMD is the result of strong recombination via surface traps, which causes a depletion of carriers near the crystal surface; the depleted bands of the active region become more absorbing at the laser wavelength, while the heat generated in the non-radiative recombination process raises the local temperature; the increase in temperature shrinks the bandgap, increasing further the absorption. At a critical optical flux density, the facet temperature becomes sufficiently high (in the order of 120-160°C) to initiate a thermal runaway, which leads to the melting of the facet; the facet damage absorbs the laser light and rapidly propagates inside the cavity. The whole runaway process can be extremely rapid, lasting less than a millisecond. To keep the device operating far from COD/COMD conditions, several design strategies have been developed, which tend to minimize the optical power density, the electrical and thermal resistances, and to maximize the efficiency (to reduce the fraction of electrical power dissipated as heat) [91, 92]. To increase the robustness of the facets, these are subject to treatments aimed to reduce defect density and surface recombination velocity, and to protect them from oxidation. There are several techniques to do this, for example the facets can be obtained by cleavage under vacuum and then passivated with a thin (a few nm) evaporated amorphous Si layer (in the so-called IBM E2 process). If the cleavage is done in air, the facets can be subject to a deoxidation procedure based on ion irradiation under UHV, followed by the passivating deposition. The deposited material can even be a thin epitaxial layer of a high band-gap semiconductor, as ZnSe. After the passivation, the facets are usually further protected with a dielectric coating, which can at the same time absolve the role of optical coating.

It is also common to avoid current injection in the region near the facet, in order to minimize the current-induced facet heating; the region where the current is not injected is sometimes referred to as non-injected mirror (NIM). To reduce the optical absorption (and consequently the heating due to non-radiative recombination) at and near facets, large non-absorbing mirrors (NAM) can be introduced, extending up to tens of microns from the facet inwards [88-90, 93-99].

For quantum well lasers, it is possible to create NAM using quantum well intermixing (QWI) techniques, which produce a local increase in the QW bandgap energy; these techniques can rely on the diffusion of impurities (mainly Zn), on the diffusion of surface vacancies (typically generated using a SiO_2 layer that reacts with Ga) and on ion implantation. A side-effect of the QW bandgap increase is a decrease of the refractive index at the not-intermixed QW photon energy and below.

A more radical alternative to create large NAMs is to remove the active layers and replace them with high-bandgap epitaxial material using a variant of the classical buried mesa structuring process, as proposed for high-power GaAs/AlGaAs/InGaAs lasers by Ungar et al. [99]. The process requires a multi-step epitaxy: the n-side of the structure, the active and part of the p cladding are grown in a first step, then the material near the facets is etched away down to the n-side; a higher band gap material is regrown in the etched regions, protecting the un-etched surface during the regrowth with a dielectric mask. The regrowth material can include a lateral reverse junction or a lateral semi-insulating layer to provide better electrical confinement. After removal of the mask, a third growth step completes the p side. One critical aspect of this strategy can be expected to be the oxygen contamination at the regrowth interface at and near the active layers.

6.2.3 Design aspects

One elementary consideration [91] that explains several characteristics of multi-mode high power lasers design is rooted in reliability constraints: if $\bar{P}_{COD}$ is the optical power density at which a given active *material* is subject to COD, the maximum output power is limited by:

$$P_{max} = \bar{P}_{COD} \cdot W \cdot d/\Gamma \qquad \text{E6.1}$$

where *W* is the active stripe width, *d* is the active thickness and *Γ* the confinement factor, which takes into account the overlap between photons and injected active volume. Consequently, higher power can be sustained reducing the confinement factor; since this implies a reduction of the gain Γg_{th} - where g_{th}=material gain at threshold - it is necessary to increase at the same time the device length *L* to satisfy the threshold condition (gain compensates the optical losses):

$$\Gamma g_{th} = \alpha_i + \alpha_m = \alpha_i + \frac{1}{2L}\ln\frac{1}{\mathcal{R}_1\mathcal{R}_2} \qquad \text{E6.2}$$

where α_i is the internal optical loss, the second term α_m represents the mirror losses and $\mathcal{R}_1$, $\mathcal{R}_2$ are the mirror power reflection coefficients. Increasing *L* is beneficial even to reduce the device electrical and thermal resistance, but to use large values of *L* it is necessary to have low values of α_i. High values of *L* make the differential efficiency η_d (output photons/injected carriers above threshold) very sensitive to the internal optical loss α_i because η_d is given by:

$$\eta_d = \eta_i \cdot \frac{\alpha_m}{\alpha_i + \alpha_m} = \left(\eta_i \cdot \frac{1}{2L}\ln\frac{1}{\mathcal{R}_1\mathcal{R}_2}\right) \Big/ \left(\alpha_i + \frac{1}{2L}\ln\frac{1}{\mathcal{R}_1\mathcal{R}_2}\right) \qquad \text{E6.3}$$

where η_i is the internal differential efficiency (fraction of the current above threshold that generates stimulated emission). To reduce α_i, the vertical structures are usually designed in order to have at the same time a *low* and *asymmetric* vertical optical confinement (LOC-ALOC); the asymmetry consist in having the resonant modes displaced towards the n-side of the junction, were the optical losses are lower (with respect to the p-doped material). For the same purpose, the doping is kept as low as possible – compatibly with the need of keeping the resistance low to reduce the energy efficiency loss due to resistive dissipation, but especially to avoid joule heating of the active, which leads rapidly to lower internal efficiency due to the increase in the number of carriers that escape from the QW and increased non-radiative processes. It can be added that the low optical confinement is beneficial in decreasing the fast-axis far-field divergence. High power lasers with low losses can have *L*=4-6 mm, the length being limited by the internal optical losses or by nonuniformity in the longitudinal optical density, which can lead to spatial hole burning and reduced efficiency [100, 101].

Based on E6.1, even increasing the width *W* and the active thickness *d* can lead to an increase of the maximum COD-limited output power. As in the case of *L*, increasing *W* is of advantage even from the point of view of the electrical/thermal resistance reduction. A typical width for high power lasers is about 100 μm, although devices with width in the mm range have been fabricated. A fundamental limit in indefinitely enlarging the devices is related to the onset of "parasitic" resonant modes in the cavity, increased ASE and modes instabilities; at the same time, the beam quality (BPP_{lat}) worsens with increasing *W*, as more modes become allowed.

Finally, the active thickness *d*, which translates in the number of QWs, can be beneficial not only to reach higher COD-limited power but even to increase the differential efficiency, because associated to a reduced threshold carrier density in each quantum well, which in turn ensures lower carrier escape [101]; the main disadvantage is a higher threshold current – which scales approximately as the number of wells. Moreover, the number of QWs impacts the vertical optical confinement, and – since the wells are normally strained – poses increasing problems of material relaxation with increasing wells number, so the optimal number of QWs must be determined by overall design and technological feasibility considerations.

6.2.4 Strategies to introduce lateral confinement

In BALs the lateral optical and electrical confinement is, in the simplest approach, defined only by the p-side electrical contact, whose shape is normally a stripe. In this case, the lateral optical confinement is due to gain-guiding, and the lateral beam is strongly sensitive to variations in the carrier density and in the temperature (and consequently in the refractive index) of the active region. With increasing injection of current, there is an increasing lateral temperature gradient, from the centre to the edge of the chip, which causes the so-called thermal lensing effect, thus broadening the lateral far field [86, 102]. To mitigate this effect, a refractive index step can be introduced at the sides of the injection stripe, for example via shallow trenches etched in the p-side of the semiconductor structure; this particular technique has the drawback of worsening the degree of polarization of the emitted light (ratio transverse electric/transverse magnetic - TE/TM). In general the introduction of a built-in lateral optical confinement stabilizes the far field, but at the same time increases its divergence already at low current, so a compromise must be found.

The lack of buried electrical confinement structures to prevent lateral carrier spreading causes an increase in threshold current density and reduces the efficiency [103]. Another related aspect is that the lateral current spreading and the carrier accumulation in the active layer at the sides of the injection stripe contribute significantly to determine the lateral broadening in the far field with increasing current, because of the carrier-induced refractive index reduction and because of gain increase at the edges, which favours high order lateral modes [104-106].

Removal of the highly-doped contact layer from the sides of the laser stripe can significantly reduce the lateral current spread; a similar effect can be obtained by damaging the contact layer using ion implantation: this makes the semiconductor material insulating. In both cases, the carriers can still diffuse laterally in the cladding, in the waveguide and in the active layers.

An implantation at the stripe edges, deep enough to reach the active layers, can suppress the lateral carrier diffusion and accumulation, but leads to a reduction in the efficiency, probably because the carriers recombine in the (defective) implanted region [105]. Moreover, the damage introduced in the active region might cause reliability problems. A modification of this deep implantation approach will be presented in chapter 7.

The already mentioned 3-step epitaxy buried-mesa approach of Ungar et al. can allow the simultaneous realization of lateral current-blocking layers, NAMs at facets and built-in lateral index guiding.

Another approach to create lateral electrical insulation, based on a 2-step epitaxy, is the so-called self-aligned structure (SAS): first the n-side, the active, a first p-cladding and an n-blocking layer are grown; then an injection window is opened in the n-blocking layer, and finally a second growth completes the p-side [87]. This approach does not extend the confinement down to the active zone, but has proved to be quite successful in the realization of very efficient BALs[4].

6.3 Structure and process

The simple two-step epitaxy strategy for the realization of InGaAs/AlGaAs BAL diodes used in this work allows the simultaneous introduction of the following features:

- non-absorbing mirrors
- lateral index guiding
- lateral electrical confinement

---

[4] At the time of writing a variant of the approach is investigated in FBH; it is not included in the present work, but preliminary results have been published in Ref. [107] (the writer is co-author).

The strategy consists in growing first the n-side, the active region and a very thin p-layer on top. The material is then etched away near the facets and at the sides of the laser stripe: the etching is stopped a few nm below the active layer, creating in this way a very shallow mesa. A second epitaxy step completes the p-side. With respect to the previously mentioned 3-step epitaxy strategies, here no reverse-junction or other blocking layer is introduced in the second epitaxy, and no mask protection is used during the regrowth.

6.3.1 The vertical structure

The laser devices were based on the epitaxial structure schematically shown in Fig. 6.2. In down-top (growth) order, it comprises:

- the n-doped $Al_xGa_{1-x}As$ cladding layers numbered 1 and 2, where layer 1 has higher Al content than 2 (and consequently lower bandgap and lower refractive index);
- the three $Al_xGa_{1-x}As$ waveguide layers numbered 3, 6, 8, having a further lowered Al content; these layers are interleaved with the etch-stop layers 4 (GaAs) and 5 (GaInP) and with the single GaInAs QW 7; except for the QW, these layers are all (slightly) n- or p-doped as indicated;
- the p-doped $Al_xGa_{1-x}As$ cladding layers 9 (same Al content as layer 2) and 10 (very high Al content);
- the p-doped GaAs subcontact layer 11, and the very highly p-doped GaAs contact layer 12.

The doping level in the etch-stop layers is comparatively high, $\approx 1\times10^{18}$ cm^{-3}, while the doping level in the waveguide AlGaAs layers is an order of magnitude lower. The GaInAs well has an In content of about 13% and a compressive strain of 9000 ppm, while the In content in the GaInP layer is about 48% with a strain < 500 ppm.

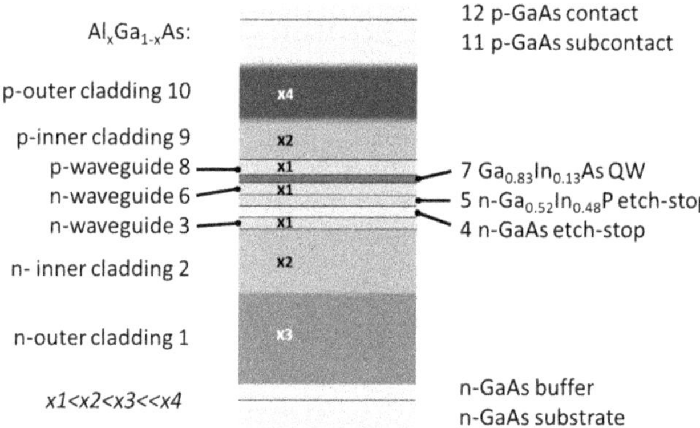

Figure 6.2 Laser epitaxial structure VS2: different shades of colour indicate different compositions, black horizontal lines between layers indicate abrupt interfaces between different compositions or different doping levels, otherwise the transition is graded.

This design, to which we will refer as "VS2" (**V**ertical **S**tructure with **2** stop etch layers) corresponds to an asymmetric large optical confinement approach, the optical fundamental mode is broadened along the vertical direction and strongly shifted towards the n side (Fig. 6.3): the predicted fundamental mode confinement factor Γ and internal absorption loss α_i, calculated with the in-house software QIP2 [108], are $\Gamma = 0.60\%$ and $\alpha_i = 0.6\ cm^{-1}$.

If the etch-stop layers are omitted, without any other change to the structure, the modal profile does not change significantly. We will refer to this simpler structure as "VS0" where the 0 is a reminder of the absence of etch-stop layers 4 and 5. Actually, structure VS0 was the starting point in developing structure VS2, and does constitute a natural benchmark for evaluating the effect of the introduction of the two etch-stop layers. Compared to VS2, the predicted optical losses are 10% lower, while Γ and the effective refractive index are almost unperturbed. The latter point is due to a mutual compensation of the refractive index values of the additional layers.

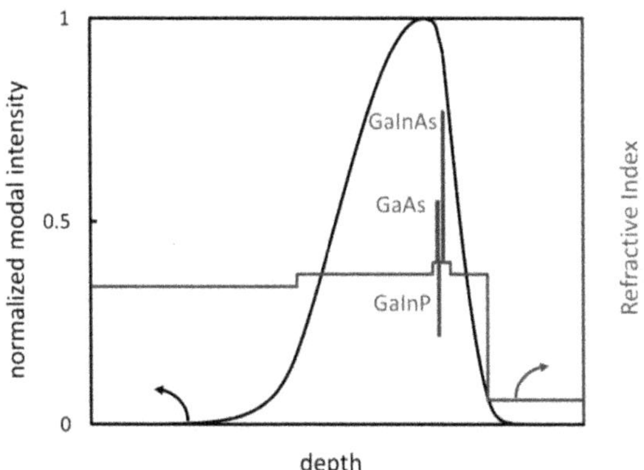

Figure 6.3 Simulated modal intensity in the vertical direction and material refractive index profile.

In a preliminary phase, both VS0 and VS2 structures have been grown in a *single* step epitaxy: then standard BALs have been fabricated and measured, in order to assess a possible detrimental impact of the etch-stop layers. Actually, a slight worsening of the laser characteristics was noticed. The results of this comparison are presented and discussed in section 6.4.

In the buried mesa (BM) lasers, the layers 4-8 are removed from the sides of the active stripe. Furthermore, in order to realize the NAM, the same layers are removed at front and back facets of the lasers. According to simulation, the change in the effective refractive index between the active and the passive sections is 3×10^{-3}, which leads to a negligible longitudinal reflection at the active-passive interface. On the other hand, it provides a non-negligible lateral optical confinement.

6.3.2 The process with two-step epitaxy

The semiconductor material was grown epitaxially in the planetary MOVPE reactor G3. The carrier gas was H_2, precursors where AsH_3, PH_3, TMIn, TMGa, TMAl. For p-type doping CBr_4 and DMZn were used, while Si_2H_6 was used for n-type doping. Substrates were 3" (100) n-GaAs wafers.

With the exception of the preliminary comparison between VS0 and VS2 mentioned in the previous section, the VS2 epitaxial structure was realized in two separate runs. The process steps comprised from the first epitaxy to the second are depicted in Fig. 6.4.

The first epitaxy included all the n-side, the quantum well, 20 nm of p-AlGaAs (*first part* of the layer 8 in Fig. 6.3) and finally a 10 nm GaAs cap layer to protect the underlying AlGaAs from oxidation.

After the first growth, a pattern was transferred onto wafers with optical lithography and subsequent selective wet-etching, in order to remove the active layer from certain parts of the surface, creating the passive sections of the devices. Then the rest of the epitaxial structure was grown over the whole surface. This procedure allowed for the realization of three different kinds of devices on the same wafer:

- standard laser diodes (VS2-STD) without any etched parts;
- lasers with non-absorbing mirrors (VS2-NAM), having the active region etched away only from the front and the back of the laser cavity;
- buried-mesa lasers (VS2-BM) with non-absorbing mirrors, having the active region etched from the front, the back and the sides of the laser cavity.

The laser stripes were aligned along the [011] direction. The etching procedure was the same used for the tunable lasers in Chapter 5: a first selective wet etchant was used to attack GaAs, AlGaAs, and GaInAs, stopping on the GaInP layer, then a second to selectively remove the GaInP. This left most of the surface covered with 10 nm of GaAs, the cap layer on the active section and the GaAs layer 4 on the passive sections (Fig. 6.4b). The vertical step between etched and not-etched areas (layers 5-8) remained unprotected. Immediately after the selective wet etching, the wafer surface was passivated with an ammonium sulfide treatment, after which the wafers were loaded in the MOVPE reactor within a few minutes.

The wafers where heated in the reactor under a mixed AsH_3/PH_3 atmosphere, and then the surface was etched in-situ using CBr_4, in order to reduce the residual oxygen contamination and to remove the GaAs upper layers. The wafer-temperature used for the etching was T_w=613°C; at these conditions, the GaAs is etched much faster than AlGaAs, and this allowed to use an overetch time in order to ensure the complete removal of GaAs without significantly attacking the AlGaAs layers. A drawback is that this etching temperature is probably not optimal for the complete removal of oxygen (a lower one would be preferable); moreover, the exposed sides of waveguide-AlGaAs below (6) and above (8) the QW, did presumably remain partially oxidized; this was estimated to be acceptable in consideration of their low Al content.

After the in-situ etch, having exposed the upper surface of the AlGaAs layers 3 and 8, the temperature was increased to T_{sp}=760°C for the growth of the subsequent layers, the first being again waveguide-AlGaAs (Fig. 6.4c).

After the second epitaxy, the p-contact stripes were defined using He^+ implantation [109] and metallized. The metal stripes were aligned with the pattern previously etched into the active region: in the case of BM devices, both length and width of the metal stripes did correspond to those of the underlying etch-defined active stripe, so that the buried active region remained exactly under the contact metal while the surrounding passive area was implanted. In this way, the width of the *active* region coincides with that of the *p-contact* stripe. Similarly, the non-absorbing mirror regions (present in NAM and BM devices) are at the same time without active layer *and* p-contact.

In STD devices the active layer is left in place everywhere, but there is still the non-injected region at the two facets of the devices. NAM and STD lasers have a cavity width defined only by the contact stripe. Fig. 6.5 schematizes the three cases.

Finally the wafers were thinned to 130 μm thickness and metallized on the back side with a sputtered multilayer of Ni/AuGe.

Facets were fabricated by cleavage. Part of the devices were measured without any coating, part were coated HR(98%)/AR(1.5%) immediately after cleavage, with no special cleaning or passivation treatment on the cleaved surface.

The overall length L of the devices spanned from 1 to 6 mm, the width of the p-contact stripes W from 10 to 100 μm, and the length L_{NAM} of NAM regions was 50 μm, identical on front and back facets. The length L_{ni} of the non-injected (implanted) regions at the facets was 50 μm too.

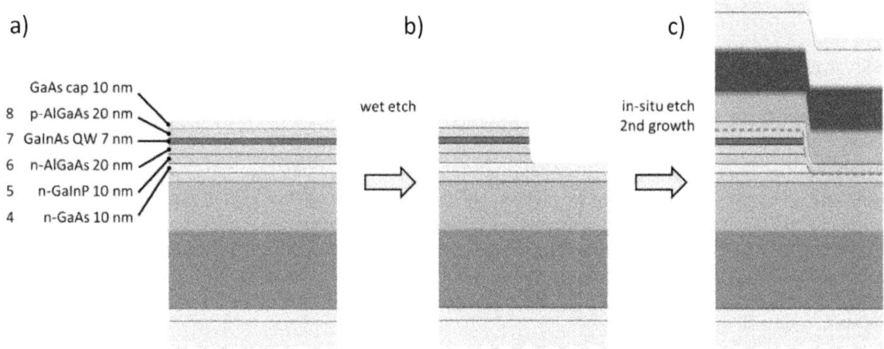

Figure 6.4 Structure VS2 after: **a)** the first epitaxy; **b)** after the wet etch; **c)** after in-situ etch and the second epitaxy. The dashed line represents the position of the regrowth interface.

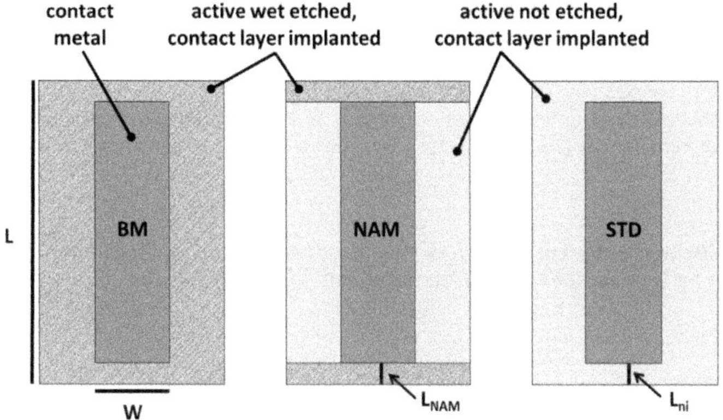

Figure 6.5 Scheme of BM, NAM and STD broad area devices.

In section 6.3.1 it has been mentioned that BAL were realized with structures VS0 and VS2 grown in a *single* epitaxial step, for the evaluation of the impact of the two etch-stop layers: these lasers will be referred to as VS2-STD0 and VS0-STD0. The process of STD0 devices was similar to the one just described and, as in STD devices, there was no wet etch of the semiconductor material. The results on these devices will be presented first, in section 6.4.1, followed by the description of the results obtained on two-step epi VS2 in the following sections, focusing on the material quality in section 6.4.2 and on device characterization in sections 6.4.3 to 6.4.5.

6.4 Results and discussion

6.4.1 Comparison of VS0 and VS2 vertical structures

The electro-optical performance of the VS2 structure was first evaluated, and compared to that of the VS0 structure, using standard broad area lasers. The facets were left as-cleaved and the devices were measured unmounted. Spectra, vertical far field angle and optical power-current-voltage (PIV) characteristics were measured in pulsed mode (1 µs, 5 kHz) for different laser lengths and widths (L from 1 to 6 mm, W=100 and 200 µm).

The emission wavelength was ~936 nm for both structures, VS0 and VS2, with a slight length-dependent variation. The vertical far field angle at FWHM was 25° in both cases. Threshold currents were slightly higher and the slope efficiency η (slope of optical power vs current) was slightly lower for VS2 compared to VS0. The internal differential quantum efficiency η_i and the internal optical losses α_i were extracted from the intercept and the slope of a linear fit of the reciprocal of external differential quantum efficiency η_d (related to the slope efficiency by $\eta_d = \eta \cdot e/h\nu$) vs L [58], using the following relation (equivalent to E6.3):

$$\frac{1}{\eta_d(L)} = \frac{\alpha_i}{\eta_i} \cdot \frac{2L}{\ln\left(\frac{1}{\mathcal{R}_1 \cdot \mathcal{R}_2}\right)} + \frac{1}{\eta_i} \qquad \text{E6.4}$$

Since the linear fit assumes that the internal efficiency is independent of L, and this hypothesis becomes weak at low L, shorter devices must be excluded from the fit.

Further characteristic parameters are obtained using the following approximate logarithmic form of the gain per unit length:

$$g_{th} = g_0 \cdot \ln\left(\frac{j_{th}}{j_{tr}}\right) \qquad \text{E6.5}$$

where g_{th} is the gain per unit length at threshold, j_{th} is the threshold current density, j_{tr} is the transparency current density (transition from absorption to gain when α_i=0). Combining E6.5 and E6.2:

$$\Gamma g_0 \cdot \ln\left(\frac{j_{th}}{j_{tr}}\right) = \alpha_i + \alpha_m \qquad \text{E6.6}$$

Defining a transparency current density for the (lossy) waveguide j_∞, as the limit of the threshold current density for $L \rightarrow \infty$, Eq. E6.6 gives in the same limit:

$$\Gamma g_0 \cdot \ln\left(\frac{j_\infty}{j_{tr}}\right) = \alpha_i \qquad \text{E6.7}$$

Inserting this expression of α_i in E6.6 one obtains:

$$\ln(j_{th}) = \ln(j_\infty) + \frac{\alpha_m}{\Gamma g_0} = \ln(j_\infty) + \frac{1}{2}\ln\left(\frac{1}{\mathcal{R}_1 \cdot \mathcal{R}_2}\right) \cdot \frac{1}{\Gamma g_0} \cdot \frac{1}{L} \qquad \text{E6.8}$$

where j_{th} and j_∞ are made dimensionless dividing them by the unit current. Using Eq. E6.8 and the facet reflectivity values derived from simulations, it is possible to extract the product Γg_0 and j_∞ from the linear fit of the logarithm of threshold current density j_{th} vs $1/L$.

Having obtained α_i and η_i from the differential efficiency fit (E6.4) and Γg_0, j_∞ from the threshold current density fit (E6.8), the transparency current j_{tr} can be calculated from E6.7, but being this parameter afflicted by a higher uncertainty with respect to j_∞, it will not be used in the rest of this chapter.

The results are summarized in Tab. 6.1. The indicated uncertainties are based only on the repeatability of the measurements on different – nominally identical – devices. The comparison of the two structures shows that, with the introduction of the etch-stop layers, there is a small (≈3%) but significant increase of j_∞ and a similarly small (≈4%) but significant decrease of the internal efficiency. The difference in j_∞

could be explained by the predicted higher optical losses in structure VS2: at the same time these are not confirmed by the experiment, actually VS2 losses seem lower than those of VS0. Given the high uncertainty on α_i it is impossible to draw a definite conclusion on this point.

Comparing the two different widths, it is interesting to notice that there is a 15% increase of j_∞ for *both* structures in passing from 200 to 100 μm, which indicates a strong lateral current leakage.

device type	α_i (cm^{-1})	η_i (%)	Γg_0 (cm^{-1})	j_∞ (A/cm^2)
VS0-STD0 *W*=200μm	0.66±0.06	90±2	10.4±0.3	91±1
VS2-STD0 *W*=200μm	0.61±0.06	86±2	10.6±0.3	94±1
VS0-STD0 *W*=100μm	0.67±0.06	89±2	10.8±0.3	105±1
VS2-STD0 *W*=100μm	0.62±0.06	85±2	11.1±0.3	108±1

Table 6.1: Electro-optical parameter of *standard* BALs based on structures VS0 and VS2. In both cases the material was grown in a *single* epitaxy step. The emission wavelength (936 nm) and the vertical far field angle at FWHM (25°) were almost identical for both structures.

In order to explain the measured efficiency differences between the structures VS0 and VS2, the possible existence of an extra current leakage in VS2 caused by recombination of carriers in the GaAs etch-stop layer (layer 4) has been theoretically investigated; this layer could behave as a secondary well (see the band diagram in Fig. 6.6) but the result of a simulation done using the software TeSCA [110] indicates that the GaInP etch-stop layer (layer 5) *should* effectively prevent the diffusion of holes from the p side of the junction into the GaAs layer.

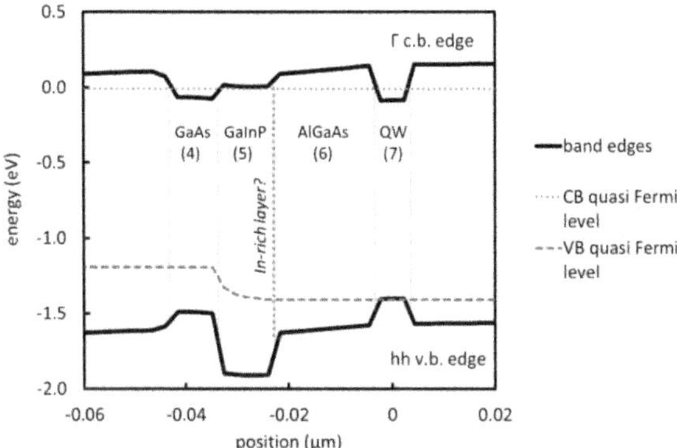

Figure 6.6. Band edges and quasi Fermi levels of structure VS2, simulated for a device with L=1 mm, W=100 μm, as-cleaved facets, at a current of 1A. In correspondence to the etch-stop layers, the quasi Fermi level of the holes lies well over the band edge, which implies that the hole density is negligible. The vertical dotted line indicates the position of an unwanted In-rich layer, possibly present in the structure as discussed in the text.

Another possible source of carrier loss in the structure VS2 could be a recombination of carriers at the staggered interface between layers 5 and 6: electrons from n-GaInP could recombine with holes diffused into the adjacent n-AlGaAs. This recombination could be promoted by an In-rich low-bandgap interfacial layer, which is possibly present at this interface [67]: this could create a spike in the band profile, which could act as a secondary well, and would not be screened by the GaInP layer. Moreover, a highly strained interface can be associated to a high surface recombination velocity. Actually, simulation on X-ray diffraction measurements indicates that strained layers are probably present at the GaAs-GaInP-AlGaAs interfaces in VS2: a tensile layer below and a compressive layer above the GaInP (Fig. 6.7a); moreover, photoluminescence measurements at 12 K show peaks around 1.54, 1.58-1.60 eV in VS2 that are not present in the structure VS0 and cannot be explained by the GaAs layer alone (Fig. 6.7b). It is then suggested that an interface-related carrier loss mechanism is active. A better interface optimization remains a possible task for future developments.

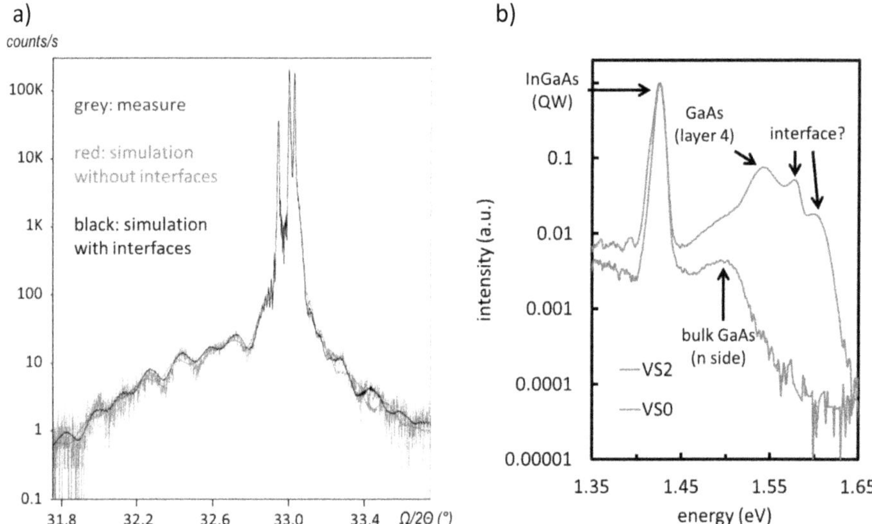

Figure 6.7 **a)** HRXRD diffractogramm of VS2 structure and simulations with (black) and without (red) strained interfaces between AlGaAs and GaInP; **b)** PL spectra of VS0 and VS2 at 12K. VS2 structure has a broad emission between 1.5 and 1.6 eV, apparently composed of 3 peaks. The lowest energy peak can be attributed to GaAs secondary well, while the other could be related to recombination at the GaInP-AlGaAs interface.

A third carrier loss mechanism is the lateral current leakage. As previously noted, this effect is expected to be strong, since the transparency current increases significantly with decreasing W. To analyse this leakage contribution, the simple analytic approach described in Ref. [103] has been used. In this model, the current flow I is ideally divided into two components: one is the transverse current I_T that flows through the active underneath the injection stripe, and one is the lateral current I_L, that spreads or diffuse laterally:

$$I = I_T + I_L \quad \text{E6.9}$$

The first component provides the pumping for the active region (albeit not necessarily with unit efficiency) while the second is entirely lost and represents the *lateral* current leakage. The lateral current leakage can be expected to be dominated by the current spread in the p side, since in the used structures this is comparatively thick and highly doped. A minor contribution should come from the lateral carrier diffusion in the QW.

Provided that the cavity is not too short, the lateral leakage current per cavity length $i_L = I_L/L$ can be considered approximately independent of L. Furthermore, under the condition that $W >> L_D$, where L_D is the lateral diffusion length of the carriers, i_L can be considered independent of W.

Based on these approximations, and considering lasers differing only in the values of W and L, operating at threshold, the following linear relation is introduced:

$$i_{th}(W, L) = j_0(L) \cdot W + i_L \qquad \text{E6.10}$$

where i_{th} is the threshold current per cavity length (i.e. the threshold current I_{th} divided by L) and the length-dependent slope $j_0(L)$ is the threshold current density which would be obtained for a stripe of infinite width. According to E6.10, the lateral leakage current per cavity length is identified with the intercept of the linear interpolation of threshold current per cavity length i_{th} versus W.

Fig. 6.8 shows the threshold currents per cavity length of devices VS0-STD0 and VS2-STD0 plotted vs W, for different values of L.

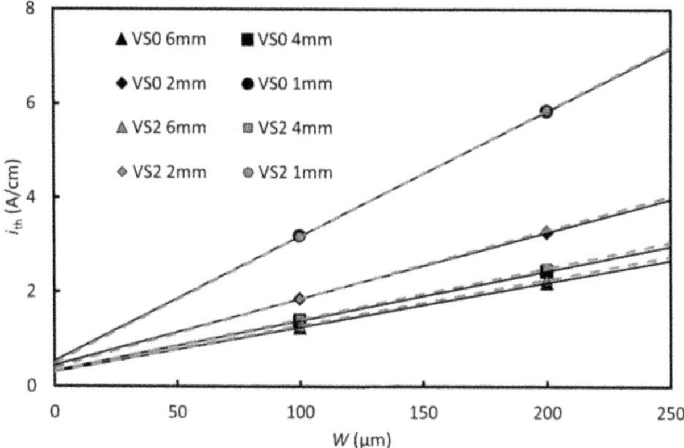

Figure 6.8 Plot of threshold current per cavity length i_{th} vs laser width W. Straight lines are linear interpolations of points related to lasers with the same value of cavity length L. The intercepts represent the lateral leakage current per cavity length at threshold, and roughly converge in one point.

As expected, the intercepts values are approximately independent of L. Moreover, i_L is almost identical for the two structures. This indicates that the introduction of the SE layers *does not* modify significantly the lateral current leakage at threshold. Nonetheless, it is important to notice that, in the considered range of L (1-6 mm), i_L (≈0.37 A/cm) represents a significant fraction of i_{th}, especially for longer and narrower devices, for example it is 23% of i_{th} for W=100 µm, L=6 mm. The lateral leakage current per cavity length can be compared with the product $j_\infty \times W$ (transparency current per cavity length): the ratio $i_L/(j_\infty \times W)$ is the fraction of the transparency current that is due to current leakage, and is 0.35 for W=100 µm and 0.20 for W=200 µm. This explains nicely the dependence of j_∞ from W shown in Tab. 6.1.

To summarize, the introduction of etch-stop layers into the epitaxial layer stack causes a 4% reduction of the differential efficiency and a 3% increase of the transparency current. Broad area lasers fabricated without lateral electrical confinement suffer from a high level of lateral current leakage, regardless which one of the two vertical structures is employed.

6.4.2 Material characterization of the two-step epitaxy

The structure VS2 was used to fabricate STD, NAM and BM broad area lasers using the two-step epitaxy process. After the second growth, the surface morphology was very smooth, with no measurable roughness, but the buried etched patterns were still clearly recognizable under the optical microscope, using interference contrast (Fig. 6.9).

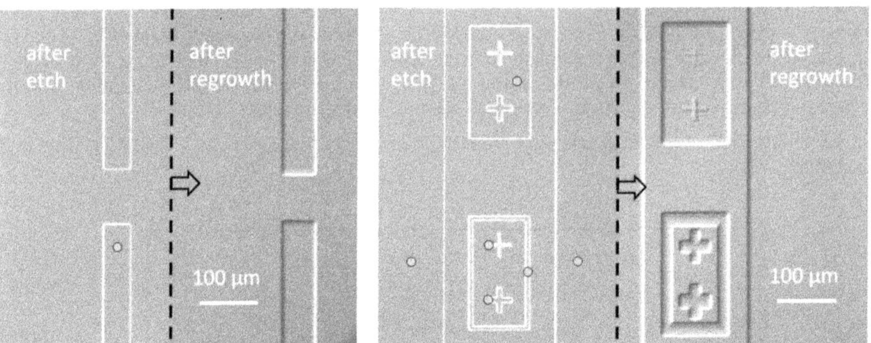

Figure 6.9 Optical microscope image of ridge stripes and markers after wet etch and after regrowth; the dots indicate the areas that have *not* been etched: after the regrowth, the corresponding shapes are larger.

The morphology is very good, with no defects on most of the samples surfaces. A comparison of the dimensions of the ridge stripes before and after the second growth shows that their size has increased by several microns, along both [011] and [0$\bar{1}$1] directions. This is due to the horizontal component of the growth on the etched steps. The samples have been studied in cross-section after regrowth with SEM, cleaving along [011] and [0$\bar{1}$1] directions. The shape of the step is very similar in both cases, and no evidence of defects has been observed at the achievable level of magnification. An example is provided in Fig. 6.10.

SIMS analysis shows that the oxygen concentration in the bulk increases with the Al content, reaching values of about 3×10^{16} cm^{3} in the worst case. At the interface between the GaAs buffer and the first AlGaAs (layer 1), the oxygen concentration has a spike, up to 6×10^{16} cm^{-3} while at the regrowth interface there is a higher peak, reaching 2×10^{17} cm^{-3} (Fig. 6.11). The first peak can be explained by some background contamination in the reactor, gettered by the introduction of aluminium; the second peak, corresponding to the start of the second epitaxy, could be in principle explained in the same way, but – since it is higher - could even indicate that not all the oxidized material has been removed by the in-situ etch. The oxygen value is in line with previous results where such concentrations did not significantly impact the efficiency of the device [42], but in the present case, given the close proximity of the regrowth interface with the quantum well and the p-n junction (20 nm), it is reasonable to expect some negative impact on device performance due to non-radiative carrier recombination.

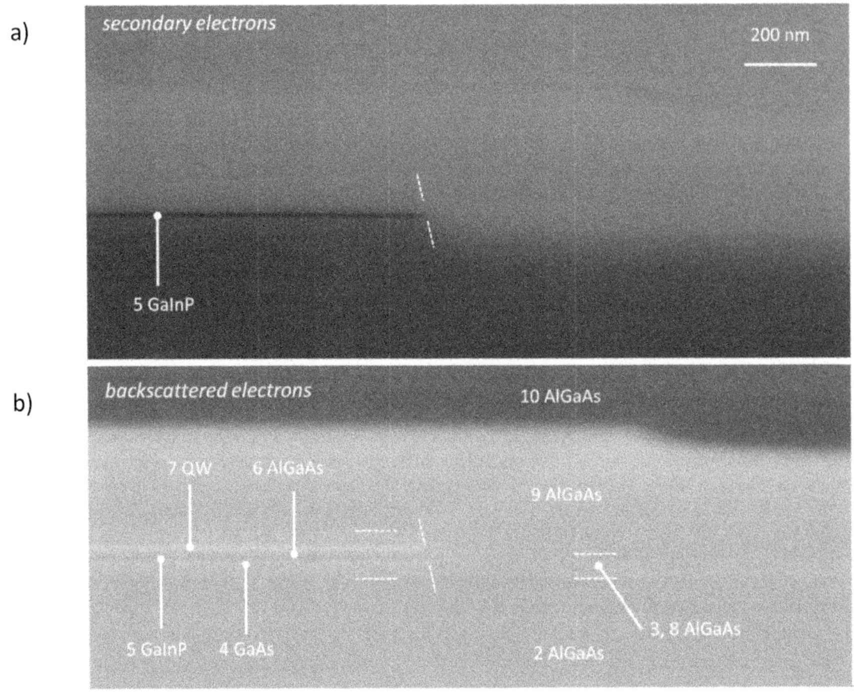

a) secondary electrons 200 nm

5 GaInP

b) backscattered electrons

10 AlGaAs

7 QW 6 AlGaAs

9 AlGaAs

5 GaInP 4 GaAs

3, 8 AlGaAs

2 AlGaAs

Figure 6.10 SEM pictures of etched step (cleavage along $[0\bar{1}1]$) after regrowth: **a)** backscattered electrons, **b)** secondary electrons. The two pictures have been taken on the same spot, the dashed lines are visual helps to locate the limits of the waveguide-layers and of the etched step.

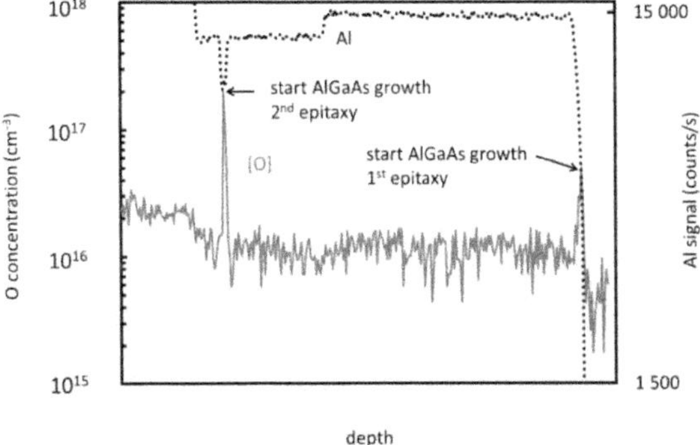

Figure 6.11 SIMS profile of oxygen concentration around the regrowth interface. The substrate is towards the right. The left peak corresponds to the regrowth interface, the right peak to the transition from GaAs to AlGaAs within the first epi run.

6.4.3 Characterization of two-step epitaxy lasers: as-cleaved devices

As in the case of the single-step epitaxy devices, a first group of lasers was characterized leaving the facets as-cleaved and without mounting, acquiring spectra and PIV characteristics in pulsed mode (1 μs, 5 kHz); The following two kinds of BALs will be compared: VS2-BM and VS2-STD. For each variant, there were 4 different values of *L* (1, 2, 4, 6 mm) and 4 different values of *W* (10, 30, 50, 100 μm). Only the STD and BM variants have been included at this level of broad analysis, since the interest in the NAM variant is mainly related to the evaluation of its reliability characteristics in comparison with those of the other two, while the performance is similar to that of STD lasers. Performance and reliability results for all the three variants, as obtained from selected coated and mounted devices, will be presented in the next sections.

The repeatability of the measurements on nominally identical devices is somewhat worse than in the case of the single-step epitaxy process, especially for STD lasers, presumably due to lower process uniformity on wafer. Moreover, due to a cleavage issue, not all of the *W*-*L* pairs for STD devices could be measured. The combination of these factors calls for more caution in the quantitative analysis with respect to the case of the STD0 lasers. Nonetheless, the matrix of experiments is wide enough to provide some interesting information, especially in terms of trends.

The most striking feature emerging from the comparison of PIV curves of BM and STD lasers is that BM devices have much lower threshold currents. The difference is particularly strong for the narrowest cavity variant, having *W*=10 μm: all the BM lasers reach the laser threshold at values of current density j_{th} in the range 200-650 A/cm^2, while STD lasers *do not* reach threshold, at least within the current density limit of 10000 A/cm^2 used in the test. Wider STD lasers (W= 30, 50, 100 μm) reach threshold condition, but at higher current values with respect to BM lasers. Fig. 6.12a shows the values of threshold current[5] I_{th} vs. *L* corresponding to different values of *W*: it can be seen that, for corresponding values of *W*, I_{th} is significantly higher for the STD devices.

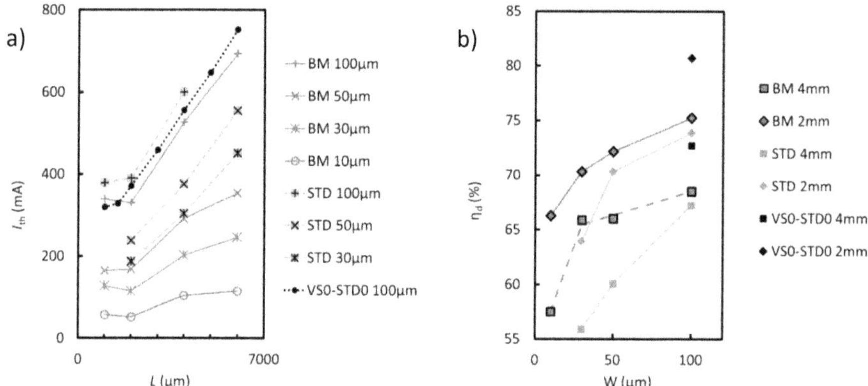

Figure 6.12 **a)** Threshold current of VS2 BM and VS2 STD lasers, realized with two-step epitaxy, plotted vs. cavity length; the values of VS0 STD lasers realized with a single epitaxy are plotted for comparison (only *W*=100 μm). The lines are an aid for the eye, and connect lasers of the same type and width. The facets are as-cleaved. STD lasers with W=10 μm did not reach threshold. **b)** Differential efficiency of the same devices (only *L*=2, 4 mm).

[5] The threshold current I_{th} is plotted here rather than the threshold current density j_{th} for graphical reasons: in the j_{th} vs *L* plot, the points are more difficult to discriminate.

The threshold current values relative to the single-epitaxy standard structure VS0-STD0 with W=100 µm are also plotted: the corresponding VS2-STD values are higher, consistent with the higher transparency current seen on the VS2-STD0 (Tab. 6.1), while the VS2-BM values are lower (except at L=1 µm), indicating that in this case the penalty related to the introduction of the etch-stop layers in VS2 is more than compensated by the increased lateral confinement. Note that defects at the sides of the mesa, in particular residual oxygen contamination, would contribute to the lateral current leakage because of non-radiative recombination at the regrowth interface of the carriers diffusing within the QW (ambipolar lateral diffusion); even this contribution is overcome - in terms of effect on the threshold current - by the extra current confinement.

Concerning the differential efficiency, VS2-STD lasers have lower values with respect to VS2-BM lasers, and the difference increases as W shrinks, as can be seen in Fig. 6.12b, where η_d is plotted vs. W for different values of L. The efficiency of the VS0-STD0 devices with W=100 µm is also plotted: in this case, the VS2 - including the BM - are *all* clearly at a disadvantage, which indicates that the compound penalties related to the introduction of the etch-stop layers and the further defects possibly introduced by the double epitaxy process at the regrowth interface are *not* compensated by the better confinement.

It must be noted that this's conclusion is valid for W=100 µm. It can be reasonably assumed that the behaviour of η_d vs. W should be similar for VS0-STD and VS0-STD0 devices, so it is quite probable that for *narrow* stripes – approximately below 30 µm – the efficiency of VS0-STD0 would become lower than that of VS0-STD (unfortunately no narrower VS0-STD0 devices were available for the comparison).

The differences in the threshold current between BM and STD devices can be further examined using the current leakage analysis already presented in section 6.4.1: the threshold current per cavity length i_{th} is plotted vs W, the intercept is identified as the lateral current leakage per cavity length at threshold, i_L. In Fig. 6.13 this plot is shown for BM and STD devices with L=2, 4 mm. It is evident that BM lasers have the lowest values of threshold leakage currents. The extrapolated leakage current can be subtracted from the threshold current, and a *transverse* threshold current density j_{Tth} (i.e. without the lateral leakage current) can be calculated for each value of j_{th}. The values of j_{Tth} are approximately (within 10%) independent of W and of the type (BM, STD) of laser: this means that the differences in the leakage current explain reasonably the differences in j_{th}.

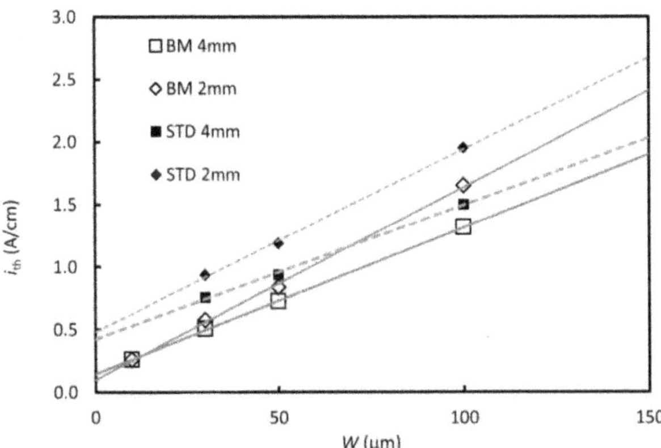

Figure 6.13 Plot of threshold current per cavity length i_{th} vs. laser width W. Straight lines are linear interpolations of points related to lasers with the same value of cavity length L. The intercept represents the lateral leakage current i_L, according to Eq. E6.10. BM lasers have lower values of leakage current with respect to STD lasers.

The lateral electrical confinement effect in BM laser has been replicated by the TeSCA simulations. Its origin can be understood comparing the turn-on potential of a simple p-n diode with that of a p-i-n laser diode (the intrinsic layer being the QW). The presence of the QW shifts the turn-on potential towards a lower voltage, so if the two diodes are connected in parallel the current will flow preferentially through the p-i-n. Moreover, the lateral diffusion of carriers in the QW should be suppressed, since the QW itself does not extend beyond the contact stripe. Fig. 6.14 shows the simulated lateral profile of hole current density for a BM device with W=10 µm, L= 6 mm.

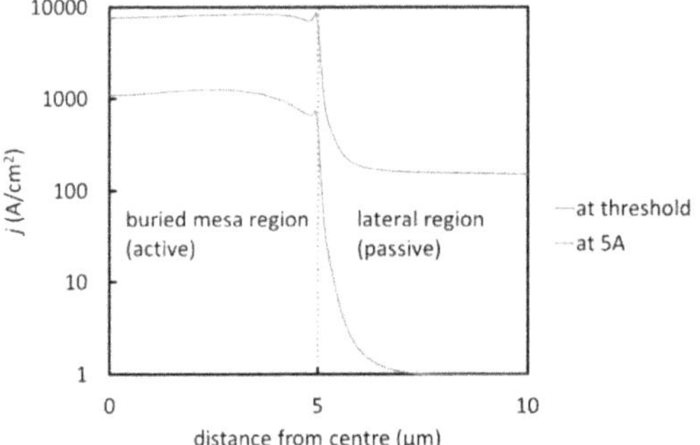

Figure 6.14 Simulated lateral profile of hole current density, slightly above the active layer, for a BM laser with W=10 µm and L=6 mm. The carrier flow is effectively confined within the active buried mesa region both at threshold and at high current (note the logarithmic scale).

Another factor causing the lower threshold currents of the BM devices, in particular of those having a smaller stripe width, is the built-in lateral index guide ($\Delta n_e = 3\times10^{-3}$) as mentioned in section 6.3.1. It is possible that without this lateral optical confinement the BM devices with W=10 µm would not reach the lasing threshold, although the relative importance of suppressing lateral current leakage and improving the lateral waveguiding cannot be directly separated.

6.4.4 Characterization of two-step epitaxy lasers: coated and mounted devices

As mentioned in section 6.3.2, part of the VS2 devices were coated AR (98%)/HR (1.5%) immediately after cleavage, with no special cleaning or passivation treatment on the cleaved surface. Selected devices with L=4 and L=6 mm were mounted p-side up on CuW-submounts. PIV curves, spectra, lateral near field (NF) and far field (FF) profiles, degree of polarization (DOP), were then measured in CW mode at 20°C (heat sink temperature) up to a current density of ~1000 A/cm². DOP is defined here as the ratio of the TE output optical power to the total optical power, and was measured at 500 mA. Experimental average values of threshold current, differential quantum efficiency, DOP, R_s are summarized in Table 6.2.

The results confirm those obtained on uncoated devices: BM lasers have significantly lower (~15%) threshold currents and higher (~5%) differential efficiencies with respect to NAM and STD lasers, while the differences between NAM and STD lasers lie within the experimental uncertainty. The degree of polarization is quite high, even in BM devices, and compares favourably with that of devices where a lateral refractive index step was introduced by means of etched trenches [86]. The STD lasers differential efficiency is ~5% lower than predicted based on the values in table 6.1 (single-step epitaxy structures) confirming the negative impact of the two-step process and indicating a need for further process improvement.

Fig. 6.15a shows the comparison of P_{out} and η_E plotted vs. I in the three cases, for L=4 mm, W=100 µm. The lower threshold and the higher differential efficiency of BM lasers combine to give to these lasers higher energy conversion efficiency.

Fig. 6.15b shows a comparison of the same parameters for different widths of BM lasers. Thanks to the reduced current leakage and improved optical confinement, even the narrower lasers have a high differential efficiency, only slightly worse than that of the larger devices. Conversely, the energy conversion efficiency η_E is higher for the narrower lasers than for the broader, because of the lower threshold current.

device type	n. chips	L (mm)	W (µm)	I_{th} (mA, ±3%)	η_d (%, ±1.5)	DOP (%, ±0.5)	R_s (mΩ, ±3%)
STD	2	6	100	972	68	100	10.4
NAM	2	6	100	953	67	100	10.7
BM	2	6	100	828	71	100	10.7
BM	2	6	50	418	70	98.5	20.5
BM	2	6	30	269	69	98	30.0
BM	1	6	10	106	67	98	90
STD	1	4	100	710	74	-	16.5
NAM	1	4	100	741	72	-	16.5
BM	2	4	100	624	80	-	15.0

Table 6.2 Electro-optical parameters measured on coated devices, mounted p-side up, T= 20°C (heat sink). The indicated uncertainty is the average reproducibility of the measure on nominally identical devices. BM lasers have lower threshold currents and higher differential efficiencies with respect to NAM and STD lasers.

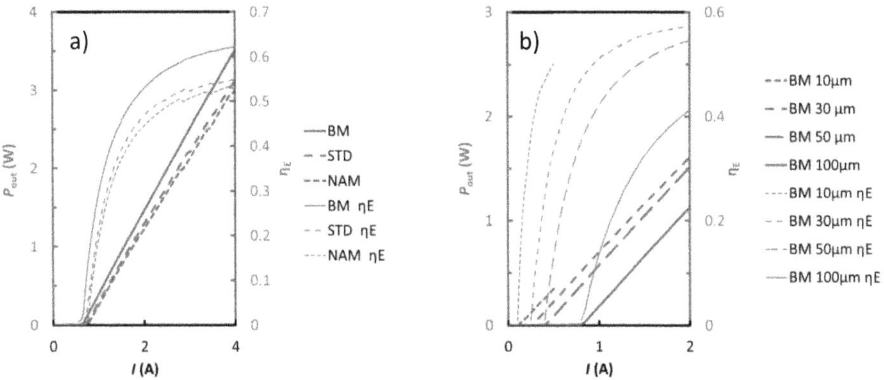

Figure 6.15 Plots of optical power and energy conversion efficiency vs. current, measured in CW mode; **a)** comparison of BM, STD and NAM lasers with L=4 mm and W=100 µm, BM lasers are more efficient both in terms of slope efficiency η and in terms of η_E; **b)** comparison of BM lasers with L=6 mm and W=10-100 µm, the narrower lasers have slightly lower slope efficiency η but higher energy efficiency conversion η_E

Near-field (NF) and far-field (FF) lateral profiles were measured at several fixed P_{out} values. STD and NAM laser behave quite similar, while BM laser have larger NF and FF distributions. In particular, BM lasers near-field is approximately as large as the laser contact stripe even at low currents, while STD and NAM lasers have a near-field width smaller than the contact stripe. Typical examples of NF and FF profiles are given in Fig. 6.16, where BM and STD lasers with the same L=6 mm and W=100 μm are compared at two values of P_{out} (the FF and NF of NAM lasers, not shown, are not significantly different from those of STD lasers).

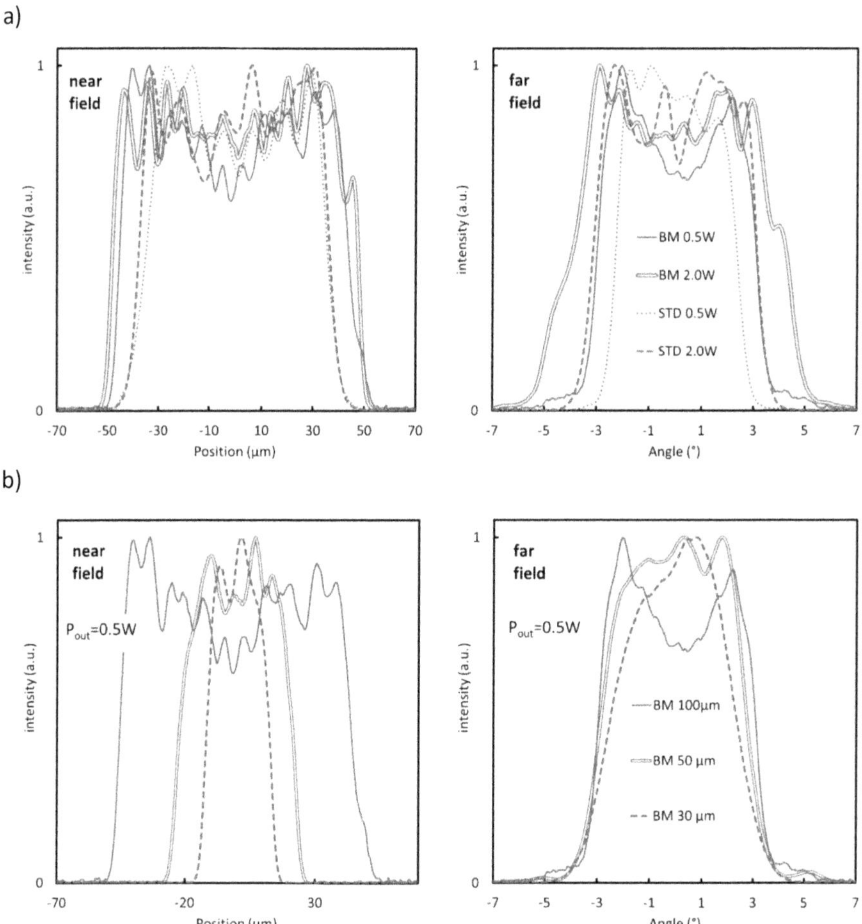

Figure 6.16 Plots of lateral near field and far field optical power, measured in CW mode at P_{out}= 0.5 and 2 W. Data are in arbitrary units and normalized setting the highest value =1. **a)** Lasers with, L=6 mm, W=100 μm: note that BM lasers have the larger values of NF width and FF angle; **b)** BM lasers with L=6 mm and different widths.

The broader values associated with BM lasers can be explained by the lateral built-in index guiding of these devices [86], that allows for the propagation of more high order modes in the laser cavity with respect to a gain-guided BAL, "filling" all the width of the buried active stripe and introducing a high-order mode related divergence in the FF.

Larger NF and FF are associated with a higher lateral beam parameter product and, as remarked in section 6.2.1, to a lower beam quality. In Fig. 6.17, the 95% BPP_{lat} of STD, NAM and BM devices is plotted versus the optical power. Comparing STD and BM devices with W=100 μm, it is not only apparent that BM devices have larger BPP_{lat} but even that they are more sensitive than the STD devices to an increase in P_{out}, at least in the considered power range. Nonetheless, it can be noted that, since BM lasers are significantly more efficient than STD lasers, and (as discuss in the following) more robust with respect to COD, it should be possible to use comparatively narrower BM devices to get the same values of P_{out} and η_E. This would reduce the apparent disadvantage in terms of BPP_{lat}.

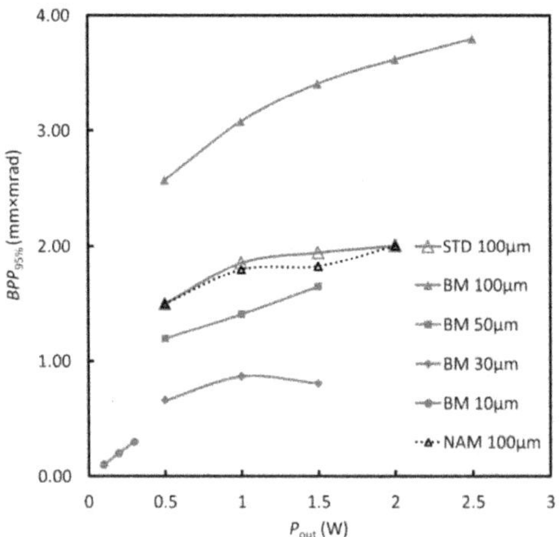

Figure 6.17. BPP_{lat} plotted vs. optical power: comparison of BM with W=10-100 μm and STD, NAM lasers with W=100 μm. L=6 mm for all devices. BM laser have higher values of BPP_{lat} than corresponding STD, NAM lasers, due to larger NF width and higher FF angle.

In published studies, gain guided devices with stripe width around 20 μm have achieved the best reported combination of power and BPP_{lat}, for a lateral brightness (defined as $B_{lat} = P_{out}/BPP_{lat}$) of more than 6 W/mm×mrad [85, 111]. However, energy conversion efficiency is there limited to less than 50%, with losses due to current spreading playing a large role. It is conceivable that BM lasers could offer a route to delivering high peak brightness, especially if the design was modified in order to reduce the lateral refractive index step. This is actually possible by increasing the thickness of the GaInP layer 5, but the modified design has not been tested yet.

6.4.5 Electrical overstress test

After the CW measurement, 9 lasers (BM, NAM and STD) were tested in QCW mode (1 ms pulse, 10 Hz repetition rate) raising the current up to the point where the output power dropped, in order to compare the maximum operation ranges and damage thresholds. After this current sweep, all the tested devices were irreversibly damaged, and did emit only a fraction of the original optical power if operated again. A visual inspection of the facets combined with the observation of the luminescence under current injection showed in 8 cases out of 9 that the active layer was damaged at the surface or near the surface of the front facet, indicating that COMD is presumably the prevalent failure mechanism of these lasers in this electrical overstress test (EOS). Still, in one case (NAM laser, L=6 mm, no facet damage observed), a different mechanism could be responsible for the damage, possibly a form of COBD (Catastrophic Optical Bulk Damage) which might be triggered either by an extended defect or by a high density of point defects acting as non-radiative recombination centers - as discussed for example in Ref. [112].

Interestingly, NAM and BM lasers did behave similarly in term of the optical power at which the damage occurred, having a higher damage threshold than STD lasers. This leads to conclude that the presence of a non-absorbing mirror is effective in enhancing the robustness of the laser, with respect to a simpler non-injected facet. At the same time, there is no evidence of any detrimental effect on the damage threshold due to the residual contamination on the etched facets and sides of the buried mesa. Obviously, this should not be considered a full reliability assessment, but only a preliminary test.

Table 6.3 reports the EOS results: j_f is the current density at which the failure occurred, P_{max} is the maximum optical output power, P_f is the output power at which the device failed and can be lower than P_{max} if a roll-off occurred before the onset of catastrophic damage; finally the parameter $\bar{P}_f$ is the front facet internal optical power density (incident + reflected) at P_{out}=P_f over the region of the QW, calculated according to [113]:

$$P_f = \frac{d}{\Gamma} W \left(\frac{1-\mathcal{R}}{1+\mathcal{R}} \right) \bar{P}_f \qquad \text{E6.11}$$

where d is the QW thickness and $\mathcal{R}$ is the reflectivity at the front facet.

	L (mm)	W (µm)	j_f (A/cm²)	P_{max} (W)	P_f (W)	$\bar{P}_f$ (MW/cm²)	facet damage
STD	6	100	2620	=P_f	10.0	8.8	yes
NAM	6	100	3580	=P_f	11.5	10.2	no
BM	6	100	4200	12.5	12.4	11.0	yes
BM	6	50	4700	=P_f	8.2	14.5	yes
BM	6	30	6390	=P_f	6.6	19.4	yes
BM	6	10	24500	3.4	2.9	25.6	yes
STD	4	100	3270	=P_f	10.1	8.9	yes
NAM	4	100	4430	=P_f	11.8	10.4	yes
BM	4	100	4250	12.0	11.7	10.3	yes

Table 6.3 Results of electrical overstress test: the meaning of the parameters is explained in the text.

Comparing the values of $\bar{P}_f$ it is possible to notice that they depend strongly on W: the narrower lasers have much higher values of $\bar{P}_f$, which could be qualitatively explained by a better heat dissipation at facets and a reduced beam filamentation[6] (and hence a reduction of transient localized spikes of optical power). $\bar{P}_f$ does not depend on L, and increases by ~20% passing from STD to NAM and BM lasers. It must here be noted that $\bar{P}_f$ is calculated using the nominal width W (contact width), but that the effective optical width, as defined by the NF intensity profile, can be different: in the case of BM lasers it

[6] Here, filamentation refers to the breakup of the optical density inside the cavity into several, temporally variable components; the theoretical treatment of filamentation is complex and partially under development, see for example [114].

is *larger* than in the case of STD lasers, at least at the comparatively low current density employed to measure it (Fig. 6.16). It could be speculated that this – and not the presence of the NAMs – is the main cause of the improved robustness, but this is in contrast to the fact that, although the NAM lasers have NF profiles almost superimposable to those of the STD lasers, they have higher COD levels, more near to those of BM lasers (Tab. 6.3). Consequently, it can be concluded that the NAMs *are* the main factor determining the increased robustness, while the NF enlargement is a secondary factor.

6.5 Chapter summary and conclusions

A comparatively simple approach for the realization of buried mesa broad area laser has been tested, based on the introduction of two etch-stop layers in the n side of the structure slightly below the quantum well, selective wet etching, in-situ etch of the lower etch-stop layer and a second epitaxy; ammonium sulphide passivation combined with in-situ etch has been used to minimize the oxygen contamination.

Compared to standard (gain guided) lasers fabricated with the same vertical structure and processing conditions, the fabricated BM lasers show a strongly reduced lateral current leakage, which in turn leads to lower threshold current, higher differential efficiency and higher energy conversion efficiency. The same technological approach allows for the introduction of non-absorbing mirrors at facets, improving the robustness with respect to catastrophic optical damage.

On the negative side, in spite of the better lateral confinement, the efficiency of large BM devices (width = 100 μm) is lower than that of standard devices realized without the additional layers and in a single epitaxial step, most probably because of increased non-radiative recombination, associated to the buried etch-stop layers under the QW and to the regrowth interface. Another drawback is that the buried mesa lasers have higher values of lateral beam parameter product than standard lasers with the same nominal width, due to the strong built-in lateral optical confinement; this issue could be probably solved with a moderate design modification.

The overall conclusion is that this approach, in its current form, is potentially interesting only for the realization of narrow broad area lasers (e.g. with width below 50 μm) where the impact of the lateral current leakage on the efficiency becomes more important and the lateral beam parameter product is anyway small.

7 Lasers with buried implantation

7.1 Chapter introduction

As discussed in the previous chapter, an effect that becomes increasingly detrimental with the reduction of the width *W* in broad-area lasers is the lateral carrier leakage, which impacts first of all the energy conversion efficiency and to some extent the beam quality. The lateral electrical confinement in BALs is realized in the simplest case by creating a stripe-shaped contact window on the top of the device, usually an opening in a dielectric deposited over the highly p-doped contact layer.

To extend the lateral confinement deeper into the structure, the contact layer and the underlying p-side layers can be insulated by ion implantation for example with He or H. This process relies on the creation of defects, substitutional, vacancies and interstitials, which reduce the free-carrier concentration and the carrier mobility. Normally this implantation is relatively shallow, so that these crystal defects do not reach the active zone, where they could act as non-radiative recombination centers, reducing the efficiency and especially the reliability of the devices. Consequently, a certain amount of current spread in the p-side is always present, along with the lateral diffusion current in the active region [115-117].

More sophisticated approaches for the realization of buried lateral current confinement, introducing elements like a reverse junction or semi-insulating layers, involve wet chemical or dry etching and multi-step epitaxy processes on a patterned surface, as in the case of the buried-mesa lasers of the previous chapter, with a resulting higher degree of technological complexity.

The purpose of the present investigation was to explore an alternative 2-step MOVPE growth strategy for the realization of lasers with a buried current aperture, by means of ion implantation outside of the active laser stripe. The implantation has been done between the first and the second growth step, without introducing any etched topology on the wafer surface and without exposing Al-containing material. Two main variants have been tested, differing in the position of the regrowth interface and implantation depth. Moreover, for each variant, two ions have been tested: $^{16}O^{+}$ and $^{28}Si^{+}$ and - for each ion - different implantation conditions.

While in the conventional implant isolation previously mentioned, the induced physical lattice damage must still be present at the end of the process, in this case the *starting* idea was to remove it completely during the second growth step, which can be equated to a long high-temperature annealing process; the isolation should have then be obtained only through "chemical" effects, i.e. the introduction of deep and shallow levels in the band gap due to the substitutional incorporation of the implanted elements in a structurally regular crystal lattice. In the actual realization, the physical damage removal was not achieved on all samples, as will be discussed later.

Potential advantages of this approach are a comparatively simple process and – assuming that the lattice damage is removed - the possibility of safely positioning the current aperture more near to the active region compared to the conventional implantation approach. In addition, there should be no need for a contact window at the p-metal/semiconductor interface or in the contact layer, i.e. the contact can be made much larger than the active width. This would result in lower electrical and thermal resistance for narrow lasers (here narrow = with buried injection window width not much larger than the p-side thickness) because the current would be less laterally constrained within the p-side. Another potential point of interest is that the buried current aperture would allow using a very thick contact layer; this is considered useful in case of devices mounted p-side down, because it can improve the uniformity of heat dissipation and more evenly distribute soldering-induced strain.

Broad area lasers emitting near 915 nm have been fabricated; with one of the variants, up to ≈12% reduction of threshold current and ≈15% increase of slope efficiency have been achieved with respect to standard lasers fabricated without the buried implantation but with the same 2-step epitaxy process. Also, in another variant, a significant improvement of the beam quality has been obtained: in terms of reduction of the lateral beam product parameter BPP_{lat} passing from 3.8 mm×mrad at 5 A for standard lasers to 2.2 mm×mrad for implanted and regrown lasers, but this happened at expense of the efficiency.

Most of the results of this investigation have been published by P. Della Casa et al. in Ref. [118] and by D. Martin et al. in Ref. [119] (here some updates and corrections are included); parts of the work have been supported by the German Federal Ministry of Education and Research contract 13N14005 as part of the EffiLAS/HotLas project.

7.2 Ion implantation

7.2.1 Interactions in the keV range and implantation profiles

When ions are implanted in a crystal using energies in the keV range, they lose their kinetic energy - transferring it to the atoms of the target material - essentially according to two mechanisms, which take the names of "electronic" and "nuclear" collisions (both due to electromagnetic interactions) [120, 121]. The first mechanism is described as inelastic collision of the incoming ion with crystal atoms. The energy is transferred to the electrons of the crystal, especially to the external electrons, promoting them to excited levels, and is ultimately transformed into heat. Normally, this kind of interaction does not lead to a large damage of the crystal structure, because the atoms retain their positions in the lattice. The second mechanism is described as elastic collisions of the incoming ions with the nuclei – or more properly with the core positive ions of the crystal. The energy is transferred to the crystal ions, exciting them vibrationally (phonon generation) or displacing them from their equilibrium position, possibly with cascade effects, forming interstitials and vacancies (Frenkel pairs) while the impinging ions change their trajectory. The displacement is not necessarily permanent: vacancies that lie at a short distance from interstitials tend to recombine with them, and this process is enhanced by increasing lattice temperature during the implantation ("dynamic annealing").

The derivative of the transferred energy with respect to path length is called stopping power, and a nuclear and an electronic stopping power are distinguished. The nuclear stopping power is low at very low energies (or ion velocities) because the nuclear charge is screened by the electrons, it reaches a maximum for intermediate energies and then drops due to the decreasing interaction time with the nuclei. The electronic stopping power is similarly low at low energies, and increases roughly linearly with the square root of the kinetic energy (at least in the keV range). The result is that the electronic interactions prevail at the beginning of the ion trajectory, and the nuclear interactions – with the associated defect formation - at the end. The final distribution of the implanted ions is shifted deeper inside the implanted sample with respect to the damage distribution, since in the very last part of the ion trajectory the nuclear collisions generate mostly phonons. In the case of $^{28}Si^{+}$ implanted in GaAs, the nuclear stopping power is higher than the electronic stopping power at low energies, it reaches its maximum at ≈25 keV, and is overcome by the electronic stopping power above ≈110 keV; in the case of $^{16}O^{+}$ implantation the corresponding values are ≈12 keV and ≈26 keV (values according to TRIM simulation software, Fig. 7.1a).

The ion implantation profile $N(x)$ (ion density at depth x) corresponding to a single implantation energy value, can be approximately described at low energies with a Gaussian distribution of the form:

$$N(x) = \frac{Q}{\Delta R_p\sqrt{2\pi}} exp\left[\frac{-(x-R_p)^2}{2\Delta R_p{}^2}\right] \quad \text{E7.1}$$

where Q is the implanted dose (ions/cm^2), R_p is the average depth (called projected range) and ΔR_p (called straggle) is the standard deviation of the distribution. As the implantation energy is raised, the profile becomes increasingly asymmetric, as qualitatively shown in Fig. 7.1b, and is better described as a skewed Gaussian distribution, the Pearson IV distribution being often used as a convenient approximation (it contains 2 extra parameters, skewness and kurtosis).

Actually, the implantation in a crystal is influenced by the relative directions of the impinging ions and the lattice structure: if an ion enters the crystal with velocity parallel to a "open channel" among the atoms, as for example along the <110> directions, it can penetrate much deeper, avoiding high-angle nuclear collisions. This channeling phenomenon is exploited for the evaluation of the implantation damage using the Rutherford backscattering technique (RBS), but is generally unwanted during the implantation (because difficult to control) and for this reason, a small angle (≈7°) is used between the implantation direction and the normal to the (100) substrate surface; this expedient allows to obtain a final distribution comparable to that obtainable in an amorphous material.

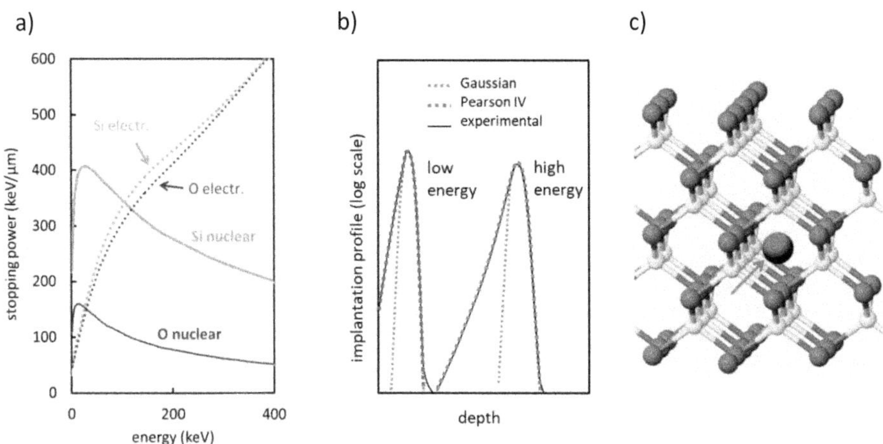

Figure 7.1 **a)** TRIM simulations of electronic and nuclear stopping power for Si and O implantation in GaAs; **b)** qualitative scheme of implantation profiles at different energies, compared with a simple Gaussian approximation and a skewed Gaussian; **c)** representation of an ion channeling along the [011] direction.

When implantation is done through a mask, the profile is broadened laterally (lateral straggle) by an amount comparable to ΔR_p; consequently, there is no lateral abrupt interface between the implanted and the not-implanted regions. A further factor potentially leading to a vertical and lateral profile broadening is the diffusion during the thermal treatment that normally follows the implantation.

7.2.2 Damage, damage removal, damage-isolation and doping

The amount of damage introduced in the crystal can be defined differently according to which properties - structural, electrical, optical - are used to evaluate it; considering the structural damage as primary criterion, a critical dose for amorphization can be defined as the dose Q_{am} at which the implanted region loses its crystalline characteristics. A direct evaluation can be done with transmission electron microscopy (TEM), while an indirect evaluation is typically done comparing the RBS channeling signal with that of amorphous and crystalline reference samples. This critical dose depends, for a given material, on several factors, including the mass of the ion, the implantation temperature and the implantation rate; the implantation energy is comparatively less important, because the damage is anyway introduced as the ions have sufficiently slowed down. As a rule of thumb, the critical dose for amorphization *decreases* with: increasing ion mass, decreasing crystal atom mass (group III of V), increasing implantation rate and decreasing temperature. Moreover, the damage caused by heavy ions tends to concentrate in clusters, while light ions cause a more evenly distributed damage. Comparing GaAs and AlGaAs, the critical dose for amorphization has been reported to increase with the Al content, although the reason for that is not entirely clear [122].

The physical damage of the crystal causes the formation of deep levels inside the bandgap, which trap the carriers reducing the conductivity; nonetheless, the as-implanted materials show a form of hopping-conductivity, with the carriers jumping from one defect to the other. The higher resistivity is obtained after a calibrated thermal annealing, that partially repairs the damage and leads to a more tight spatial localization of the deep levels. On the other hand, if the annealing is done at sufficiently high temperature, the physical-damage related trap density will fall, together with the resistivity.

In the case when the aim of implantation is to introduce electrical dopants, these must be electrically activated by a high-temperature annealing that allows, as much as possible, to eliminate the lattice damage and promote the substitutional incorporation of the implanted atoms; this is more difficult for III-V semiconductors than – for example – in the case of silicon, because of poorer recrystallization properties. In general, to this purpose, a high degree of lattice amorphization should be avoided. Ultimately, the effectiveness of the annealing depends on the amount of damage that has to be healed. According to Ref. [120], in the case of amorphized GaAs, temperatures in the order of 200°C are sufficient to obtain a coarse recrystallization, but the material remains highly defective, containing extended defects as twins and stacking faults; these defects starts to anneal out at temperatures in the order of 500°C, but still leaving behind a high density of dislocation loops, which in turn grow and annihilate above 700°C; optimal dopant activation requires generally temperatures ≥ 750°C, which annihilate (to some extent) point defects and point-defect clusters. In general, in GaAs, effective donor activation requires higher temperatures (≥ 850°C) as acceptor activation. It must be said that the annealing is usually done in rapid thermal annealing (RTA) systems, and the time spent at the highest temperature is of the order of 30 seconds; a prolonged annealing under group V protection (as in a MOVPE reactor) might partially reduce the temperature needed for structural damage removal.

7.2.3 Oxygen in GaAs-AlGaAs

Oxygen can be effectively used to obtain damage-related isolation in GaAs; it has been evaluated [120] that each atom can remove 10 to 50 carriers from n-doped GaAs. This isolation effect is largely removed by annealing above 600°C, although some thermally stable compensation remains, attributed to the formation of deep acceptor centers. The nature of these centers is not precisely understood: it has been proposed that they should be "off-center substitutional" defects, with the oxygen taking the place of arsenic but bonding only two out of four neighboring Ga atoms, and displaced towards what would be normally an interstitial position; the two remaining Ga atoms, which have (in a localized description) a total of 2 dangling bonds and one unpaired electron, could "relax" into a negatively charged defect, capturing an electron and forming a Ga-Ga bond (this model is fundamentally analogous to the DX center described in appendix 1). Nonetheless, alternative models have been proposed [123-125] and specific donor-neutralizing defects are also possible, for example – in the case of Si doping – with formation of Si-O bonds.

When implanted in GaAs p-doped with Zn, Mg or Cd, the isolation effect is more stable, up to ≈700°C, and in the case of Be-doped GaAs even up to 900°C, due to a specific, strong Be-O interaction. Vice-versa, the isolation obtained in C-doped GaAs is reported to drop already at 500°C.

In n–doped AlGaAs, a thermally stable isolation effect can be achieved by oxygen implantation; it has been attributed to the formation of deep acceptor centers during the annealing, in analogy to GaAs, but in this case the trap concentration seems not to be limited to low values. The nature of these centers is even less precisely known than in the case of GaAs [123, 126].

In p-doped AlGaAs, the effects of oxygen implantation on the electrical properties have been scarcely reported, although they might be *expected* to be similar to those obtained in p-doped GaAs; a thermally stable isolation has been confirmed only in the case of Be-doped AlGaAs.

In particular, to the author's knowledge, no studies on the effect of oxygen implantation specifically in C-doped AlGaAs have been published, but a thermally stable isolation effect due to O implantation has been reported by Bryan et al. [127] in a p-doped GaAs-AlAs multilayer and - although not specified in their paper – it is presumable that the p-dopant was carbon. Moreover, a compensation effect of oxygen contamination in MOVPE-grown intrinsic-carbon doped AlGaAs has been determined by Kakinuma et al [128] and attributed to formation of deep hole traps by oxygen.

7.2.4 Silicon in GaAs-AlGaAs

Although the implantation of silicon can be used, as that of any other ion, to create a damage-related isolation, it has been studied mostly as a means to introduce n-doping. The activation efficiency of Si in GaAs is reported to be comparatively insensitive to the implantation temperature, but requires high annealing temperatures: according to [129], the activation starts at T>700°C but T≥850°C is required to bring it to completeness (saturation) using RTA. According to [130], T=750°C should be sufficient to achieve the activation saturation, provided that the annealing time is of the order of one hour. While at low doses ($<1\times10^{13}$ cm^{-2}) the activation after annealing can be high (≈80%), at higher doses ($\approx1\times10^{15}$ cm^{-2}) it reaches a saturation value corresponding to a concentration of about 2×10^{18} cm^{-3}, possibly because of silicon amphoteric character and/or because of the formation of Si-Si neutral complexes [130-132]. Further increasing the dose, the post-annealing carrier concentration tends to decrease rather than increase, because of increased residual defectivity [133].

In AlGaAs, the activation efficiency has been reported to be lower than in GaAs, and to require higher (≈100°C) annealing temperatures [120, 129].

7.2.5 Quantum-well intermixing effects of implantation

Ion implantation near or across the quantum well, followed by a thermal annealing, can cause a perturbation in QW compositional profile, which is usually indicated as quantum-well intermixing (QWI) [88, 134, 135]. Similar effects can be obtained with several other techniques, an example being the in-diffusion of vacancies generated at the semiconductor surface by means of chemical reactions with a deposited SiO_2 layer.

In its simplest form, the QWI can be explained with the help of Fig. 7.2a; the original QW, with abrupt interfaces, is "smeared out" by implantation/vacancy diffusion, resulting in a band-energy profile which is narrower towards the center of the well; the VB and CB levels within the well drift away from each other. This causes a blue-shift in the QW radiative recombination wavelength.

The effect can be used to realize NAMs, but has even an impact on carrier transport, because of the small energy barriers (indicated by the red segments in the picture). In an edge-emitting laser, the ambipolar diffusion within the QW in the direction of the intermixed region will be reduced. Another effect is expected to be a reduction of the refractive index - at the original QW wavelength - in the implanted region, due to the blue-shift of the QW absorption peak (can be deduced from the Kramers-Kronig relation).

Fig. 7.2b shows an example of blue-shift obtained with He implantation followed by regrowth on the same structures used in this work (although He buried implantation has been tested, and is potentially interesting for introducing NAMs, it has not been integrated into working devices and will not be further discussed).

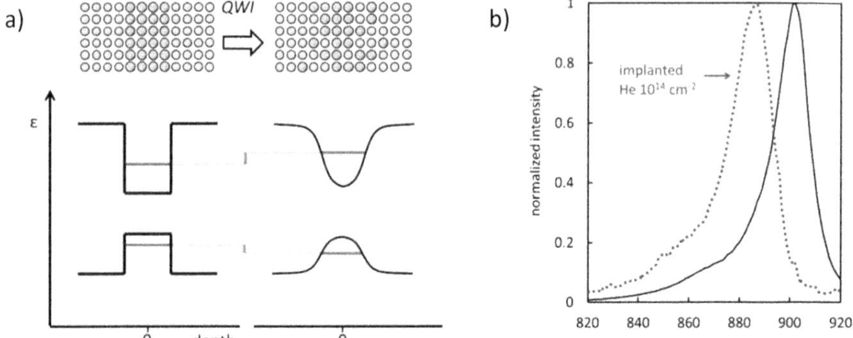

Figure 7.2 **a)** Scheme of quantum-well intermixing showing the potential energy for the electrons (black), the electron energy levels (blue) and their difference between implanted and not-implanted zones (red); **b)** example of blue-shift obtained with He implantation (electroluminescence spectra).

7.3 Device description and fabrication procedure

7.3.1 Vertical structure

The realized laser devices were based on an epitaxial structure whose simplified schematic is shown in Fig. 7.3, consisting in a down-top (growth) order of the following layers:

- n-doped $Al_xGa_{1-x}As$ cladding and waveguide layers numbered 1 and 2 where layer 1 has higher Al content – and higher bandgap, lower refractive index - than 2;
- active layer 3, consisting in a single GaInAs quantum well;
- p-doped $Al_xGa_{1-x}As$ waveguide and cladding layers numbered 4 (same Al content as layer 2) and 5 (very high Al content);
- the p-doped GaAs subcontact layer 6, and the very highly p-doped GaAs contact layer 7.

The AlGaAs n-cladding is doped $\approx 2\times10^{18}$ cm^{-3}, the AlGaAs n-waveguide has a graded doping, gradually decreasing to nominally undoped near the QW; the doping of the AlGaAs p-waveguide is graded from nominally undoped to 2×10^{18} cm^{-3}, and the following p-cladding and subcontact layers retain a similar value of $\approx 2\times10^{18}$ cm^{-3}, the contact is doped $\approx 2\times10^{19}$ cm^{-3}.

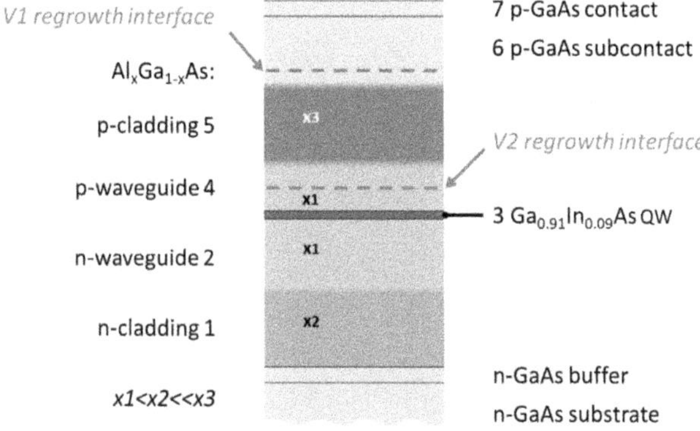

Figure 7.3 Laser epitaxial structure: different shades of color indicate different compositions, black horizontal lines between layers indicate abrupt interfaces between different compositions or different doping levels, otherwise the transition is graded. The dashed lines indicate the position of the regrowth interface according to the 2 process variants.

The well has an In content of about 9% and a compressive strain of about 6500 ppm. The room temperature electroluminescence of the QW has a maximum near 905 nm. The total thickness of contact + sub-contact is ≈850 nm, the p cladding is ≈650 nm and the p waveguide is 250 nm thick. On the n side, waveguide and cladding are much thicker than on the p side. The design corresponds to an asymmetric large optical confinement approach: the optical fundamental mode is broadened along the vertical direction and strongly shifted towards the n-side, as in the structures of the previous chapter.

The structures have been grown in two epitaxial steps, with the current confining ion implantation done before the second step. In a first process variant **V1** the regrowth interface is located just above the upper cladding, within the GaAs sub-contact.

In a second process variant **V2** the regrowth interface is located within the upper waveguide layer, about 200 nm above the QW. The positions of the interface in the two cases are indicated in Fig. 7.3 by dashed red lines.

No GaAs or GaInP etch stop layers remain below or above the active zone, which – based on the results presented in the previous chapter - represent a possible advantage in terms of device performance.

7.3.2 Process with 2-step epitaxy and intermediate implantation

The layer structures were grown in the planetary MOVPE reactor G3. The carrier gas was H_2, precursors where AsH_3, PH_3, TMIn, TMGa, TMAl. For p-type doping CBr_4 and DMZn were used, the latter only for the second growth step, while Si_2H_6 was used for n-type doping. Substrates were 3" (100) n-GaAs wafers.

The first epitaxial step for variant V1 is depicted in Fig. 7.4 left. It includes all the p cladding, 20 nm of the GaAs sub-contact (etch-stop 2), an GaInP layer (etch-stop 1) and a GaAs cap, the last two being sacrificial layers.

In the case of variant V2, Fig. 7.4 right, the first epitaxial step includes 200 nm of the p-waveguide (out of a total of 250 nm), a GaAs layer (etch-stop 2), an InGaP layer (etch-stop 1) and a GaAs cap, the last three being sacrificial layers.

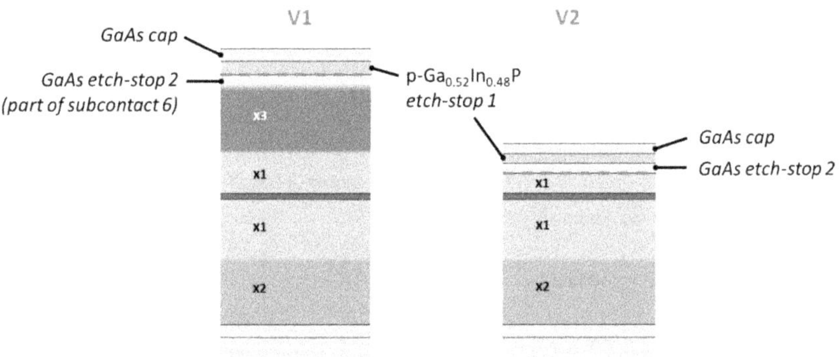

Figure 7.4 Structure after first epitaxy, process variant V1 and V2. The dashed lines indicate the position of the regrowth interface, the layers above it are removed by ex-situ or in-situ etching before regrowth.

In the first process step, WSiN markers were created on the surface by means of sputtering and lift-off lithography, to allow the alignment of all the subsequent steps. A photoresist mask was then defined by optical lithography, creating protective stripes over the sections designated to become the active part of the lasers; the unprotected areas between the stripes were implanted at room temperature (the implanter did not allow to heat the samples) using $^{28}Si^{+}$ or $^{16}O^{+}$. The implantation profiles were simulated using TRIM software, and the results were confirmed by secondary ion mass spectrometry (SIMS) measurements. The simulated profiles are shown in Fig. 7.5. The energy-dose combinations used in the experiments, along with the simulated peak concentration values and peak positions with respect to the QW are listed in Table 7.1.

It must be specified that, before starting this matrix of experiments, preliminary tests were done fabricating with a simplified process – but with the same epitaxial structures – diodes which were either not-implanted (reference) or implanted on the whole contact area. Their I-V characteristics were then compared, to estimate the minimum value of the implanted dose that could be used obtaining a significant impact on the turn-on voltage and series resistance.

Moreover, in the case of the variant V2, laser diodes were fabricated with the complete process and oxygen implantation having the energy/dose combinations: 30 keV/1×10^{14} cm^{-2},120 keV/1×10^{14} cm^{-2}, 65+95 keV/1+2×10^{14} cm^{-2}. These devices, which are not listed in Tab. 7.1, had extremely poor characteristics: very high threshold current and very low slope efficiency. Observing the emission spot size with and infrared camera, it was evident that the light was emitted from a width much larger than that of the buried current aperture *W* (see later Fig. 7.6), indicating that the dose was insufficient to obtain the desired lateral isolation. This is an important point, because it clarifies that the dose range contained in Tab. 7.1 could not be significantly extended to lower doses.

sample #	variant	atom	energy (keV)	dose (10^{14} cm^{-2})	*d (nm)*	*peak conc. (10^{18} cm^{-3})*
1	V1	Si	250	0.8	*655*	*3.3*
2				1.6	*655*	*6.6*
3				10	*655*	*41*
4				50	*655*	*207*
5		O	250	4	*495*	*14*
6				10	*495*	*35*
7	V2	Si	45	0.4	*185*	*5.6*
8				2	*185*	*29*
9		O	30	6	*175*	*75*
10			65+95*		*75*	*36*
11			120		*5*	*28*
12			240		*-195*	*19*

**2.4×10^{14} cm^{-2} at 65 keV, 3.6×10^{14} cm^{-2} at 95 keV*

Table 7.1 List of samples with ion-implantation conditions, the last two columns give the simulated values of peak concentration and peak distance d from QW (negative values indicate peak concentration below the QW); note that in each variant and ion, either the dose or the energy have been varied.

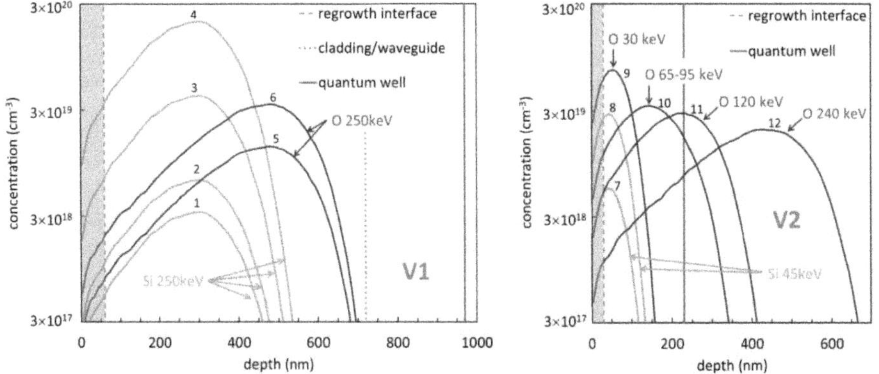

Figure 7.5 Simulated implantation profiles (in logarithmic scale) of the two process variants; the depth corresponding to the sacrificial layers, removed after implantation, is shaded in grey.

From the simulations it can be seen that:

- for variant V1, both ions stop in the p-cladding and do not reach the p-waveguide;
- for variant V2, silicon is implanted in the upper part of the waveguide and does not reach the QW, while oxygen is implanted at different depths: its distribution peaks above, at or below the QW.

After the implantation, the GaAs cap and the GaInP etch-stop layers were removed with selective wet etchants. No passivation treatment (e.g. ammonium sulfide) was used in this case after the etching; within half an hour after the second etch, the wafers were loaded into the MOVPE reactor.

In the case of variant V1, the second epitaxy did start with GaAs (subcontact, remaining part of layer 6) over the underlying GaAs etch-stop 2. In the case of variant V2, the etch-stop 2 was removed in-situ using CBr_4, uncovering the underlying AlGaAs inside the reactor, in order to minimize oxygen contamination, at T_{sp}=600°C; at this temperature, the etching is not selective and a few nm of waveguide AlGaAs were removed. Subsequently, the growth did start with AlGaAs (p-waveguide, remaining part of layer 4).

To ensure comparable (although not identical) annealing conditions, the regrowth did start in both cases with a high temperature (T_{sp}=760°C) step of similar duration. For V1, this was the growth of the GaAs sub-contact, which was followed by lower temperature growth of the contact layer. To grow the highly p-doped GaAs at 760°C it was necessary to use Zn as dopant, because intrinsic carbon incorporation (i.e. carbon originating from TMGa) would have been impossible, and doping with CBr_4 resulted in a very poor morphology, due to competition with strong etching. For V2, AlGaAs waveguide and cladding were grown at 760°C, while sub-contact and contact layers were grown at lower temperature.

The regrowth duration (considering the time spent at T>400°C) was approximately 3600 s in the case of V1 (1700 s at 760°C) and 4900 s in the case of V2 (1550 s at 760°C).

In this temperature range, the electrical activation of silicon can be expected to be low - estimated no more than 30%, even though it is difficult to make a precise prediction based on literature data. Moreover, with doses $\geq 1\times10^{15}$ cm^{-2} and room temperature implantation, the healing of the physical damage can be expected to be incomplete, especially in the case of silicon. Nonetheless, it was decided not to try to use higher regrowth temperatures.

After completing the layer structure, p-contact stripes were defined using evaporated Ti/Pt/Au. A conventional He^+ insulating implantation in the contact and subcontact layers was used to create the current aperture in standard lasers (i.e. lasers without the buried implantation) that were fabricated on the same wafers along with the buried-implanted lasers for comparison. The wafer process was then completed by electroplating gold on the p-metal stripes, thinning and n-side metallization. Facets were fabricated by cleavage. The devices were measured first without any coating, then selected devices were coated with high-reflective (HR) and antireflective (AR) layers HR (98%)/AR (1.5%) and mounted p-up on CuW carriers for more extensive characterization.

7.3.3 Device types

Two main kinds of lasers were present on each wafer of either variant: lasers with lateral buried implantation (LBI) and standard lasers (STD) without the lateral buried implantation. The laser types are depicted in Fig. 7.6.

In the case of STD lasers, the width *W* of the injection window corresponds to the width *W1* of the p-metal, while in the case of LBI lasers it corresponds to the aperture created by the buried implantation. For STD lasers, two values of *W* were used: 100 µm and 200 µm. For LBI lasers the current apertures *W* were 10, 30, 50, 100 and 200 µm, while the width *W1* of the metal was always 400 µm. The lengths *L* of the devices were 1, 4 and 6 mm. No lateral optical confinement (as for example index-guiding trenches) was introduced, so the lasers were gain-guided unless the implantation itself did create some degree of optical confinement.

With respect to V1, structure V2 has the potential advantage of introducing the lateral confinement only in close proximity to the active cavity, which should minimize the current spreading, but has the disadvantage of a regrowth interface positioned in the waveguide layer, where the presence of non-radiative recombination centers can impact the laser performance. Moreover, there might be non-radiative recombination at the transition between implanted and not-implanted regions near the pn junction. Finally, part of the optical power overlaps the implanted region; the optical properties of the implanted material are not well known, but can be reasonably expected to be significantly different – especially in terms of absorption – with respect to the not-implanted.

A further sub-variant of STD and LBI lasers had both ends completely buried-implanted: the buried implantation extended along the whole width of the device and from the facet inwards up to 50, 100 or 200 µm; consequently, in these devices the current injection was prevented near the facets, creating two short passive sections. These devices have been used to evaluate the impact of the implantation on the optical absorption (section 7.5.5) and the possibility to realize NAMs simultaneously with the buried current aperture.

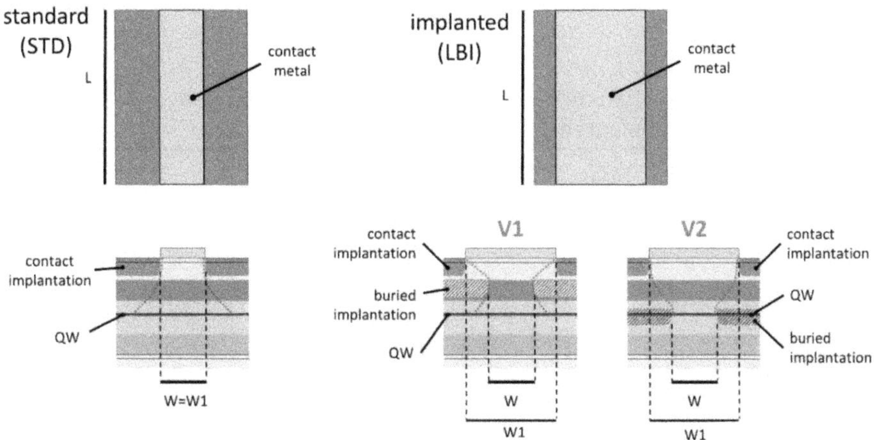

Figure 7.6 Top and front views of: standard laser (left) and lasers with buried lateral implantation V1 and V2 (right); V1 and V2 differ only in the position of the buried implantation. The dotted lines show qualitatively the different expected current spreads in the p-side.

7.4 Material characterization

7.4.1 Residual implantation damage after regrowth

Preliminary tests were done to assess the effects of the implantation and subsequent regrowth on the QW and more generally on the crystal structure. After the first epitaxy, Si or O was implanted into wafers with epitaxial structures (corresponding to the first epitaxy of V2 but without the GaInP etch-stop layer) using doses in the range from 1×10^{13} to 1×10^{15} cm^{-2}, and energies in the range 30-250 keV. The implantation was done only into parts of the wafers, leaving the rest as a reference. High-resolution X-ray diffraction (HR-XRD) Omega-2Theta scans of the implanted areas show additional features not present in the reference areas (Fig. 7.7a), due to compressive strain in the implanted layers. Main source of compressive strain are the interstitial atoms, associated with the formation of Frenkel pairs [136]. After the regrowth, the XRD diffractograms in the implanted and not-implanted areas became identical (Fig. 7.7b) except for the higher tested dose (Fig. 7.7c-d), where small differences are still present[1]. These differences can be simulated introducing a residual compressive strain in the implanted layers of approximately 500 ppm. The results indicate that the lattice has – at least at this level of resolution – largely recovered from the implantation damage although not completely for doses $\geq 1\times10^{15}$ cm^{-2}.

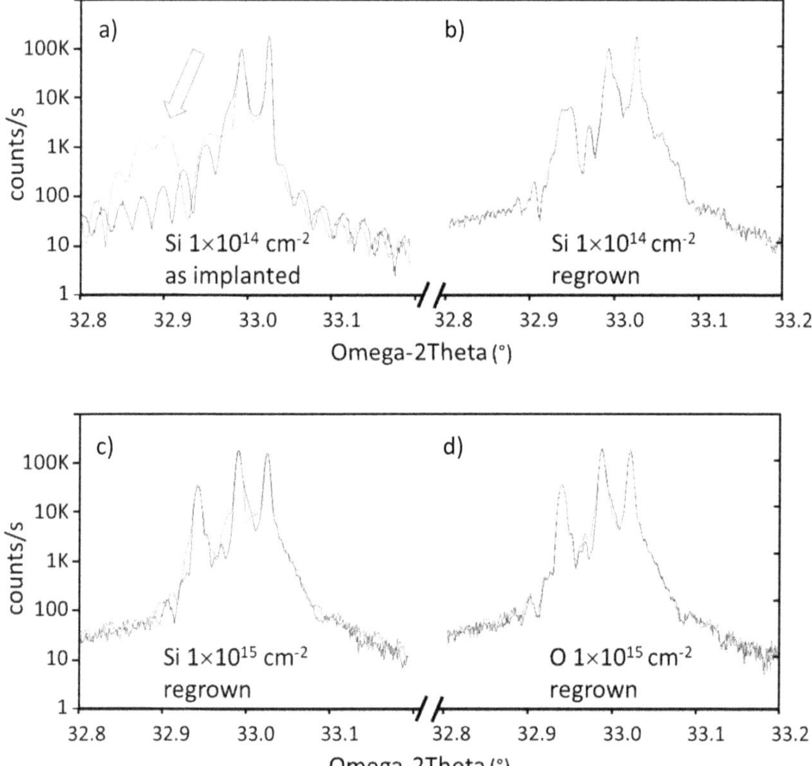

Figure 7.7 XRD diffractograms of partially implanted V2 test structures (Si at 45 keV. O at 120 keV, doses as indicated): **a)** after the first epitaxy, **b-c)** after the second epitaxy; red lines correspond to the implanted area, the blue lines to the not implanted, the arrow indicates the features introduced by the implantation. It can be seen that the lattice distortion disappears in the regrown sample with the lower dose, but not completely in the samples implanted with the higher dose – especially in the case of Si implantation.

---

[1] In the test samples of Fig.7.6 the composition of the AlGaAs cladding was slightly different in b) with respect to c) and d): this explains the different shape of the leftmost peak.

Additionally, low-temperature (80 K) cathodoluminescence (CL) spectra were taken in implanted and non-implanted areas as exemplarily shown in Fig. 7.8. In the case of V1 and Si implantation at 250 keV in the cladding (Fig. 7.8a), the cathodoluminescence intensity in the implanted region was almost unperturbed but for doses $\geq 4\times10^{14}$ cm^{-2} the wavelength was slightly ($\approx$ 5 nm) blue shifted. In the case of V1 and oxygen implantation at 250 keV in the cladding (Fig. 7.8b), the intensity was reduced and the blue shift was more pronounced (up to 15-20 nm).

The blue-shift is presumably due to quantum well intermixing (QWI), caused by the diffusion of vacancies from the implanted volume into the QW during the second growth step. In this case, NAMs should be present in the lasers that had the oxygen implantation at the facets.

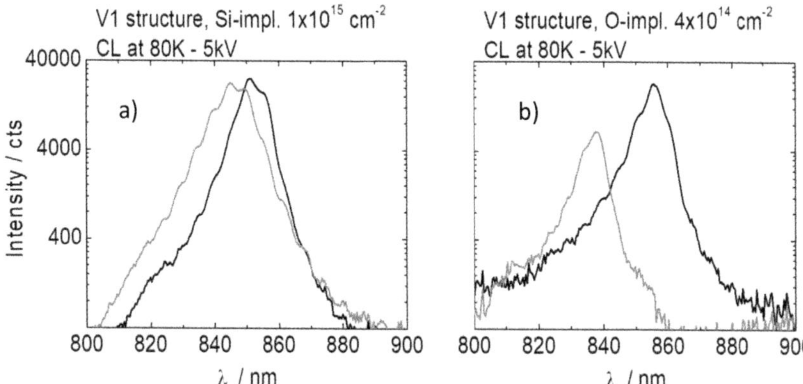

Figure 7.8 80 K cathodoluminescence spectra of V1 structures after regrowth: not implanted region (black lines) and implanted region (red lines); **a)** silicon implanted (250 keV, dose 1×10^{15} cm^{-2}) has 5 nm blue shift **b)** oxygen implanted (250 keV, dose 4×10^{14} cm^{-2}) has 17 nm blue shift and intensity reduction.

In the case of V2 and Si (implantation at 45 keV in the p-waveguide), the CL intensity of the QW was reduced and showed no or very small blue shift in the implanted areas. In the case of V2 and oxygen, at E_{imp}=30 keV (implantation in the p-waveguide) the QW luminescence was almost unperturbed in intensity and wavelength, indicating a negligible amount of damage to the QW, while with energies > 30 keV (implantation peak in the QW or below), the QW emission in the implanted area was completely suppressed after regrowth. The different damage caused by low and high energy oxygen in V2 structures is interpreted as being due to the formation of oxygen-related non-radiative recombination centers in the active zone when the atoms are implanted directly in the QW. Concerning QWI, since with E_{imp}>30 keV no luminescence was detected, we have no information for these cases, while the absence of wavelength shift for E_{imp}=30 keV indicates that no appreciable QWI happened at this energy.

The lack of QWI in Si-implanted V2 samples, in spite of the implantation being more near to the QW with respect to V1, might be explained by a combination of factors: a comparatively low dose ($\leq 2\times10^{14}$ cm^{-2}) and a reduced number of vacancies per ion produced at low energy (by a factor of 4 according to TRIM simulation). Moreover, since the implantation is very shallow, almost one half of the damage is generated in the sacrificial layers which are removed before the regrowth, further reducing the number of vacancies available for QWI. Finally, again because of the very shallow distribution of the damage, it can be expected that in V1 a considerable amount of vacancies will diffuse upwards and be lost at the surface during the heating and the initial phase of the regrowth.

The lack of QWI in the V2 sample implanted with O at 30 keV might be explained by similar considerations.

7.4.2 Electrical effects of the implantation

The depth of the implantation was selected on the basis of the expected electrical effects of the implanted species after annealing, in conjunction with the preliminary tests previously mentioned. On the assumption that silicon would have been at least partially activated, it was always implanted in the p-side of the devices, to compensate the p doping and possibly to create a reverse junction (in V1 structures) or to shift the position of the junction away from the QW (in V2 structures). Moving the junction away from the QW can have an isolation effect, increasing the turn-on voltage of the diode.

Implanted and annealed O has a compensating effect in n-doped material, and in one case (sample 12, V2) it was implanted with energy high enough (240 keV) to have a projected range below the QW, into the n-side of the waveguide.

As discussed in section 7.2.3, it is not well known from literature how effective an oxygen implantation in carbon-doped AlGaAs could be in creating a thermally stable isolation. The measurements done on test diodes with and without the implantation (section 7.3.2) did show that the current-voltage curves of the implanted diodes were shifted towards higher voltage values with respect to the not-implanted, provided a sufficiently high dose was employed. The causes of this behavior are not precisely known, it can be suggested that they could include three factors: neutralization of the original p-doping via specific O-C interactions in the AlGaAs matrix, formation of oxygen-related deep donor levels or hole traps, and reduction of hole mobility via scattering at lattice defects and ionized impurities. Based on these preliminary test results, O implantation in the p-side or in both p- and n- sides was included in the matrix of experiments (samples 9-11).

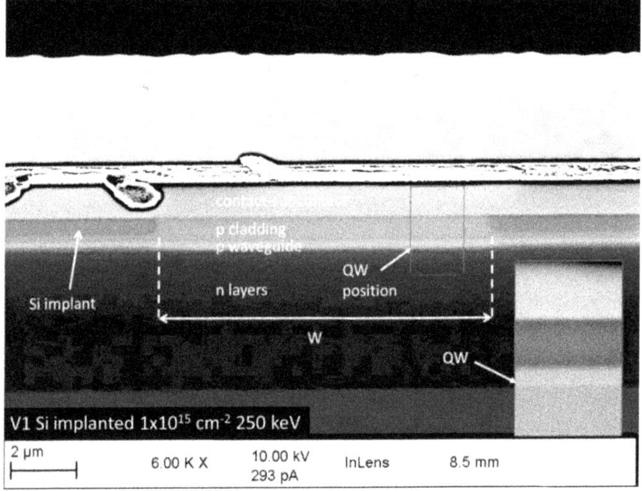

Figure 7.9 Secondary electron SEM images of the facets of V1 lasers with W=10 µm, implanted with Si. The small insert is a backscattered electron image taken in the corresponding position, where the QW can be identified as a thin white line. In the case of Si implantation, a pnp reverse junction is formed within the top cladding layer, and the buried current aperture is clearly visible. In the case of O, no contrast due the implantation was visible.

To gain a qualitative insight into the electrical effect of the implantation, the as-cleaved facets of several devices have been studied by scanning electron microscopy. Using secondary electrons, it is possible to identify the position of a pn homojunction, since the emission from the p-doped material is more intense than that coming from the n-doped part, due to different effective electron affinities and secondary electron escape depths of p- and n-doped areas [137].

Samples 1-4 were all implanted with Si in the p-cladding; Fig. 7.9 shows the secondary electron image of a V1 laser from sample 3, implanted with dose 1×10^{15} cm^{-2}: a dark region is visible in the cladding, indicating a change to n-doping.

In this case a pnp junction is formed, acting as an effective current barrier. To cross-check this conclusion, capacitance-voltage measurements have been done on a simplified test structure with the same Si implantation, confirming the shift to weak n-doping of the implanted cladding ($\approx3\times10^{16}$ cm^{-3}) after analogous thermal processing. A similar reverse junction was visible even for the lower Si doses, but the dark n-region was thinner. This is in line with the expected n-doping effect of the implanted Si, but at the same time indicates that either the electrical activation is very low or that the electrically activated Si is compensated, either by self-compensation or by the deep traps created by the implantation process that have survived the annealing.

When the ion implanted in the p-cladding of V1 lasers was oxygen (samples 5, 6), no reverse junction was detected with the secondary electrons, confirming that the cladding is still predominantly p, as could be expected based on the previous discussions.

In the case of V2 lasers, Si implantation in the upper part of the weakly doped p-waveguide (samples 7, 8) changes its character to n-type, and the junction moves upwards (Fig. 7.10a).

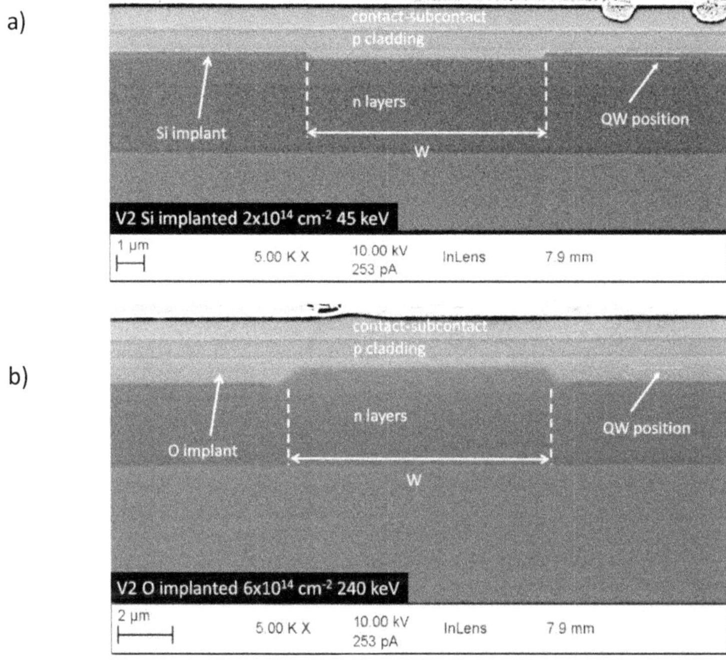

Figure 7.10 Secondary electron SEM images of the facets of V2 lasers with W=10 µm: a) implanted with Si, 45 keV: the junction moves up in accordance to the donor character of silicon; b) implanted with O, 240 keV: the junction moves down, indicating acceptor character of oxygen.

Actually, according to TRIM simulations, there should still be a thin (≈80 nm) p-doped layer embedded between the n-doped layers, and a reverse junction similar to the case of V1 should be present. It is not visible in the secondary electron images either because of depletion of the p-doped layer or because the calculated Si profile is not sufficiently accurate, possibly widened by channeling and/or diffusion during the second growth step.

Oxygen implantation in the p-waveguide (sample 9, E_{imp}=30 keV) does not impact the junction position, while a deeper implantation reaching the n-waveguide (samples 10-12) moves the junction downwards. Fig. 7.10b shows this for the case of sample 12 (E_{imp}=240 keV), but a similar effect was visible on samples 10 and 11. This indicates that O compensates the n-doping and – presumably - that the material becomes predominantly p-type.

7.4.3 Surface morphology and regrowth interface

From a morphological point of view, the surface of the samples after the second epitaxy appeared generally smooth, with a low but not negligible defect density and, occasionally, rough areas, especially near the WSiN markers; it is probable that these defects were originated by the lift-off lithography process used to create the markers, based on positive resist AZ5214 image reversal technique. The defect density did vary from sample to sample, showing some instability in the preparation of the surface before the regrowth; an example of "good" and "less good" morphology is shown in Fig. 7.11.

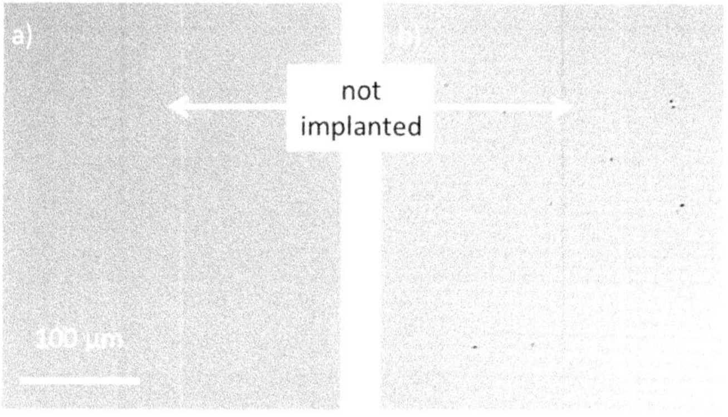

Figure 7.11 DIC-microscope images taken on a V2 structure implanted with oxygen after regrowth: **a)** area with low defect density; **b)** area with higher defect density; in the second case, the surface is slightly rough.

Using the optical microscope with interferometric contrast, the not-implanted stripes were visible on V2 structures after the regrowth, especially when high implantation doses were used; AFM measurements showed that these stripes were higher than the surrounding implanted areas by up to ≈2 nm. Note that this is the opposite of what one would expect based on the fact that the implantation introduces new atoms into the lattice (about 1.6 monolayers for a dose of 10^{14} cm^{-2}). A possible explanation is that the implantation damage – including the compositional change due to GaAs/AlGaAs intermixing - has increased the in-situ etch rate of the AlGaAs material lying just below the GaAs etch-stop 2.

The presence of oxygen contamination at the regrowth interface in the V1 variant was not considered important, given its position above the p-cladding, but in the case of V2 variant its impact could be more detrimental to laser performance. SIMS analysis, done on dedicated not-implanted areas of several processed wafers, did actually show comparatively high oxygen peaks - about 10^{18} cm^{-3} - and variable from sample to sample. These values are much higher than what previously obtained with in-situ etching/regrowth processes; this is most probably the result of three factors: the impact of the additional WSiN markers definition process, the lack of ammonium sulfide passivation after the wet-etching, and a higher than usual background oxygen contamination in the MOVPE chamber at the regrowth start.

7.5 Characterization of as-cleaved devices

The as-cleaved unmounted devices have been characterized acquiring the emitted optical power versus current (PI curves) using a current generator and pulsed linear ramps up to 2 A, with ramp time=500 µs, repetition frequency=10 Hz.

7.5.1 Comparison of 2-step and single-step growth

STD lasers coming from the V1 and V2 samples listed in Table 7.1 were compared with similar STD lasers fabricated in a single growth step, indicated with STD0. Note that STD0, V1-STD and V2-STD vertical structures are nominally identical, and differ only for the presence and position of the regrowth interface. As in the previous chapter, their electro-optical parameters α_i, η_i, Γg_0 and j_∞ (optical losses, internal differential efficiency, confinement factor times gain coefficient and transparency current density for $L\rightarrow\infty$) are extracted from the length dependence of the threshold current I_{th} and of the differential quantum efficiency η_d; the lateral leakage current per unit length at threshold (i_L=I_L/L) is extrapolated from the intercept of a plot of the threshold current per unity length i_{th} vs W. The results are summarized in Table 7.2 for devices with W=200 µm. The indicated uncertainties are based only on the dispersion of the values measured on different devices and they are larger for the 2-step growth process, partially because less values of L were available and partially because of wafer-to-wafer process-related variability. It must be added that the process used to fabricate the STD0 devices was not identical to that used for the V1-STD and V2-STD, because it used a different p-metallization.

device type	W (µm)	α_i (cm^{-1})	η_i (%)	Γg_0 (cm^{-1})	j_∞ (A/cm^2)	I_L/L (A/cm)
STD0 *single step*	200	0.5±0.2	80±2	10.2±0.3	90.5±2	0.40±0.06
V1-STD *2-step*	200	0.6±0.2	79±3	10.3±0.6	87±3	0.38±0.06
V2-STD *2-step*	200	0.6±0.2	77±3	10.3±0.6	95±3	0.40±0.06

Table 7.2 Electro-optical parameters of standard lasers based on structures V1 and V2: comparison of devices made using single step growth and 2-step growth.

The threshold lateral leakage is directly evaluated by the parameter I_L/L, the value is approximately the same in all cases, single and two-step epitaxy; it impacts significantly the threshold current (it is very similar to the values obtained in the standard structures of the previous chapter).

Comparing the 2-step growth with the corresponding single step growth values, one can see that the threshold current leakage and the optical losses are almost the same in all cases. For variant V1, the efficiency is 1% lower and the transparency current is 4% lower, although these differences are comparable with the uncertainties. For variant V2, the differences are somewhat larger: j_∞ is higher (5%) and η_i is smaller (4%) with respect to the single epitaxy; even these differences are comparable with the estimated uncertainties and must be taken with some caution, but in this case they both point to the same direction, i.e. worsening of laser performance. These results strongly suggest the presence of additional vertical current leakage caused by non-radiative carrier recombination in the p-side of V2, presumably due to the defects and residual oxygen contamination previously mentioned, confirming the necessity of improvements in the process.

7.5.2 PI curves of LBI and STD lasers

V1-STD and V2-STD lasers (W=100, 200 µm) have linear optical power–current (PI) curves up to 2 A (an example is given in Figure 7.12, black curves); the slope efficiency η is systematically lower for W=100 µm, consistent with the lower internal efficiency already shown in Table 7.2 for the single-epitaxy STD lasers. I_{th} is higher and η is lower in V2-STD than in VS1-STD lasers, even in this case consistent with the behavior of j_{∞} and η_d seen on single-epitaxy lasers.

V1-LBI lasers with buried current apertures W=200, 100, 50, 30 µm, have in general linear PI curves and the slope is almost independent of W. The narrower stripes with W=10 µm have very irregular PI curves or they do not lase at all. All V1-LBI lasers have higher slope efficiency and similar or lower threshold current than V1-STD lasers with the same nominal current aperture (Fig. 12a), indicating a general improvement due to the buried implantation.

V2-LBI devices have a much less uniform behavior across different samples. Even in the best samples, the PI curves are linear up to 2A only for the broader devices (100-200 µm) with the slope decreasing with increasing W. LBI devices with W=10 µm are mostly not lasing. On the positive side, the best V2-LBI samples have higher slope efficiency and slightly lower threshold current than V2-STD lasers with the same nominal current aperture (Fig. 12b).

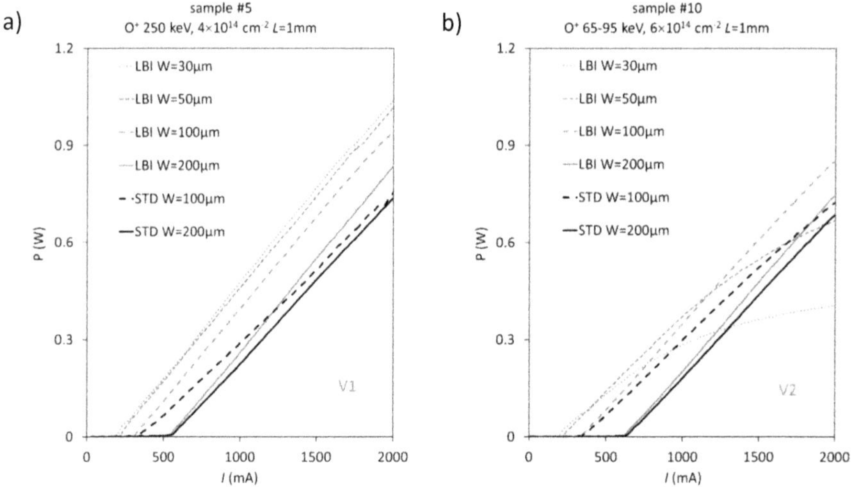

Figure 7.12 PI curves of STD and LBI lasers for different values W of the current aperture, L=1 mm: **a)** variant V1, **b)** variant V2; the indicated power is per facet.

7.5.3 Effects of implantation on I_{th} and leakage current

To provide a concise comparison of the relative performance of LBI lasers with the corresponding STD lasers (both fabricated with the 2-step epitaxy), two dimensionless parameters are introduced:

1) the ratio of their threshold currents $r(I_{th})=I_{th}(LBI)/I_{th}(STD)$;

2) the ratio of their leakage currents $r(I_L)=I_L(LBI)/I_L(STD)$.

Note that, in both cases, values < 1 indicates an improvement of LBI with respect to the STD.

The parameters $r(I_{th})$ and $r(I_L)$ have been determined independently for each length L (1, 4, 6 mm) but to simplify the analysis only the average over the three lengths is plotted in Fig. 7.13 (for graphical reasons a different scale is used for the parameters $r(I_{th})$ and $r(I_L)$.

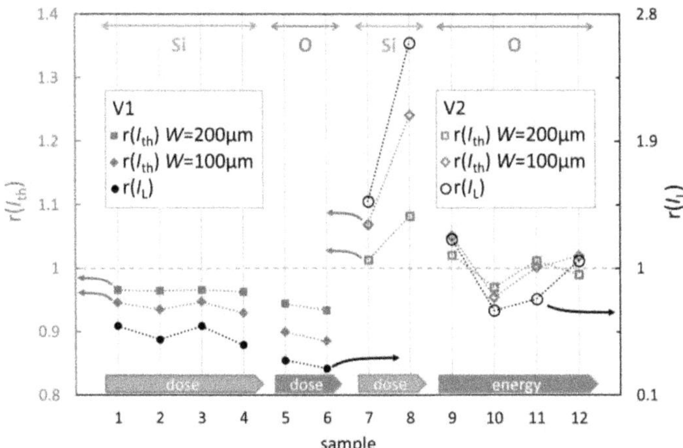

Figure 7.13 Comparison of LBI lasers vs STD. Left vertical axis: ratio of threshold currents $r(I_{th})$ for W=200 µm (squares) and W=100 µm (diamonds); the uncertainty is ≈5%. Right vertical axis: ratio of lateral leakage currents $r(I_L)$ (circles); the uncertainty is ≈20%. The dotted lines are only a guide for the eye. For V1, the values are <1, indicating that LBI lasers have reduced threshold current and reduced leakage current with respect to STD lasers. For V2 this happens only for sample 10 and partially 11, 12.

It can be seen that $r(I_{th}) < 1$ holds for all the V1 samples, indicating that LBI lasers have reduced threshold current with respect to the STD lasers, and that this improvement is more pronounced for W=100 µm than for W=200 µm. The lateral confinement has naturally a stronger impact on the lasers with smaller W, since there I_L is a larger fraction of the total current. The leakage parameter confirms this interpretation: the values of $r(I_L)$ for V1 samples are all < 1, and follow a trend similar to that of $r(I_{th})$.

The O-implanted samples 5, 6 have lower values of $r(I_{th})$ and $r(I_L)$ compared to the Si-implanted samples 1-4. The effect of the increasing dose seems to be a very small reduction of $r(I_{th})$ for both Si and O (but the variations are too small to be considered significant).

In the best case (sample 6, W=100 µm) I_L is reduced by ≈30% and I_{th} by ≈12% with respect to the value in STD lasers; this is true even when compared to STD lasers realized with a single-step epitaxy.

The same comparison gives a higher diversity of results for V2 samples. The Si-implanted samples 7 and 8 have high values of both $r(I_{th})$ and $r(I_L)$, indicating a poorer lateral current confinement than in STD lasers. Increasing the dose (7→8) strongly worsens the result, possibly because a buried reverse junction which might be present with the lower dose is wiped out by Si diffusion with the higher dose (see section 7.4.2). The remaining samples 9-12 (all of them O-implanted with the same dose but different energies) have better values, although only sample 10 seems to be consistently better than the STD (5% lower threshold current). Since in the case of V1 the threshold current of STD lasers fabricated with 2-step epitaxy is 5% higher than that of STD lasers fabricated with the single-step epitaxy, there is no significant improvement with respect to the latter.

7.5.4 Effects of implantation on the slope efficiency

In a similar way to what has been done for the threshold current, the slope efficiency η of LBI and STD lasers have been compared using the ratio parameter: $r(\eta)=\eta(LBI)/\eta(STD)$. Note that a value >1 indicates now an *improvement* of LBI over STD. Also in this case, to simplify the analysis $r(\eta)$ is averaged over the three cavity lengths; $r(\eta)$ values are plotted in Fig. 7.14.

In the case of V1, $r(\eta)$ is always >1 and is higher for W=100 μm than for W=200 μm, confirming the improvements on all samples and especially on lasers with smaller W. Surprisingly, even in the case of V2, $r(\eta)$ is >1 in almost all cases (except sample 8) indicating an improvement of η with respect to the STD even for samples (in particular 7 and 9) that did not show any improvement according to the threshold current ratio parameters examined in the previous section. In other words, the criteria for the evaluation of the quality of the current confinement based on I_{th} or η lead to different conclusions for some V2 samples. The highest improvement is comparable to that of V1 variant (both up to ≈10%) for large stripes, but is less strong for the narrower stripes (≈10% vs ≈20%).

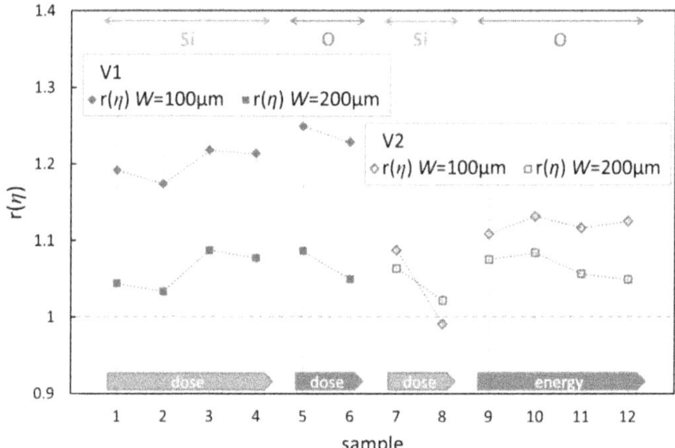

Figure 7.14 V1 and V2 variants, comparison of LBI vs STD. Ratios of slope efficiencies r(η); the uncertainty is ≈5%. The dotted lines are only an aid for the eye. Values >1 indicate that LBI lasers have improved slope efficiency with respect to STD lasers.

A qualitative interpretation of the discrepancy between the quality of the current confinement as inferred from current-related and power-related parameters is that while $r(I_{th})$ and $r(I_L)$ provide an evaluation of the effectiveness of current confinement *at threshold*, $r(\eta)$ provides a *beyond-threshold* evaluation. In general, only the leakage currents that do not clamp at threshold can contribute to reduce the slope efficiency [58, 117]. Calling $I_L(I)$ the lateral leakage current at a given total current I flowing in the device, and I_{Lth} its value at threshold, the component that can impact η is: $\Delta I_L(I)=I_L(I)-I_{Lth}$. Based on the experimental results, it appears that V2-LBI lasers have – in comparison to STD lasers – higher or similar I_{Lth} but lower ΔI_L. Factors that cannot be simply reduced to lateral current leakage with respect to the current aperture width W can play a role in determining the observed trends, in particular the fact that at threshold the optical power is concentrated towards the center of the stripes, so that the carriers injected in the active region within the current aperture but near the edges will be anyway partially lost (not contributing to the amplification of lasing modes); the lateral electrical confinement can then be expected to become more important as the current increases, because the optical power distribution broadens towards the stripe edges, and the carriers injected near the edges can better contribute to light amplification.

7.5.5 Effects of implantation on optical absorption

It is also of interest to determine the impact of the lateral buried implantation on the internal optical losses α_i due to variations in the complex refractive index n+iκ in the implanted part. The values of α_i of LBI devices extracted from the length dependence of η_d are near 0.6 cm^{-1} for W=200 µm, similar to the STD lasers, and more variable but ≤ 1 cm^{-1} for narrower devices, except the LBI-V2 lasers with W=30 µm coming from samples 10, 11, 12, that had values between 1.5 and 2 cm^{-1}. Unfortunately, since only 3 lengths were available, one of which was too short for the fit (see chapter 6, section 6.4.1) the uncertainty on α_i was particularly high.

In the following, an evaluation of the optical absorption *in the implanted region* based on the optical losses of devices with passive sections is provided. As mentioned in section 7.3.3, part of the devices (both STD and LBI) was implanted near the facets creating two short (max 200 µm) not electrically pumped passive regions. One effect of these passive sections is that a variation of the loss α_{ip} in the passive sections with respect to the loss α_{ia} in the active section impacts the slope efficiency, and α_{ip} has been evaluated from the experimental values of η of lasers with and without the implantation at facets, making use of an approximate model described in appendix 4. The resulting average values are given in Table 7.3 (STD and LBI had the same losses in the passive sections). Note that the uncertainty becomes comparable to the values for low optical losses.

<table>
<tr><td colspan="5">variant V1</td></tr>
<tr><td>sample</td><td>element</td><td>energy (keV)</td><td>dose (10^{14} cm^{-2})</td><td>α_{ip} (cm^{-1})</td></tr>
<tr><td>1</td><td rowspan="4">Si</td><td rowspan="4">250</td><td>0.8</td><td>0.5±0.5</td></tr>
<tr><td>2</td><td>1.6</td><td>0.5±0.5</td></tr>
<tr><td>3</td><td>10</td><td>0.5±0.5</td></tr>
<tr><td>4</td><td>50</td><td>3±1</td></tr>
<tr><td>5</td><td rowspan="2">O</td><td rowspan="2">250</td><td>4</td><td>0.5±0.5</td></tr>
<tr><td>6</td><td>10</td><td>0.5±0.5</td></tr>
<tr><td colspan="5">variant V2</td></tr>
<tr><td>7</td><td rowspan="2">Si</td><td rowspan="2">45</td><td>0.4</td><td>5±1</td></tr>
<tr><td>8</td><td>2</td><td>5±1</td></tr>
<tr><td>9</td><td rowspan="4">O</td><td>30</td><td rowspan="4">6</td><td>4.5±1</td></tr>
<tr><td>10</td><td>65-95</td><td>31±6</td></tr>
<tr><td>11</td><td>120</td><td>19.5±4</td></tr>
<tr><td>12</td><td>240</td><td>18.5±4</td></tr>
</table>

Table 7.3 Optical losses in the implanted passive sections, α_{ip}; in devices without passive sections, the contribution to the optical loss of the implantation is limited to the fraction of the optical power that propagates beyond the sides of the buried current aperture.

In the case of V1, all samples except #4 have $\alpha_{lp} \approx \alpha_{la}$. Given the experimental uncertainty, a possible loss *reduction* in the O-implanted samples due to the blue-shift of the QW wavelength mentioned in section 7.4.1 can be neither confirmed nor excluded. The Si-implanted sample 4, that has significantly higher losses, is the one with the highest implantation dose: the increase in α_{lp} is most probably due to strong residual lattice damage.

In the case of V2, α_{lp} is always significantly higher than α_{la}, especially in the samples that were oxygen-implanted with the higher energies (10-12). This is in line with the higher values of α_l noted on the narrow devices from the same three samples. One explanation can be the presence of a high concentration of non-radiative recombination centers in the active zone. This can be understood considering that the QW is not removed from the passive sections and, being thus part of the laser cavity, it is optically pumped. The gain of a passive section in absence of electrical or optical pumping is strongly negative (in the order of -40 to -50 cm^{-1} for these structures) but it will increase with optical power asymptotically approaching 0 (transparency). A reduced non-radiative recombination lifetime of carriers will hinder the approach to transparency.

In LBI lasers without facet implantation, the extra optical loss is limited to the sides of the active stripe; the impact of this *lateral* optical absorption on the total optical losses in BALs is relatively small, since most of the optical power is confined in the not-implanted region, but it becomes larger for smaller values of W, because the fraction of the optical power at the edges becomes higher.

In addition to the effect on optical absorption, the presence of non-radiative centers at the edges might introduce a further source of lateral current leakage, similar to the lateral interface recombination typical of buried-mesa lasers, also in this case more relevant for smaller values of W; in other words, part of the lateral current spread reduction due to implantation is probably compensated by lateral carrier recombination, occurring near the pn junction in the transition region defined by lateral straggle between the not-implanted and the fully-implanted regions. It can be noted that both optical absorption and lateral "interface" recombination should cause an increase of the temperature at the edges of the injected stripe, while the reduction of lateral current spread should have the opposite effect.

Based on the results and considerations of this and of the previous sections, we can gain a better understanding of the PI curves of narrow LBI lasers: it is suggested that the reason why the PI curves in narrow V2-LBI lasers bend more strongly than those of V1-LBI lasers, is a combination of the following factors: poorer lateral electrical confinement - as can be seen by the comparison of the parameters $r(I_{th})$ and $r(I_L)$, non-radiative carrier recombination at the regrowth interface, recombination in the lateral partially implanted regions in or near the QW, and - especially for the V2 samples 10-12 – significantly higher optical losses caused by the implantation.

7.6 Characterization of coated and mounted devices

7.6.1 PI curves

Selected LBI and STD devices, both fabricated with the 2-step epitaxy process, with L=4 mm, W=100 µm were coated HR (98%)/AR (1.5%) and mounted p-up on CuW carriers. Two to six lasers were taken from each of the samples 1, 2, 4, 5, 6 (V1) and 9, 11, 12 (V2). PVI curves, spectra, lateral near field (NF) and far field (FF) profiles, were then measured in CW mode at 25°C (heat sink temperature).

Qualitatively, the relative variations in I_{th} and η of LBI lasers with respect to the corresponding STD lasers seen on the uncoated-unmounted devices are roughly confirmed on the coated-mounted devices measured in CW mode. Quantitatively, the reduction of I_{th} and the increase of η are in most cases less pronounced (although the differences lie mostly within the estimated uncertainty). The reason is not completely clear, but might be related to the fact that STD devices have a narrower p-metal stripe (Fig. 7.6), and might be penalized by a poorer current injection when contacted only in one point – as happens during the measurements on unmounted devices. At any rate, the results obtained on mounted devices must be considered more reliable. The average values of the ratio parameters r(I_{th}) and r(η), from devices grown in two steps, are given in Table 7.4.

The best results are those of sample 6, (V1, oxygen implanted, 250 keV, 10^{15} cm^{-2}) with 12% reduction in the threshold current and 15% increase in the slope efficiency.

In the case of V2 process, the STD lasers fabricated with the 2-step epitaxy do not show improvements in I_{th} (rather a worsening) but they have higher η with respect to the corresponding two-step grown STD lasers, especially sample 11 (oxygen implanted, 120 keV, $6{\times}10^{14}$ cm^{-2}).

Since both the VSB and the STD devices fabricated with the 2-step epitaxy did suffer – especially in the case of V2 – of process-related penalties on threshold current and slope efficiency, the obtained improvements on these parameters should be considered in a relative sense.

sample (atom)	*element*	*energy (keV)*	*dose (10^{14} cm^{-2})*	*coated CW r(I_{th})*	*coated CW r(η)*
variant V1					
1	Si	250	0.8	0.96	1.12
2			1.6	0.94	1.12
4			50	0.93	1.11
5	O	250	4	0.92	1.14
6			10	0.88	1.15
variant V2					
9	O	30	6	1.08	1.07
11		120		0.99	1.16
12		240		1.03	1.08

Table 7.4 Parameters r(I_{th}) and r(η_d) determined with CW measures on coated and mounted LBI lasers with L=4 mm and W=100 µm; the uncertainty is ≈5%.

An "absolute" improvement evaluation can be done with respect to STD0 references fabricated with the same vertical structure but with a single-step epitaxy, as that reported in Tab. 7.2 for unmounted and uncoated devices. Moreover, the "best conventional" process solutions can be introduced (making the process more complex). To this purpose, single-epitaxy grown lasers, fabricated with the same vertical structure and dimensions (L=4 mm, W=100 µm) were HR/AR coated, mounted p-up and measured in CW mode; the process, more sophisticated than that used for the STD0 devices of Tab. 7.2, introduced a different p-metallization with large pads to get a better injection uniformity, improved thermal dissipation and more uniform material strain, a different and deeper He implantation scheme, which extended the isolation damage down to the upper half of the cladding layer 5, and double lateral trenches, with inner trenches for controlling the optical confinement and external trenches for reducing the reflection of high-angle modes; all this did lead to improved devices (indicated as STD0+), which did constitute a most challenging comparison for the LBI devices.

In the case of variant V2, and considering sample 11 which had the lowest threshold current and the highest slope efficiency, the results were as follows (quoted values are averages): sample 11, I_{th} =690 mA, η_d =1.07 W/A; STD0+ references, I_{th} =680 mA, η_d =1.13 W/A. While the result on I_{th} confirms that of Tab. 7.4 (no improvement) the result on η_d is very different, and is due to the fact that STD0+ references are much more efficient than V2-STD references: there is no "absolute" efficiency improvement with respect to an optimal standard approach, but rather some worsening.

In the case of variant V1, considering sample 6 (again the one with lowest threshold current and highest efficiency) the results were: I_{th} =580 mA, η_d =1.13 W/A; there is then an improvement in the threshold current, but even in this case, no "absolute" improvement of the slope efficiency[2].

7.6.2 Near-field and far-field

Comparing the shape of the NF and FF profiles of STD[3] and LBI lasers, a common trait is that, in the tested current range (up to 5A), the NF profile of LBI lasers is more top-hat shaped and with minor tendency to develop side lobes than that of STD lasers. Correspondingly, the FF profiles have more sharp edges without side lobes. The values of NF and FF widths, and the 95% BPP_{lat} measured at 5 A for V1 and V2 lasers are listed in Table 7.5.

sample (atom)	*element*	*energy (keV)*	*dose (10^{14} cm^{-2})*	*NF (µm)*	*FF (°)*	*BPP_{lat} (mm×mrad)*
			variant V1			
1	Si	250	0.8	100	6.6	2.9
2			1.6	89	7.6	2.9
4			50	70	9.0	2.7
5	O	250	4	101	6.8	3.0
6			10	107	7.2	3.4
			variant V2			
9	O	30	6	90	5.5	2.2
11		120		73	7.1	2.3
12		240		72	9.4	3.0
STD	-	-	-	106	8.3	3.8
STD0+	-	-	-	79	8.7	3.0

Table 7.5 NF, FF and BPP_{lat} 95% of coated lasers (LBI lasers when not otherwise indicated) measured at 5 A.

A further - and more testing - comparison is made with STD0+ lasers, which included index-guiding trenches as in reference [86] to improve the beam quality. The trenches have a depth of about 1.5 µm, ending within the p-waveguide, and they have an offset of 5 µm from the stripe edge, defining a ridge of width=110 µm that provides a lateral effective refractive index step Δn_e of approximately 1×10^{-3}.

[2] In Ref. [118, 119] an improvement on efficiency is reported, based on a comparison with less performant reference lasers (although realized with the same vertical structure).

[3] There was no significant difference in NF, FF and *BPP* values of STD lasers between V1 and V2 variants.

V1. The BPP_{lat} in all cases is smaller for LBI lasers, due to reduced NF and/or FF width. It can be seen that in V1-LBI the near-field width shrinks with increasing Si dose while the far-field becomes larger and the resulting BPP_{lat} remains roughly constant; increasing the O dose, the effect on NF and FF is less strong, but both slightly increase resulting in a larger BPP_{lat}. The effect of the increasing dose can be tentatively related to a corresponding increase in current confinement and in optical confinement, the latter is due to indirect effects (thermal lensing and lateral current accumulation as discussed in the following) and possibly to the direct modification of the material refractive index of the implanted region; the impact of the (annealed) implantation on the refractive index is actually not known, but if the carrier density reduction in the cladding layer were the dominant effect, an increase of the refractive index would take place.

An exhaustive explanation of NF and FF behavior would require a more extensive characterization, extended to a statistically significant number of devices and to higher currents, investigating the evolution of the optical modes in conjunction with the onset of new modes as the current increases; the results should be reproduced with a model taking into account the multiple effects introduced by the implantation on material optical properties (complex refractive index) electrical properties (carrier traps/non-radiative recombination centers) thermal properties (thermal conductivity of the implanted region). This was beyond the limits of the resources dedicated to the investigation, but a limited analysis has been done specifically on Si-implanted devices (corresponding to sample 2 but fabricated in a second iteration), using currents up to 12 A, and simulating the device behavior with the software BALaser [138]; the details can be found in C. Goerke master's thesis [139]. Some simple conclusions can be summarized: the lateral implantation in the cladding is effective in confining the current, with consequent reduction of the lateral gain outside the nominal current aperture; the clamping of the carrier density at threshold is particularly weak towards the stripe edges, and actually the carrier density increases there with increasing current more in the LBI laser than in the STD, because the onset of new lateral modes (which could cause lateral carrier clamping) is delayed (with respect to current increase) by the strong gain reduction outside the current aperture. The lateral carrier accumulation depresses locally the refractive index, adding-up to the thermal lens effect in increasing the lateral refractive index gradient and the slow-axis far-field divergence. Since the lateral current confinement does not extend down to the waveguide, the lateral carrier accumulation will increasingly promote the carrier diffusion away from the cavity with increasing current, spoiling to some extent the advantage of introducing the confinement above the active layer. Moreover, at high currents the gain near the edges will increase more rapidly than in a STD lasers, favoring the onset of lateral modes in spite of the lower gain outside the current aperture. The overall result is that the beam quality of the analyzed LBI laser was not significantly improved, especially at high current, with respect to STD lasers.

V2. In the case of V2-LBI, increasing the energy of O implantation reduces the NF width and increases the FF widths. The lowest values of BPP_{lat} are obtained in V2-LBI lasers from samples 9 (narrowest FF) and 11 (narrowest NF); in both cases, they are lower than the value of the STD0+ laser. The near-field and far-field of samples 9 and 11 at 5 A are shown in Fig. 7.16, their BPP_{lat} as a function of current in Fig. 7.17. The values of STD and STD0+ lasers are also shown for comparison.

For V2 structures, the analysis is further complicated by the probable presence of significant non-radiative recombination at the edges of the buried current aperture, which should cause a local temperature increase and reduce the carrier accumulation, especially in sample 11, where the implantation profile peaks in the QW. The non-radiative recombination will reduce the lateral gain, delaying the onset of lateral modes, and both the reduced lateral carrier accumulation and the lateral temperature increase should mitigate to some extent the growth of lateral refractive index confinement with increasing current. Nonetheless a non-radiative recombination can be hardly considered as a positive factor. Another effect contributing to reduce the BPP_{lat} is probably the strong optical absorption in the lateral implanted regions, which would inhibit the lasing of high order lateral modes: in this case, the effect could be considered positive, *provided* that the impact on the optical losses is small, i.e. that only the "unwanted" modes are strongly impacted. This could be the case of sample 9 (compare α_{ip} in Tab. 7.3); the narrow FF shown in Fig. 7.16 might be an indication that a "sweet spot" to position the implantation has been found – at least from the *optical* point of view.

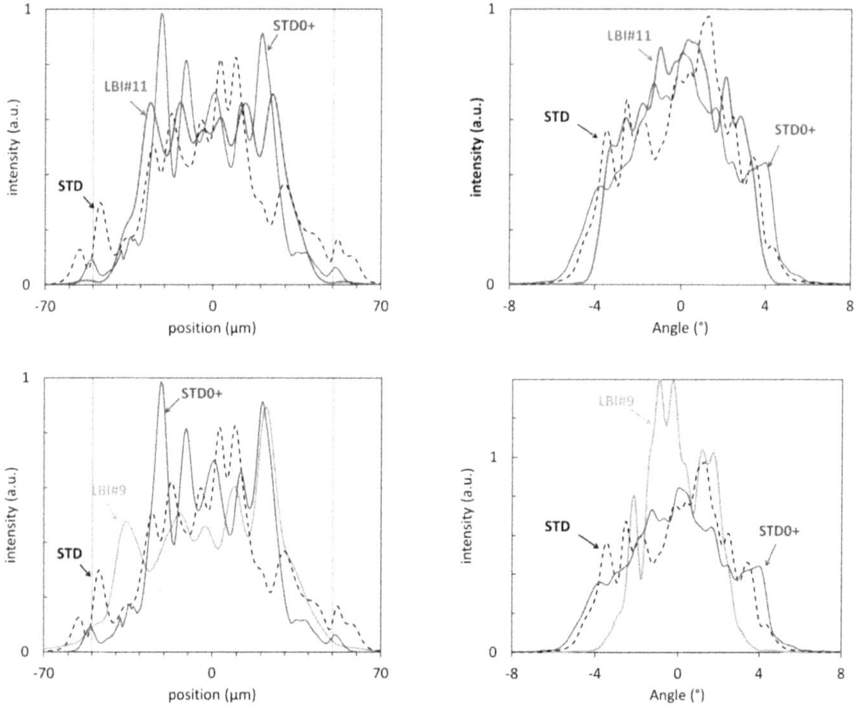

Figure 7.16 Comparison of lateral NF and FF profiles @ 5 A (CW): red/orange V2-LBI lasers, samples 11/9, blue STD lasers, black STD0+ with index trenches (5μm offset). All lasers have W=100 μm (indicated by the vertical lines).

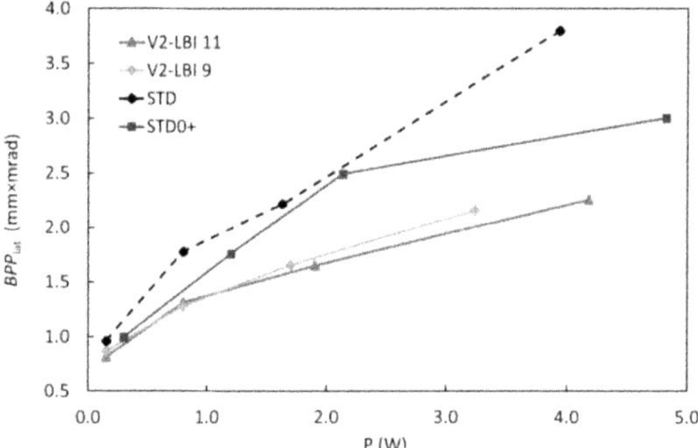

Figure 7.17 BPP_{lat} at 95% power content vs. emitted optical power: V2-LBI sample 11 (triangles) and 9 (diamonds), STD laser (black dots) and STD0+ laser with index trenches (blue squares). The lines connecting points are a guideline for the eye.

7.7 Step-stress tests

In section 7.4.1 it has been noticed that, since the oxygen implantation in V1-LBI lasers causes a blue-shift of the QW luminescence, the buried implantation at the facets is expected to create NAMs. To test this effect, V1-LBI lasers with and without facet implantation were subjected to a step-stress life-test at 25°C, increasing the cw current in steps of 1 A/week till catastrophic failure occurred. The selected devices were taken from samples 1 and 2 (Si-implanted) and 5 (oxygen-implanted with dose 10^{14} cm^{-2} at 250 keV). V1-STD devices were also tested. For a simple comparison, a mean time to failure normalized to I=4 A has been calculated in each case, assuming a proportionality MTTF$\propto I^{2.3}$.
The result was that the devices implanted with oxygen at the facets had a MTTF much higher than *all* the other devices (35% higher than the second best). At the same time, SEM+CL post-mortem failure analysis did show the presence of COMD on *all* the devices (an example is shown in Fig. 7.18). The conclusion is that oxygen implantation is effective in creating NAMs in V1 structures, although the residual optical absorption is still sufficient to make COMD (as opposite to "bulk" catastrophic damage) the main source of failures.

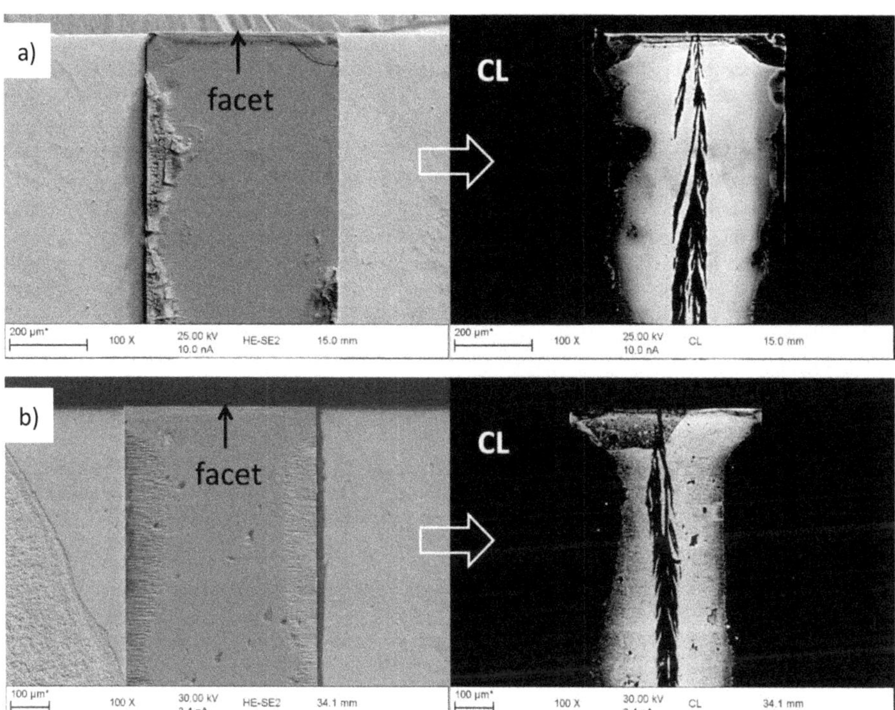

Figure 7.18 COMD observed with CL on two V1-LBI lasers, both including 50 µm near-facet implantation: **a)** Si-implanted, **b)** O-implanted. The lasers are bonded with the p-side over a metal holder, the n-side metal has been removed and the n-side GaAs semiconductor selectively etched.

7.8 Chapter summary and conclusions

A comparatively simple approach for the realization of broad area lasers has been tested, based on a two-step MOVPE growth process, where lateral current confinement is created by means of an intermediate ion implantation; the second growth step has the double role of completing the structure and of annealing the implantation damage. Two main variants have been tested: the first (V1) with the implantation in the p-cladding layer, the second (V2) with the implantation in the waveguide and in - or very near - the QW. In both cases, two (alternative) ions were implanted: Si^+ and O^+, testing different energies and doses.

The performance of the devices has been compared with that of standard lasers, fabricated in the same 2-step growth process (STD), or in a single-step growth using the same vertical structure and a more sophisticated design (STD0+). It has been possible to realize an effective lateral current confinement with both approaches V1 and V2: the best results in terms of threshold current reduction and increased efficiency were obtained with oxygen implantation. Silicon implantation appears to be anyhow a feasible alternative for the first approach, while it has given less promising results with the second.

The first approach has given, in the best case (oxygen implanted with dose 10^{15} cm^{-2}) ≈12% reduction of threshold current and ≈15% increase of slope efficiency with respect to STD lasers, although the efficiency improvement disappears when the comparison is done with STD0+ lasers. The BPP_{lat} could be reduced with respect to the STD lasers, but remained comparable to the values that can be obtained by the more sophisticated design of STD0+ lasers.

An interesting result is that oxygen implantation allows introducing NAMs simultaneously with the creation of the buried current aperture; this does not happen with silicon implantation.

The second approach – implantation in the waveguide layer and active zone - is technologically more challenging, because the regrowth interface is positioned very near to the active zone and contamination or other interface defects lead to a reduction in the laser efficiency with respect to lasers realized in a single-step growth. Besides, the comparatively high implantation doses needed to obtain an effective lateral isolation introduce non-radiative recombination centers that cause an increase in optical losses - and probably in carrier losses - which becomes more severe for narrow-stripe lasers. With this approach, using a 120 keV oxygen implantation, where the implanted profile encompasses the QW, a maximum ≈15% increase of the slope efficiency and an equivalent threshold current have been achieved with respect to the corresponding STD lasers (which suffer from the same problems), but both parameters remain worse with respect to those of STD0+ lasers.

A significant reduction in the BPP_{lat} parameter has been demonstrated using 120 keV or 30 keV oxygen implantation; the second case is particularly interesting, because of the reduction of far-field divergence, although the current confinement obtained with 30 keV is rather poor.

The following conclusions are drawn:

- the first approach (implantation in the cladding) might be considered a competitive alternative to the current-confinement technologies currently employed for broad-area lasers, *provided* that the positive results in term of threshold current reduction can be reproduced and that the reliability is not negatively impacted.

- the second approach (implantation in the waveguide) is problematic, having significant drawbacks in terms of energy conversion efficiency that make it unsuitable for high power applications; while some of them could be solved by process improvements, others are probably unavoidably related to the residual implantation damage at high doses. Nonetheless, the results suggest that a carefully positioned implantation in the waveguide *might* be exploited to introduce an optical-filtering element to improve the far-field of devices having otherwise a different current-confinement solution.

8 Summary and outlook

The in-situ etching with CBr_4 has been investigated, and most of the results are condensed in **chapter 4**. In the case of GaAs, the etching is an activated process, the etch rate increases exponentially with temperature and decreases with increasing group V partial pressure; a limited investigation on GaInP shows a similar behavior. In the case of AlGaAs the etch rate decreases with increasing temperature, indicating a different mechanism: this allows selective etching based on temperature.

The possibility to obtain a stable etching condition and a good final morphology are largely determined by the quality of the substrate *and* the surface conditions. The etching on GaAs behaves as a defect-delineation etch, revealing the dislocations and possibly other defects present in the material. Using low-quality substrates, several different kinds of surface defects are formed, and occasionally a catastrophic roughening occurs, while using high-quality substrates a good morphology can be obtained after the etching *and* after a subsequent regrowth. Nonetheless, even on high quality substrates, the formation of morphological defects is observed after in-situ etch + regrowth when the surface has been preliminary processed by means of lithography and wet etching (chapters 5 and 7). A passivation treatment with ammonium sulfide appears to be beneficial for the reduction of the problem, and it is suggested that the defects might be related to the presence of oxide islands.

A semi-empirical model has been used to interpret the experimental kinetic data relative to GaAs etching: it is proposed that the activated step in the reaction is the desorption of GaBr, for which an activation energy of 159 kJ/mol is estimated. It is further proposed that arsenic species inhibit the reaction, competing with bromine species for the same surface sites, and that the extent of arsine pyrolysis influences this competition; only at high temperatures the etch rate is expected to be limited by mass transport.

The effect of adding TMGa or TMAl during the etching has been also investigated. While in the case of TMGa the results can be explained in terms of simple competition between GaAs growth and etch, in the case of TMAl two anomalies have been noticed: an increase in the etch rate in presence of a moderate quantity of TMAl and inhibition of the defect-delineation effect.

CBr_4 does not attack or de-oxidize an air-exposed $Al_{0.5}Ga_{0.5}As$ surface. The effectiveness of CBr_4 etch in removing oxygen contamination from GaAs has been evaluated analyzing etched and regrown samples with SIMS. The results indicate that a moderate oxygen contamination on GaAs surface *can* actually be eliminated: this opens the possibility to use a thin GaAs sacrificial cap to protect the underlying aluminum-containing layers during the wafer process, removing it in-situ before the regrowth. The optimal condition to do this appears to be an etching done at the lowest possible temperature.

In spite of the potential interest of its effects on morphology, the addition of TMAl during the etching has the serious drawback of stabilizing – rather than removing – the oxygen contamination on the surface.

The investigations on in-situ etching paved the way to devices based on 2-step epitaxy, combined with in-situ etching for surface-contamination removal. Thermally-tuned SG-DBR lasers operating around 975 nm have been realized (**chapter 5**). The GaInP buried grating, embedded in a GaAs waveguide, was well defined and no indication of compositional perturbation was noticed within the waveguide, although small perturbations were present in the AlGaAs cladding at the transition between the active area and the passive area where the active layers have been removed. A satisfactory device performance, in terms of tuning characteristics, has been achieved. Nonetheless, a comparatively low efficiency indicates that further optimization of the vertical structure, of the process and of the general design is advisable, in order to reduce the sources of current leakage, non-radiative recombination and optical losses.

Electronic (current-injection) tuning would allow changing the lasing wavelength on a nanosecond timescale, compared to the millisecond timescale needed for thermal tuning; such devices have been already realized with InP-based materials (so operating at longer wavelengths, 1300-1550 nm). When the possibility of obtaining an electronic tuning was investigated, the results were disappointing: the wavelength shift resulting from the current injection in a passive section of the devices was too small for tunable-laser applications; increasing the tuning current leads, beyond a certain value, to the inversion of the wavelength shift due to the heating effect. Increments of the electronic wavelength shift were obtained reducing the oxygen contamination and with other process and design modification, but still remaining well below the desired result. Most probably, the introduction of a lateral current confinement, the improvement of thermal dissipation and the minimization of the residual sources of non-radiative recombination would allow further improvement of the results. Nonetheless, a comparative analysis of the properties of GaAs-based and InP-based materials indicates several limitations of the former, including a weaker change of the refractive index with increasing carrier density. It is suggested that in the case of GaAs an effective electronic tuning *might* be possible, but only at short wavelengths ($\lesssim$900 nm).

High-power, broad-area lasers having a buried-mesa structure have been fabricated using 2-step epitaxy combined with ex-situ and in-situ etching (**chapter 6**). The process allowed for the introduction of lateral current and optical confinement and non-absorbing mirrors (NAMs) at the facets. In this case, the in-situ etch was necessary not only to "clean" the surface, but even to remove completely the protective GaAs cap from the underlying AlGaAs layer without exposing this to air or other oxidizing agents.

The results could be defined as encouraging, especially considering that only one process iteration was done. The devices show performance and reliability improvements with respect to reference standard devices, realized with the same 2-step epitaxy process but not including the ex-situ wet etch used to define the mesa and the non-absorbing mirrors. One drawback is an excessive built-in lateral optical confinement, which causes an increase in the far-field divergence, but this aspect could be addressed with an appropriate design change. A more challenging problem is that the efficiency of the buried-mesa BALs is inferior to that of standard BALs fabricated with the same vertical structure but with a single epitaxy. The effect has been attributed to an increased non-radiative recombination at the regrowth interface. To which extent process improvements could address this problem remains an open question A dissatisfactory aspect that is unavoidable with this approach is the fact that the etched sides of the mesa, which in the used structures include AlGaAs layers (with low Al content), remain exposed to oxidation. An ideal process design should completely avoid this. Using the in-situ etch this is *in principle* possible (compare for example the scheme in Fig. 4.18b) but the technological feasibility remains to be proved.

A different strategy to introduce buried current aperture in BALs has been presented in **chapter 7**, based on ion implantation (Si or O, different doses/energies) followed by a second epitaxial step.

In a first approach, the first epitaxy includes the upper cladding, and the implantation extends in this layer without reaching (or barely reaching) the waveguide. Given its position, the regrowth interface does not introduce problems due to non-radiative recombination. When compared to reference devices grown in one epitaxial step followed by a deep He implantation, the lasers with the buried current aperture do show a significant advantage in terms of threshold current; moreover, buried oxygen implantation can be used to create NAMs simultaneously with the buried current aperture.

In a second approach, the first epitaxy includes only part of AlGaAs waveguide above the QW, and protective sacrificial layers. The implantation is done in the waveguide and – depending on the energy used – across the QW. The in-situ etch is necessary to remove the protective layers before the start of the regrowth. This process has proved to be problematic, both because non-radiative recombination can occur at the regrowth interface and laterally at the transition between the implanted and not implanted region and because of significant optical losses in the implanted region, attributed mostly to deep levels created within the QW. Nonetheless, it is interesting that a device having oxygen implantation in the waveguide – but not extending to the QW – has shown a sensible reduction in its far-field width; this effect – possibly related to filtering of high-order modes -might deserve further exploration.

A scheme of the relations among process elements, to which purpose their combinations have been used, and the devices realized in this work is shown in Fig. 8.1 (the in-situ pattern transfer has been added for its potential importance but was not part of the present work and is greyed-out).

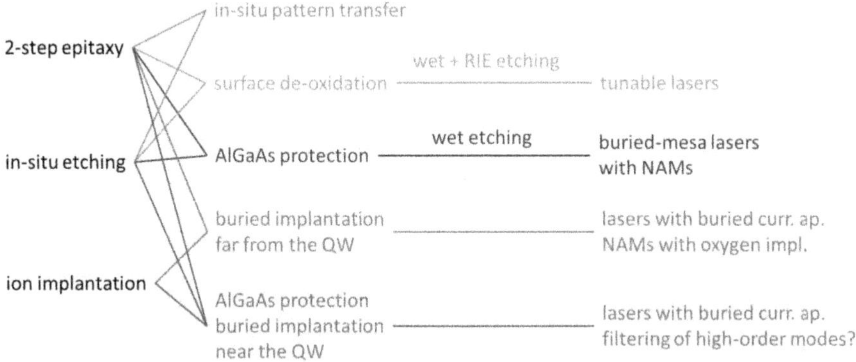

Figure 8.1 Scheme of process elements combinations, resulting effects and realized devices.

In conclusion, it is the hope of the writer that the investigations summarized above can provide usable information about the combination of 2-step epitaxy, in-situ etching and ion implantation for the realization of laser diodes: in particular, in terms of what these "technological strategies" give – or reasonably promise to give - and what probably requires a different approach.

A1 Zincblende III-V semiconductors and related properties

A1.1 Appendix content

This chapter summarizes several characteristics of III-V Zincblende semiconductors; structural, electrical and optical properties, especially those that are more relevant to the realization of optoelectronic devices, are presented for bulk materials, layered structures and surfaces. Pertinent definitions and concepts of solid state physics and chemistry are outlined.

A1.2 Composition, bonding and related properties

III-V semiconductors are solid crystalline chemical compounds, composed by at least one element of group III (boron group) and at least one element of group V (nitrogen group) of the periodic table.[1]

Their ideal stoichiometric composition has a 1:1 ratio between group III and group V elements, but small deviations (excess of group III or group V) can be present in real crystals, either unintentionally or deliberately introduced in order to modify the properties of the material. Similarly, small[2] amounts of foreign atoms are always present as unwanted impurities and/or deliberately added - in the latter case typically as electrical dopants.

For technological applications in the fields of electronics and optoelectronics, the most commonly employed constituent elements are Al, Ga, In from group III and N, P, As, Sb from group V. The compounds are called nitrides, phosphides, arsenides and antimonides (or stibnides). The simplest III-V compounds are the binaries, like GaAs and InP. Ternary compounds have two group III or two group V elements, as in $Al_xGa_{1-x}As$ and $GaAs_yP_{1-y}$ respectively. Quaternary compounds can have 1, 2 or 3 group III elements and correspondingly 3, 2 or 1 group V elements. Higher order compounds are also possible, although less commonly used.

A selection of III-V binary compounds properties is reported in table A1.1 and the data are plotted in Fig. A1.1 to better visualize the trends. Group V elements have higher electronegativity values than group III elements, and the III-V bonds are covalent with a partial ionic character (sometimes group V atoms are called anions and group III cations, even if the bonding is not properly ionic). To quantify the amount of ionic character, at least four different ionicity parameters have been introduced according to different models (Phillips, Harrison, Coulson, Pauling) [1, 140]. Pauling electronegativity differences $\Delta\chi$ and Phillips ionicity parameter values for several binary compounds are listed in columns 3 and 4. Phillips ionicity and $\Delta\chi$ both increase in the group V sequence Sb < As < P < N, but they have different

---

[1] The roman numbers III and V refer to the old numbering scheme of the periodic table groups, following either old IUPAC or CAS convention. Since 1990, IUPAC recommends a simpler scheme with groups numbered from 1 to 18, where the boron group is 13 and the nitrogen group is 15. Since most of the semiconductor community continues to use the old conventions, the old IUPAC scheme is used throughout this work.

[2] Depending on the specific material, synthesis technique and application, the amount of foreign atoms present in technologically useful III-V semiconductors can vary by several orders of magnitude, with unwanted impurities from the ppb to the ppm range and dopants usually in the ppm range; ppb-level impurities are often difficult to determine experimentally.

trends with respect to group III: Phillips ionicity – which is expected to be a more accurate representation of the III-V bond character – increases in the sequence Al < Ga < In.

III-V semiconductors' most common crystal structures are Zincblende (or Sphalerite) and Wurtzite. When the bonding has a significant ionic component, the Wurtzite structure is slightly more energetically stable than the Zincblende, because of a different long-range configuration which favors polar interactions [141]. The more stable structure for nitrides is Wurtzite, while it is Zincblende in the other cases (column 2).

binary	crystal Wurtzite, Zincblende	$\Delta\chi$ V-III (Pauling)	ionicity (Phillips)	lattice param. a (Å)	atom density $\times 10^{22}$ (cm^{-3})	cohesive energy (kcal/mol)	melting point (K)	band gap at T_{amb} (eV)
AlN	w	1.43	0.449	3.110	9.6	210	3487	6.0 d
GaN	w	1.23	0.500	3.199	8.9	203	2791	3.42 d
InN	w	1.26	0.578	3.585	6.5	175	2146	0.8 d
AlP	zb	0.58	0.307	5.4635	4.91	198	2823	2.49 i
GaP	zb	0.38	0.327	5.4508	4.94	174	1730	2.27 i
InP	zb	0.41	0.421	5.8690	3.96	159	1335	1.35 d
AlAs	zb	0.57	0.274	5.6614	4.41	179	1740	2.17 i
GaAs	zb	0.37	0.310	5.6533	4.43	155	1513	1.43 d
InAs	zb	0.4	0.357	6.0583	3.60	144	1210	0.36 d
AlSb	zb	0.44	0.250	6.1355	3.46	165	1338	1.62 i
GaSb	zb	0.24	0.261	6.0959	3.53	139	991	0.72 d
InSb	zb	0.27	0.321	6.4794	2.94	128	797	0.17 d

Table A1.1 Selected III-V binary compounds properties, data taken from [1, 142-143]; in the last column, d = direct bandgap, i = indirect bandgap. The bandgap energy value of InN is still somewhat debated because of the difficulty in synthetizing good-quality material.

In both crystal structures each atom has four nearest neighbors of the other species, arranged in a tetrahedron, and the bonding can be interpreted in terms of valence bond theory assuming sp^3 hybridization. In comparing the lattice parameter a (column 5) it must be considered that it is differently defined for Wurtzite (a_w) and Zincblende (a_{zb}): while a_w is the distance between two atoms of the same species in a tetrahedral unit, in Zincblende this distance is given by $a_{zb}/\sqrt{2}$ (see below Fig. A1.3). In the corresponding plot of Fig.A1.1, the lattice parameter a_w has been scaled accordingly to allow a consistent comparison, and with this scaling the binary lattice parameters increase in the sequence N << P < As < Sb and Al ≈ Ga < In. These trends correspond to those of the atomic covalent radii. The atomic density (column 6) is proportional to $1/a^3$.

Cohesive energy and melting point (columns 7-8) have trends similar to each other, decreasing in the sequence Al > Ga > In and N > P > As > Sb. This can be seen as a reflection of the corresponding polar-covalent single-bond energy [144]. The same trend is present for similar reasons - as better explained in section A1.7.2 - on the bandgap (column 9), with the exception of the inversion of InN with respect to InP.

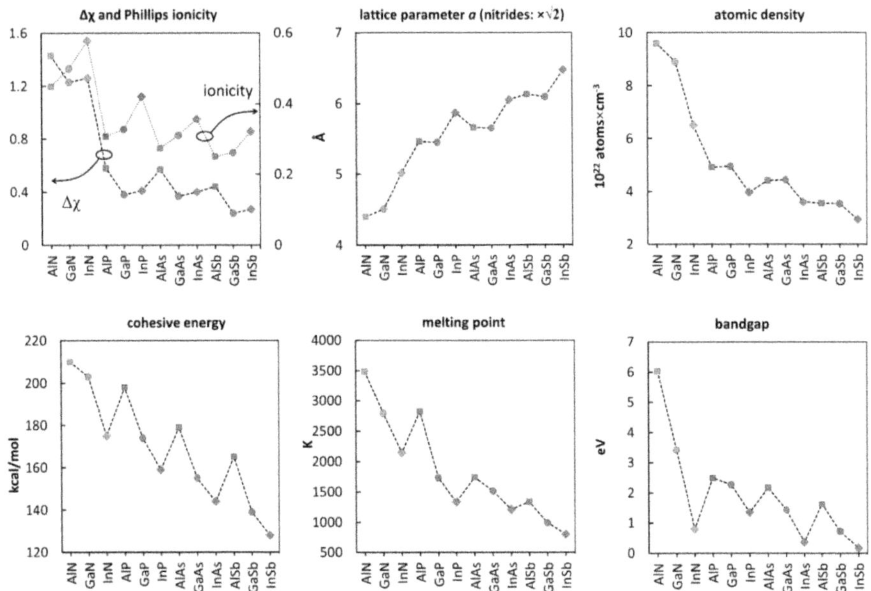

Figure A1.1 Selected V-III binary compounds properties (from table A1.1). In the case of the lattice parameter a, the value has been multiplied by $\sqrt{2}$ for consistency of comparison, as explained in the text.

A1.3 Crystal structure

A1.3.1 Zincblende and Wurtzite crystalline structures

Zincblende is characteristic of antimonides, arsenides and phosphides, and it is usually described in the close-packing approach [145] (depicted in Fig. A1.2) as being composed by two interpenetrating *cubic close packed* (ccp) sub-lattices, each one with one type of atom: this can be done taking the group V atoms to form a first ccp lattice, and occupying the positions corresponding to one of the two equivalent families of tetrahedral interstitial sites (T+, T-) with the group III atoms[3].

Actually, although this construction is topologically correct, it does not give a realistic picture of the actual atomic sizes (based for example on covalent radii): group V atoms must be pushed apart to enlarge the interstitial sites enough to accommodate the group III atoms.

It can be noted that, in Zincblende, the empty tetrahedral sites of the group V sublattice correspond to the octahedral sites of group III sublattice and vice-versa.

Wurtzite is characteristic of nitrides, and it can be described in the close-packing approach as composed by two interpenetrating hexagonal closed-packed (hcp) lattices, each one with one type of atom. As in the case of Zincblende, taking group V atoms to form the main hcp lattice, the group III atoms will occupy the positions corresponding to one of the 2 families of tetrahedral interstitial sites.

---

[3] Closed-packed lattices have 2 tetrahedral interstitial sites and one octahedral interstitial site for each atom.

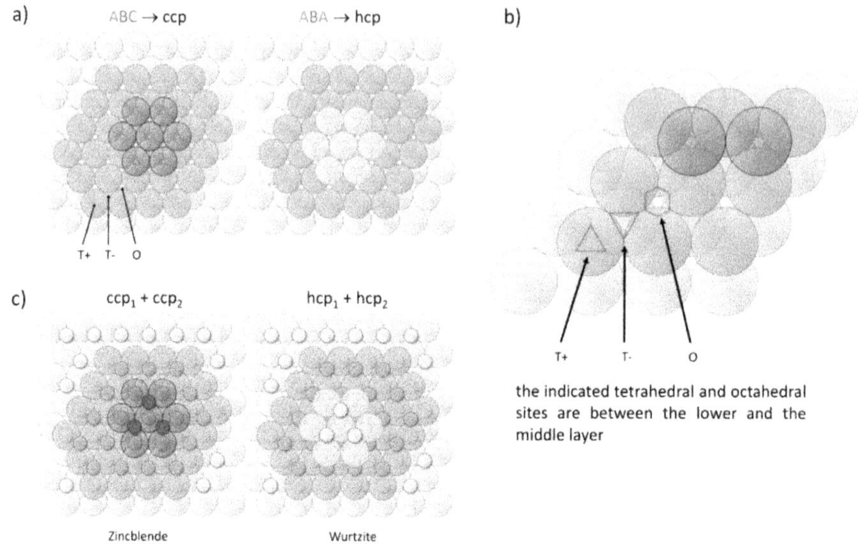

Figure A1.2 **a)** the simplest closed-packed stacking sequences of closed-packed layers of spheres: ABC (cubic closed packed, ccp) and ABA (hexagonal closed packed, hcp); T and O are the tetrahedral and octahedral interstitial sites. **b)** close-up view of the interstitial sites. c) a second ccp (hcp) lattice of small spheres is nested in the T+ sites of a first ccp (hcp) lattice, resulting in Zincblende (Wurtzite) structure.

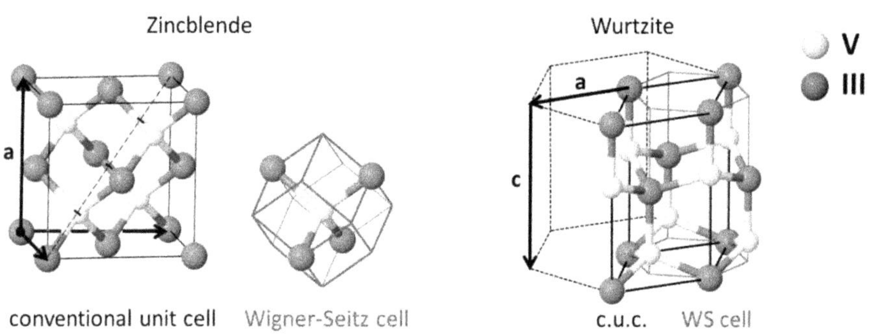

Figure A1.3 Unit cells of Zincblende and Wurtzite. Black lines contour conventional unit cells, red lines contour Wigner-Seitz cells, black arrows indicate the basis vectors $\mathbf{a}_1$, $\mathbf{a}_2$, $\mathbf{a}_3$ of the unit cell; the Miller index (hkl) denotes the crystal planes parallel to the geometric plane which intercepts the points $\mathbf{a}_1/h$, $\mathbf{a}_2/k$, $\mathbf{a}_3/l$. For Zincblende, the (100) planes can be taken as parallel to any of the faces of the fcc conventional unit cell, and the ccp closed-packed atom layers are on (111) planes; one cube diagonal is drawn, showing that the group III and V sublattices are shifted by ¼ of its length. For Wurtzite, the hexagonal (001) planes – or (0001) in the Miller-Bravais notation - are parallel to the hexagonal faces of the unit cell and correspond to hcp closed-packed atomic layers.

While in the hcp lattice the stacking sequence of the close-packed atomic layers is ABAB... - meaning that the third layer is superimposed to the first - in the ccp lattice the sequence is ABCABC...

The ccp lattice is equivalent to a face-centered cubic (fcc) *Bravais* lattice, with the ccp closed-packed layers parallel to the (111) fcc planes, and the Zincblende lattice can be seen as two interpenetrating group III and group V fcc sub-lattices, shifted along a body diagonal of the conventional fcc unit cell (shown in Fig. A1.3) by ¼ of its length. Since the basis vectors of the conventional unit cell have cubic symmetry, the vectors [hkl] indicating a direction in the lattice basis $\mathbf{a}_1$, $\mathbf{a}_2$, $\mathbf{a}_3$ are normal to the planes defined by the Miller indices (hkl). The unit cell is completely specified by the lattice parameter a.

The Zincblende crystal structure is not in itself a Bravais lattice, but it can be mapped into one (fcc) using a *lattice base* [146, 2]. The conventional unit cell can be used as base, but in connection with the description of the electronic structure a more useful choice is that of the smallest possible (or *primitive*) cell, and in particular of the Wigner-Seitz cell, defined as the region of space that is closer to a selected Bravais lattice point than to any other[4]. While the conventional unit cell is a cube containing 8 atoms, the primitive Wigner-Seitz cell is a rhombic dodecahedron containing 2 atoms (Fig. A1.3).

As already mentioned in chapter 2, Zincblende Bravais lattice belongs to a symmorphic space group ($F\bar{4}3m$ Hermann-Mauguin, T_d^2 Schoenflies), which contains 24 point symmetry operations and does not include an inversion center.

Wurtzite Bravais lattice is an hexagonal lattice: it can be generated using as base a primitive cell containing 4 atoms, two of each kind (III, V), its space group is the non-symmorphic $P6_3mc$ Hermann-Mauguin (C_{6v}^4 Schoenflies) which contains 12 point symmetry operations; as in the case of Zincblende there is no inversion center and, moreover, there is a net permanent polarization along the c axis. To completely specify the unit cell, 3 parameters are needed: the lattice parameters a, c and the internal parameter u (anion-cation bond length divided by c). In the ideal close-packing description only one parameter must be specified, because in this case c/a = 1.633 and u = 0.375 (real values can differ by some percent from these ideal values).

A1.3.2 Zincblende crystal facets: thermodynamics and surface reconstructions

Zincblende lowest-index crystal planes and facets have been already described in chapter 2. Here, some *additional* information is provided concerning crystal facets.

At thermodynamic equilibrium, the ideal tridimensional crystal shape is a polyhedron, whose facets correspond to crystal planes, which minimizes the total Gibbs *excess* surface energy for the given volume; the excess energy is defined with respect to bulk values, and is mainly due to the missing bonds at the surface. The polyhedron can be derived from Wulff's construction, which is based on the distribution of the surface energy per unit area γ_{hkl} over the facets: for each possible crystallographic direction [hkl] a vector of module γ_{hkl} is drawn from a common origin, and the plane normal to the vector intersecting its tip; the inner envelope of all such planes define the equilibrium crystal shape. Only the facets with the smaller values of γ_{hkl} do actually contribute to this shape, and according to available experimental data and theoretical evaluation, they correspond in Zincblende to the lowest-index planes {100}, {110} and {111}, resulting in a polyhedron similar to those shown in Fig. 2.1 and 2.2 of chapter 2, but with the relative size of the facets modified by the selected conditions, e.g. with $(111)_B$ facets much broader than $(111)_A$ facets in a group V rich environment [147, 148].

Constructions analogous to this thermodynamic Wulff's plot have been used to predict - or at least to rationalize - the time-evolution of the crystal shape during etch or growth; in this case the empirical etch/growth rates on different planes take the place of the surface energy, and the problem is made computationally much more complex because the construction cannot be limited to a single origin, but

[4] There is only one type of Wigner-Seitz cell for a given Bravais lattice, and it contains exactly one lattice point; using the WS cell, the full symmetry of the Bravais lattice underlying the crystal structure is obtained.

must be extended to a two or tri-dimensional starting surface [149]. It can be added that the assumption of etch/growth rates depending only on the crystal planes is a simplification, because the specific geometry can play an important role when there can be significant surface diffusion of species from one inert or masked area to a more reactive one.

The surface energy is always lower than what could be expected based on the assumption that the crystal's surface retains the same configuration and interatomic distances as in the bulk: bond lengths and angles of the external atomic layers change with respect to bulk values, (a phenomenon indicated as *relaxation*) and even changes in the bonding topology of surface atoms can occur, often - but not necessarily - leading to the reduction of the number of dangling bonds; such rearrangements are called *reconstructions* [147, 148, 150, 151] and have been experimentally observed with a variety of techniques, among which scanning-tunneling microscopy (STM) has given the most direct information about the arrangement of the atoms (but still with limits in the achievable resolution). The reconstructed surfaces exhibit a certain degree of regularity in the bi-dimensional patterns originated by the atomic displacements, and are ideally described as bi-dimensional crystals.

The theoretical prediction of the thermodynamic stability of reconstructions involves the case-by-case quantum-mechanical computation of their impact on the electronic structure (and in principle even the impact on vibrational modes should be considered). An additional complication is that in most cases of technological interest, reconstructions take place under conditions far from thermodynamic equilibrium, for example during the growth in a MBE reactor (and in a MOVPE reactor, but information about reconstructions is - in comparison to MBE - extremely limited).

For tetrahedrally coordinated compound semiconductors, an "electron-counting rule" has been proposed. Formulated for III-V compounds, the rule states that any reconstruction must redistribute the electrons in bonds and dangling bonds leaving no net charge on the surface, all the residual dangling bonds of group III empty, and those of group V full. The rule can be justified assuming that for group V, dangling bonds correspond to energy levels below the Fermi energy, and that the opposite holds for group III; no partially filled (i.e. conductive) surface bands are expected in this model and the rule is sometimes formulated in the form: surfaces tend to be autocompensated.

At least in the case of GaAs, the prevalent reconstruction of (110) facets is found in UHV experiments to be essentially a relaxation, with charge transfer from Ga to As: rehybridization of the involved atomic orbitals occurs, and As atoms shift outwards while Ga shift inwards (this kind of rearrangement is called "buckling").

The $(111)_{Ga}$ facets have been found in UHV experiments to be stabilized by a reconstruction involving the formation of surface Ga *vacancies* and As dangling bonds; according to theoretical models, and to satisfy the electron-counting rule, 25% of Ga atoms should ideally be missing. It is interesting to note that without these vacancies, and considering an even distribution of the electrons in all the bonds and valence orbitals, the Ga dangling bonds would be fractionally filled. The reconstructions of $(111)_{As}$ facets are somewhat more debated, but have been reported to involve the presence of *extra* arsenic in the form of trimers adsorbed on the terminating As layer.

On (100) facets, pairs of neighboring surface-atoms of the same species can use the orbitals not already involved in bonds with the atoms of the other species of underlying layer to form bonds between each other; these pairs are called *dimers*. Several reconstruction have been experimentally found depending on the experimental conditions used (e.g. temperature and ratios of As and Ga fluxes to the surface in a MBE reactor); the most common involve alternation of dimers with dimer *vacancies* – again satisfying the electron-counting rule.

A1.4 Ternary and higher order compounds

Ternary and higher order compounds [152] can be seen as random mixtures (solid solutions) of binary compounds with the same crystal structure. Thermodynamically, not all the potential III-V alloys are stable in the entire composition range, several of them exhibiting some miscibility gap region; the instability is determined by the the fact that, fixed the system overall composition, its Gibbs free energy does not lie in a minimum when only a single phase exists.

Gibbs free energy G is composed by an enthalpic (H) and an enthropic (S) terms: G=H-TS. When a multinary alloy is formed starting from two binary compounds, the values of these functions are modified by terms called enthalpy and entrophy of mixing. The gap region is largely due to the enthalpy of mixing, which in III-V compundes is always positive (unfavourable) because of the strain introduced by different atomic sizes within the cation and (especially) within the anion sublattices [153]. The entropy of mixing is viceversa always favourable. Fig. A1.4 shows the calculated temperature-dependent miscibility gap of the quaternary $Ga_xIn_{1-x}As_yP_{1-y}$. When the system composition lies within the gap, the decomposition of the single phase A in two phases B and C (identified by the end points of the isothermal tie-line passing through the system composition) is favoured because $G_{(B+C)} < G_{(A)}$.

The miscibility gap shrinks at higher temperatures, because the entropic contribution TS to the free energy becomes more important. When the material is grown epitaxially, metastable homogeneous phases within the miscibility gap can be obtained; this can happen either because the growth conditions are far from the thermodinamic equilibrium or because the formation of separate phases would actually induce locally an extra strain, and this introduces a kinetic barrier against formation of (or decomposition into) different phases; the latter effect, present even in bulk crystals, is enhanced in the case of epitaxial layers due to the stabilizing effect of the underlying substrate.

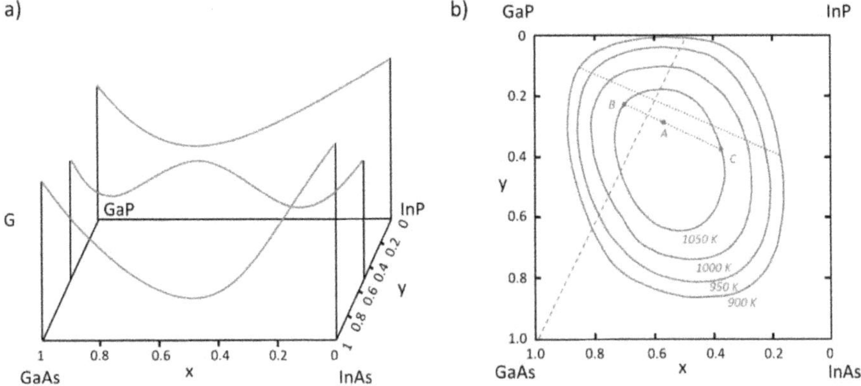

Figure A1.4 **a)** Tridimensional representation of Gibbs energy G, in the quaternary $Ga_xIn_{1-x}As_yP_{1-y}$ at fixed temperature (qualitative); **b)** solubility gap at different temperatures, according to [153]. The dashed line indicates the compositions lattice matched to GaAs, the dotted lines are examples of tie-lines at 1050 and 900 K.

To a good approximation, the lattice parameter of multinary alloys can be obtained as the correspondent linear interpolation of the binary lattice parameters (Vegard's law). Taking as example the ternary $Al_xGa_{1-x}As$, the lattice constant $a(x)$ is expressed by:

$$a(x) = x \cdot a_{AlAs} + (1 - x) \cdot a_{GaAs} \qquad \text{E A1.1}$$

For a quaternary with 3 atoms of the same group as $Al_xGa_yIn_{1-x-y}As$:

$$a(x,y) = x \cdot a_{AlAs} + y \cdot a_{GaAs} + (1-x-y) \cdot a_{InAs} \qquad \text{E A1.2}$$

For a quaternary with 2 atoms of each group as $Ga_xIn_{1-x}As_yP_{1-y}$:

$$a(x,y) = x \cdot y \cdot a_{GaAs} + x \cdot (1-y) \cdot a_{GaP} + (1-x) \cdot y \cdot a_{InAs} + (1-x) \cdot (1-y) \cdot a_{InP} \qquad \text{E A1.3}$$

Only small deviations from this linear model must be taken into account to accurately fit the experimental values. It must pointed out that the interatomic distances resulting from Vegard's law represent only averages, the real distances are typically more close to those of the individual binary compounds than to the averaged values. A similar linear interpolation approach can be successfully used to evaluate some other properties, in particular the elastic constant, the thermal expansion coefficient and the specific heat, but in general deviations from linearity are significant and more complex interpolation schemes are required. For example, a linear interpolation can be used as first approximation for the bandgap, but an accurate representation is usually obtained introducing quadratic terms to allow for some bowing.

The composition dependence of the refractive index, at a wavelength λ inside the transparency region (which lies between the electronic interband transitions region and the vibrational transitions or "Reststrahlen" region), is approximately linear, provided that λ does not get too close to either limit. The thermal conductivity and the carrier mobility have strongly non-linear behavior, and become much smaller than those of the binaries for intermediate compositions. Figure A1.5 shows the dependence for these parameters on composition in the case of $Al_xGa_{1-x}As$.

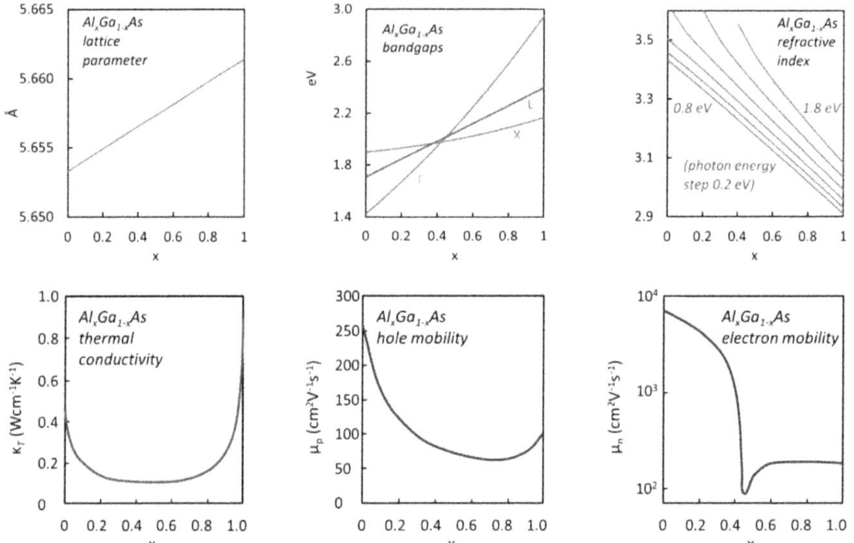

Figure A1.5 Dependence on composition of selected properties in the $Al_xGa_{1-x}As$ system: thermal conductivity, direct (Γ) and indirect (X, L) bandgaps, hole and electron mobility. The step-like trend in the electron mobility is related to the shift from Γ to X minima of the smaller bandgap (see section A1.7.4). Lines originate from models or interpolations of experimental points, and are adapted from [152, 154-156].

There are cases where the atomic distribution is *not* random: this phenomenon is called *ordering*. The driving force for ordering in III-V compounds is not expected to be the thermodynamic stability, but rather a dynamic effect of the growth conditions, correlating for example with the surface orientation of the substrate and with the presence of specific surface reconstructions. A well-known example is that of GaInP grown with vapor-phase epitaxy tecniques on a GaAs, (100) oriented wafer: depending on the growth conditions, there is a more or less pronounced formation of alternated In-rich and Ga-rich $(111)_B$ planes. The ordering impacts the properties of the material, for example the band gap, and is generally unwanted, because ordered materials tend to be inhomogeneous, with domains having different degrees of ordering [157].

A1.5 Epitaxial multilayers: mismatch, strain, relaxation

Most cases of technological interest involve the epitaxial growth of multiple layers on crystal wafer substrates, each layer having the same crystal structure[5] but different composition (heterostructures). In special cases, materials with different compositions can have the same lattice parameter, as in the case of GaAs and $Ga_{0.51}In_{0.49}P$ or of InP and $Ga_{0.47}In_{0.53}As$. More in general, even when the crystal structure is the same, a difference in the stand-alone lattice constant (mismatch) is present, introducing strain in the stack and accumulation of strain energy with increasing layer thickness: beyond a certain critical value, the strain can be spontaneously removed by formation of linear defects called misfit dislocations: this effects is called relaxation. Figure A1.6 schematizes a pseudomorphic strained layer, and a similar layer relaxed via dislocation formation at the interface.

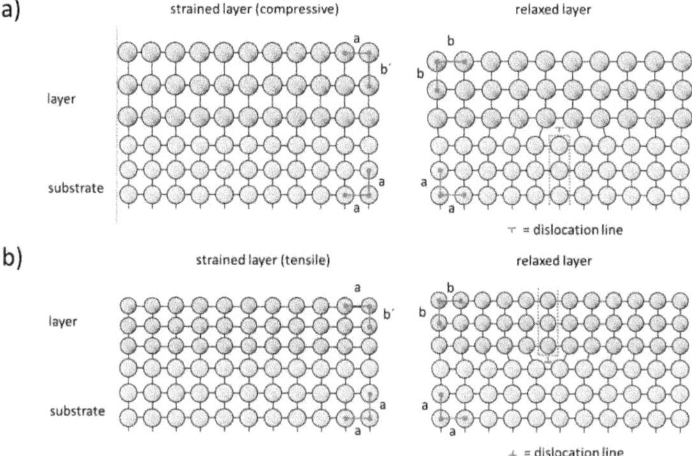

Figure A1.6 Sections of a cubic crystal with lattice parameter b grown on a substrate with lattice parameter a. The layer is represented strained on the left and totally relaxed via dislocation formation on the right.

a) $a < b$: in absence of relaxation, the layer is compressively strained, its lattice parameter in a plane parallel to the interface is reduced to match that of the substrate, while in the normal direction, due to Poisson's effect, there is an elongation ($b' > b$). In the relaxed layer, the lattice parameter is that of the stand-alone material.
b) $a > b$: without relaxation, the layer is tensively strained, its lattice parameter in a plane parallel to the interface is increased, while in the normal direction it is shortened ($b' < b$).

[5] Which is usually the same crystal structure of the substrate, at least in the cases of interest in this thesis; substrates with a different crystal structure with respect to the grown layers can be used either because of availability/cost reasons or because they have desirable properties for specific applications.

Dislocations are extended one-dimensional defects; they are generally unwanted, because they are non-radiative recombination centers, negatively impact the carrier mobility and are detrimental for the reliability and performance of semiconductor devices. They are present not only in multilayered structures but even in bulk crystals, and are described more in detail in the section A1.6.4. Several theoretical models for the determination of the critical layer thickness have been developed, the most cited being probably that of Matthews and Blakeslee [158]: an example of the model prediction is given in Fig. A1.7 [159]. In general the results of this and similar models provide rather guidelines than accurate values [160].

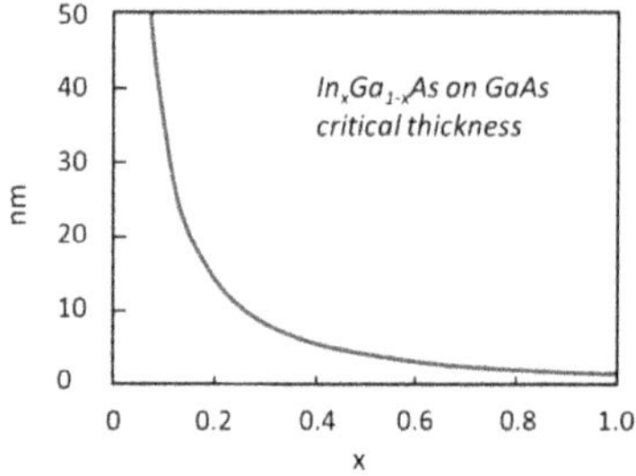

Figure A1.7 Critical thickness of $In_xGa_{1-x}As$ grown on GaAs calculated according to Matthews and Blakeslee model.

A1.6 Defects

As already mentioned in chapter 2, defects related to the presence of foreign atoms are called extrinsic, while defects that represent deviations from the regular arrangement of the lattice but are not associated to foreign atoms are called intrinsic. The presence of *intrinsic* defects in a crystal can be expected on a thermodynamic basis: the minimum value of a crystal Gibbs free energy at any T > 0 K is achieved in correspondence to a finite amount of configurational entropy, which is generated by the creation of defects. Assuming for example that a crystal is synthesized from melt near thermodynamic equilibrium conditions and then cooled down to ambient temperature, the configurational entropy "frozen" in the crystal will be that corresponding to the solidification temperature. In practice, synthesis conditions can be far from the thermodynamic equilibrium, especially in the case of epitaxial growth techniques.

A1.6.1 Point defects

To a first approximation, point defects [4, 5] can be classified based on the anomalies in the position of the atoms within the lattice, considering only the possibility that either regular lattice sites or interstitial positions can be occupied (or left empty). Taking as example GaAs, with Si as impurity, the elementary point defects that can be present in III-V semiconductor crystals, along with the corresponding Kröger-Vink symbols, are:

intrinsic point defects

- vacancies: v_{Ga} = missing Ga atom, v_{As} = missing As atom
- antisites: Ga_{As} = Ga on As lattice site, As_{Ga} = As on Ga lattice site
- self-interstitials: Ga_i, As_i = Ga, As occupying an interstitial site

extrinsic point defects

- substitutional: Si_{Ga} = Si on Ga site, $Si_{As,}$ = Si on As site
- interstitials: Si_i = Si occupying an interstitial site[6]

When the defect is electrically charged, this is indicated by a superscript: + (or •) for each elementary positive charge and - (or ') for each negative; a neutral state can be indicated with 0 (or x).

The generation of a certain type of point defect can be associated with the simultaneous generation of a second type of point defect: an example is a Frenkel pair, composed of a self-interstitial plus a corresponding vacancy, as $Ga_i + v_{Ga}$ in GaAs. This is a typical defect created by particle irradiation or ion implantation, when an atom is knocked out of its lattice position to an interstitial site; the interstitial and the vacancy are not strongly bound together; they can move apart from each other or recombine. More in general, elementary point defects can stably associate to form complex defects.

A1.6.2 Electrical characteristics of point defects

As discussed in chapter 2, electrically active defects are usually classified in shallow and deep according to the position of the energy levels that they introduce in the semiconductor bandgap; shallow levels can easily generate carriers in the conduction and valence bands, deep levels act rather as carrier traps. Some integrative information is added in this section, together with references to other related sections of this appendix.

The localization of the electrons in shallow defect states is comparatively weak, their wavefunctions extends to several nm: when the dopant concentration becomes sufficiently high, their wavefunctions start to merge generating delocalized states (impurity bands) which are conducting, albeit with low

[6] In Zincblende, there are two kinds of interstitial sites, but this aspect is not specified in the notation.

mobility. In parallel, with increasing carrier concentration the ionization energy of the dopants is reduced, because - once ionized - the defects are stabilized by electrical screening. The combination of these factors leads to a reduction of the activation energy for the conduction, the condition corresponding to zero activation energy being called Mott transition. For example, the Mott transition in n-doped GaAs occurs approximately at a dopant concentration of 1×10^{16} cm^{-3}.

Further increase of dopant concentration leads to significant perturbations of the band structure, causing narrowing of the CB-VB bandgap and broadening of the density of states profile at the band edges (band tailing) as better described in sections A1.7.2 and A1.9.5.

Deep electronic levels have energies more near the center of the band gap, and a stronger localization of the electrons. They are responsible of introducing fast non-radiative recombination paths, and negatively impact the carrier mobility. Although these effects are in many cases detrimental for the realization of electronic or optoelectronic devices, they can be useful when there is the necessity of making the material semi-insulating. For example Fe is used to this purpose in InP, and defects related to an excess of As, in particular the antisite As_{Ga} (EL2 defect) are used for GaAs: they behave both as deep acceptors, and are introduced to compensate an originally prevalently n-type conductivity. The effects of deep levels is further discussed in relation to interfaces, carrier transport and interband recombinations in sections A1.7.3, A1.8.2, A1.9.2,

A special case of extrinsic point defects often present in III-V semiconductors is related to hydrogen contamination; hydrogen can *passivate* donor and acceptor atoms forming complexes. Since it can easily diffuse interstitially, in-depth passivation of a semiconductor can happen in processes where the material surface is exposed to hydrogen plasma (as in reactive ion etching) or to hydrogen-containing molecules at high temperature (as in MOVPE epitaxy). Annealing at temperatures of the order of 500°C (in absence of sources of atomic H) causes the decomposition of the complexes and H_2 outdiffusion.

A1.6.3 Structure of deep defects; an example: silicon DX center

A point defect introduces always some degree of local perturbation in the crystal structure (i.e. in the local disposition and bonding of the atoms). Moreover, the ionization of the defect can be associated to a significant change, with formation or breaking of a bond and atom displacements. This can happen especially in the case of deep centers: in principle, a full description of such defects should involve the knowledge of their electronic states and of the corresponding nuclear configurations, but even with the interplay of several experimental techniques and advanced computational approaches, the microscopic structure models of deep defects remains often debated.

A well-studied and prototypical example of an electrically active defect, having different possible nuclear configurations and which can play the different roles of shallow and deep center, is the substitutional silicon impurity in epitaxial $Al_xGa_{1-x}As$ grown on GaAs. The different geometries and the corresponding energies are depicted in Fig. A1.8.

In one configuration a substitutional Si occupies the lattice site of a group III atom, maintaining the normal local bonding structure of the lattice (sp^3 hybridization, bonds with four As atoms). In this configuration, the defect can be either neutral or become positively charged, promoting an electron to the conduction band: Si_{III} is a shallow donor with ionization energy ≈6 meV and is easily thermally ionized:

$$\mathrm{Si_{III}} \rightarrow \mathrm{Si}_{\mathrm{III}}^{+} + \mathrm{e}_{CB}^{-} \qquad \text{R A1.1}$$

According to the so-called large relaxation model, the neutral defect can even *capture* an electron becoming negatively charged and changing its atomic configuration into a new one indicated with *DX*:

$$\mathrm{Si_{III}} + \mathrm{e}_{CB}^{-} \rightarrow DX^{-} \qquad \text{R A1.2}$$

In this reaction, one of the Si-As bonds is broken and the Si atom is displaced through a (111) plane towards an interstitial site, maintaining its bonds to the other 3 arsenic atoms and leaving behind a group III vacancy.

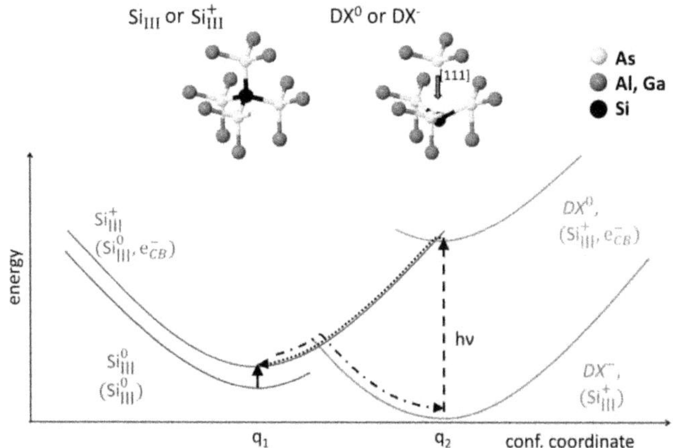

Figure A1.8. Atomic configurations of substitutional, Si and of the associated *DX* center, in $Al_xGa_{1-x}As$ having high Al content ($x > 0.2$). The configuration coordinate refers to a single Si impurity, but the energy is referred to a neutral system comprising two impurities. The parabolae in q_1 represent the shallow donor configuration, those in q_2 the DX configuration, the vertical arrow in q_1 indicates the ionization R A1.1, the dashed vertical arrow in q_2 the photoexcitation R A1.4, the bended dash-dotted arrow the capture R A1.2 and the bowed dotted arrow indicates reaction R A1.5.

In terms of elementary point defects, the DX^- center could be described as a complex formed by the association of a negatively-charged silicon interstitial and a group III vacancy ($DX^- \approx Si_i^- + v_{III}$) although its structure is rather to be considered in its entirety rather than composed by two independent elementary defects.

The DX^- configuration can relax compressive strain present in the lattice, and is stabilized with respect to the unrelaxed one at high pressure or high Al content ($Al_xGa_{1-x}As$ grown on GaAs is compressively strained, in proportion to the Al fraction). The activation energy is of the order of 150 meV for the capture reaction R A1.2, while for the inverse emission reaction it is of the order of 300 meV.

The formation of *DX* centers *reduces* the concentration of shallow donors in Si-doped $Al_xGa_{1-x}As$ according to the overall exothermic reaction:

$$2Si_{III} \rightarrow DX^- + Si^{+}_{III} \qquad \text{R A1.3}$$

The concentration reduction of free electrons due to *DX* centers is important at low (< 200 K) temperatures, but becomes rather small near room temperature.

An interesting effect associated with this defect is the low-temperature persistent photoconductivity. Upon illumination, the carrier concentration (and the conductivity) increases due to photoionization:

$$DX^- \xrightarrow{h\nu} DX^0 + e^{-}_{CB} \qquad \text{R A1.4}$$

The neutral defect DX^0 has the same nuclear configuration of DX^- but it is metastable and reverts to Si_{III} (without an energy barrier) increasing the donor concentration:

$$DX^0 \rightarrow Si_{III} \qquad \text{R A1.5}$$

When the illumination is interrupted, Si_{III} is kinetically stabilized (at low T) by the activation energy of the capture reaction R A1.2 and the conductivity remains higher than before the illumination.

A1.6.4 Extended defects

Three-dimensional defects [145] include all the accidental inclusions (particles), voids, precipitates and in general all volumes that differ in composition, structure or orientation from the rest of the crystal. In the case of MOVPE epitaxial growth, the inclusion of particles can occur because they detach from the reactor chamber, or form due to unwanted reactions in the gas phase. Precipitates can form when the thermodynamic solubility limit of a component is exceeded in the grown solid phase.

Two-dimensional defects include stacking faults, twin boundaries, antiphase boundaries and - in a broad sense – the crystal external surfaces and the internal interfaces, as those present in multilayered structures.

A stacking fault occurs when there is an alteration in the normal sequence of atomic layers: this can happen in particular along the direction perpendicular to the close-packed planes of ccp/fcc lattices – the (111) planes - perturbing the regular stacking sequence, as for example: ABCABC... → ABC**AB**ABC... The formation energy of stacking faults decreases with increasing ionicity of the III-V compounds [161].

Twin defects (also considered tridimensional defects) occur when two differently oriented regions (domains) join in one common plane (twin boundary), the two regions being related by a non-lattice point symmetry operation (as a reflection or a rotation). An example of mirror twinning in a ccp/fcc lattice is the closed-packed planes sequence ABCABC**CBACBA**, the mirror plane being parallel to a (111) plane. Zincblende crystals can have this mirror twinning and a rotational (180°) twinning, where the rotational axis is normal to a (111) plane [162].

Antiphase defects occurs when two equally oriented crystal regions join in one common plane (antiphase boundary) and the relative position of the atoms of each species in the two regions is shifted by a non-lattice translation: in III-V crystals this happens when the lattice positions of group III and V are interchanged in the two domains.

Linear defects commonly to be found are the dislocations [163]. There are two basic types of dislocations: edge and screw, as depicted in Fig. A1.9.

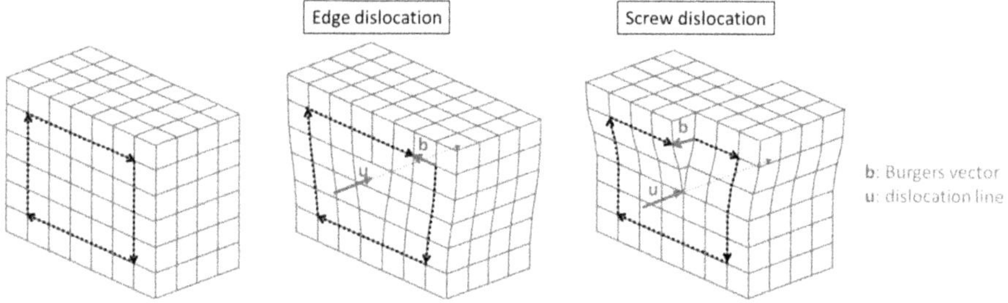

Figure A1.9: edge and screw dislocations, construction of Burgers vectors with the Burgers circuit around the dislocation lines.

An edge dislocation can be visualized as the linear edge of an incomplete crystal plane, beyond which the neighboring planes meet again. A screw dislocation can be visualized as the line, from where two crystal halves slip one from the other in a plane containing the dislocation line. Dislocations are associated to Burgers vectors **b**, which are translation vectors of the lattice and are found describing a lattice-point to lattice-point circuit, in such a way that it would be closed in a perfect crystal but remains open when a dislocation line **u** is included, as shown by the dotted arrows in Figure A1.9: the Burgers vector connects the start to the end of the circuit. Burgers vectors are perpendicular to edge dislocations and parallel to screw dislocations.

Real dislocations have often a mixed edge-screw character and can be curved. A dislocation cannot end abruptly: it must reach the crystal surface, end at a junction with other dislocations or describe a closed loop. Dislocations emerging from the substrate will extend in the epitaxial layers. The dislocation segments that extend along the interface between two epitaxial layers (or between the substrate and an epitaxial layer) are called misfit dislocations, because they can relax the strain associated with interface misfit (while a dislocation in the bulk of a crystal always introduces extra strain). The dislocation segments that extend through the epitaxial layers are called threading dislocations. Provided that sufficient thermal or mechanical energy is supplied, dislocations can move; the movement is relatively easy in a plane containing the dislocation line and the Burgers vector - this is called "glide" or "slip" movement - while movements outside the glide plane - called "climbing" - imply the absorption or creation of vacancies (or interstitials) and require more energy.

The dislocations that can potentially exist in the Zincblende lattice have been classified [164] according to the directions of **u**, **b,** and to the glide plane (if present). Commonly found dislocations are the 60° dislocations, represented in Fig. A1.10: they are mixed dislocations with a prevalent edge character, they slip in the (111) planes, and have a 60° angle between **u** and **b**. In a strained multilayer structure grown on a (100) wafer, these dislocations glide to the interfaces, where they can relax the misfit strain.

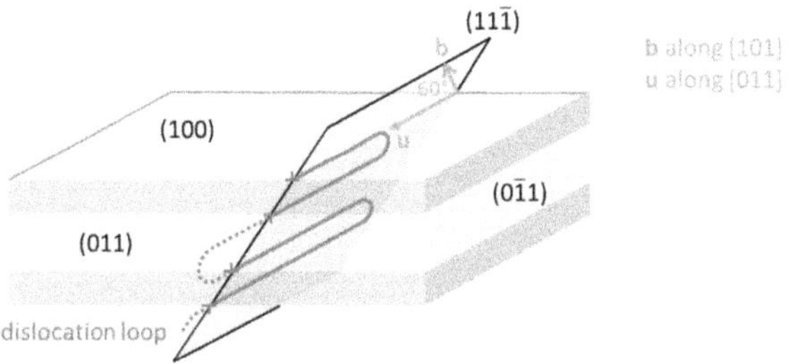

Figure A1.10 Schematic of a 60° dislocation that intersects the interfaces of a strained multilayer, originating horizontal misfit segments.

Dislocations can decompose into partial dislocations, which are linear defects similar to dislocations, but their Burgers vector are not a translation of the lattice, while the sum of their Burgers vectors is a translation of the lattice and corresponds to the vector of the total dislocation; an example of a dislocation in an fcc crystal, which splits in two partial dislocations and creates a stacking fault ribbon, is shown in Fig. A1.11.

Most of the extended defects are associated with deep electronic levels, and have similar effects of deep point defects: reduction of the carrier density, reduction of the mobility via ionized defect formation and increase of non-radiative carrier recombination.

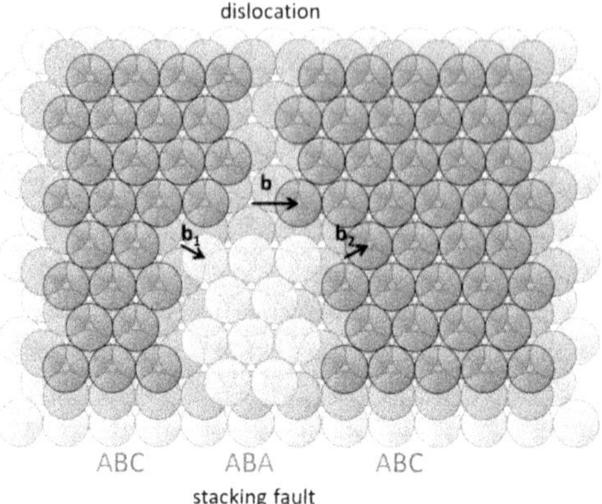

Fig. A1.11 In an fcc crystal, a dislocation with Burgers vector **b** and laying in a (111) slip plane splits into two partial dislocations, with Burgers vectors $\mathbf{b}_1$ and $\mathbf{b}_2$ (where $\mathbf{b}=\mathbf{b}_1+\mathbf{b}_2$). The region between the two partial dislocation lines is a stacking fault.

A1.7 Electronic structure and related properties

A1.7.1 Band structure of crystals [2, 146]

In absence of external time-dependent perturbations, the electrons within a solid crystal can approximately be described by stationary multi-electron wavefunctions Ψ which are solutions of the time-independent Schrödinger eigenvalue equation: $\hat{H}\Psi=\mathcal{E}\Psi$, where $\hat{H}$ is a time-independent Hamiltonian operator. This approach is based on Born-Oppenheimer approximation: the full electron-nuclear many-body wave function is expanded in terms of electronic and nuclear wave functions, the nuclear kinetic energy term – but not the internuclear repulsion - is excluded from $\hat{H}$, i.e. nuclear position are fixed; in this sense, the eigenfunctions Ψ and the eigenvalues $\mathcal{E}$ are both parametric functions of the nuclear coordinates $\mathbf{X}_i$. Solving the Schrödinger equation for different nuclear coordinates leads to the construction of a potential energy surface $\mathcal{E}(\mathbf{X}_i)$ for the motion of the nuclei, which can be used in a second step to determine the nuclear wavefunctions.

The electron wavefunction Ψ contains space and spin (intrinsic angular momentum) coordinates of all the electrons, and must be antisymmetric (change sign) with respect to electron permutations, in order to satisfy the indistinguishability principle and the generalized Pauli principle; moreover, $\hat{H}$ contains significant terms related to electron-electron coulomb interactions and to spin-orbit magnetic interactions: all these factors conjure to make in principle Ψ not separable, neither in single-electron components nor in space and spin components.

In practice, this is done in a further simplification called *independent electron approximation*, leading to a representation of the electronic structure where single electrons can "move" or "jump" in a partially quasi-continuous and partially discontinuous domain of single-electron quantum states, the electronic *bands.* Corrections can be added to take into account the terms neglected in the accumulated approximations without abandoning the fundamental picture of independent electrons.

According to the independent electron approximation, each electron is described by a spatial wavefunction $\varphi(\mathbf{r})$ which satisfies the single-electron Schrödinger equation:

$$\left(-\frac{\hbar^2}{2m_0}\nabla_r^2 + U(\boldsymbol{r})\right)\varphi(\boldsymbol{r}) = \varepsilon\varphi(\boldsymbol{r}) \qquad \text{E A1.4}$$

Where $\mathbf{r}$ is a vector representing the electron position in space, m_0 is the electron rest mass, ε is the single-electron energy eigenvalue and $U(\mathbf{r})$ is an effective potential which contains the interactions with the nuclei, and the interactions with the other electrons only in an averaged way. For each $\varphi(\mathbf{r})$ there are actually two occupation possibilities corresponding to the two possible spin states.

On general grounds, it can be proved (Bloch's theorem) that the solutions of the Schrödinger equation E A1.4 *in a Bravais lattice* can be selected to have the form of a "plane wave" or "envelope" term $e^{i\boldsymbol{k}\cdot\boldsymbol{r}}$ with wavevector $\mathbf{k}$ multiplied by a "Bloch function" $u(\mathbf{r})$ that has the periodicity of the lattice:

$$\varphi_{n,\boldsymbol{k}}(\boldsymbol{r}) = e^{i\boldsymbol{k}\cdot\boldsymbol{r}}u_{n,\boldsymbol{k}}(\boldsymbol{r}) \qquad \text{E A1.5}$$

The quantum number *n* is called band index, the set of wavefunctions with a common band index are called bands and the set of the corresponding energies $\varepsilon_n(\mathbf{k})$ are called *energy* bands.

This form satisfies the condition that the wavefunctions are simultaneously eigenstates of the Hamiltonian operator *and* of the operators representing the translations of the Bravais lattice to itself. A particle that satisfies this condition is called Bloch particle. Within a band, $\mathbf{k}$ does not vary in a strictly continuous way: the density in k-space of the allowed values of $\mathbf{k}$ is proportional to the crystal volume.

Because of the uncertainty principle, a Bloch electron has no well-defined velocity (the corresponding operator $\hat{v}$ does not commute with the Hamiltonian). Nonetheless it has an expectation-value velocity $\langle\boldsymbol{v}\rangle$, and it can be proved that $\langle\boldsymbol{v}\rangle$ is proportional to the k-space gradient of the band according to:

$$\langle\boldsymbol{v}\rangle = \langle\varphi_{n,\boldsymbol{k}}|\hat{v}|\varphi_{n,\boldsymbol{k}}\rangle \equiv \int_{-\infty}^{\infty}\varphi_{n,k}^{*}\frac{-i\hbar}{m_0}\nabla_{\boldsymbol{r}}\varphi_{n,\boldsymbol{k}}d\boldsymbol{r} = \frac{1}{\hbar}\nabla_{\boldsymbol{k}}\varepsilon_n(\boldsymbol{k}) \qquad \text{E A1.6}$$

In order to define, to a certain extent, position and velocity for the electrons, it is necessary to relax the precision with which **k** is known, making use of wave-packets which are linear superposition of the Bloch wavefunctions in a limited k-space interval around a given **k**: the expectation-value velocity corresponds to the group velocity of the wave-packet. This approach justifies a semiclassical description of motion where the electrons are treated as particles.

As for the case of velocity, the Bloch-electron momentum is not well-defined: its expectation-value is $m_0\langle \boldsymbol{v} \rangle$. The vector **k** is associated to a quantity $\hbar\mathbf{k}$ called *crystal* momentum (or quasi-momentum), and only in the limit of a zero or constant potential U(**r**) (free electron) the crystal momentum becomes properly the electron momentum. Nonetheless, since **k** is a quantum number of the stationary states, for $\hbar\mathbf{k}$ a conservation law holds similar to that of a proper momentum (more in section A1.8.1).

Due to the translational symmetry of the crystal, the energy bands can be represented as a function of **k** (dispersion relation) taking only **k** values within a limited portion of the k-space: this portion is called *first Brillouin zone* (BZ) and corresponds to the primitive Wigner-Seitz cell of the reciprocal lattice[7]. Any **k**′ outside the first BZ, is equivalent to a vector **k** inside the first BZ obtained through a translation given by a vector **K** of the reciprocal lattice. In the limit of very weak interaction of the electrons with the lattice (free-electron approximation) the band energy has a quadratic dependence on |**k**|.

A1.7.2 Carrier statistics and semiconductors bands [2, 5, 146, 165]

Each band is characterized by a density of states[8] per unit energy and per unit real-space volume $g_n(\varepsilon)$. The occupancy probability of an electronic state having energy ε obeys at thermodynamic equilibrium the Fermi-Dirac distribution (which assumes non-interacting particles):

$$f(\varepsilon) = \frac{1}{e^{(\varepsilon-\varepsilon_F)/k_BT} + 1} \qquad \text{E A1.7a}$$

where k_B is the Boltzmann constant and ε_F is the Fermi level. The Fermi level is the energy value at which the occupation probability is 0.5 at the given temperature and, from a thermodynamic point of view, is the chemical potential of the electrons. In the case of doped semiconductors, the carriers originates even from localized states in the bandgap, whose occupation follows a Fermi distribution modified inserting a pre-exponential factor $1/\gamma$ in the denominator, to take into account their ground-state degeneracy γ:

$$f(\varepsilon_d) = \frac{1}{\gamma^{-1}e^{(\varepsilon_d-\varepsilon_F)/k_BT} + 1} \qquad \text{E A1.7b}$$

where ε_d is the energy of the dopant level.

For each band, the *electron* density per unit energy and per unit real-space volume $dn(\varepsilon)$ is the product of the Fermi function times the density of states; the electron density per unit volume n is found integrating $dn(\varepsilon)$ over energy:

$$n = \int dn(\varepsilon)\cdot d\varepsilon = \int f(\varepsilon)\cdot g_n(\varepsilon)\cdot d\varepsilon \qquad \text{E A1.8}$$

For symmetry reasons, a completely filled band (and an empty one) does not contribute to the electrical conduction, while a partially filled one does, because the electron population can redistribute in the intra-band states in order to produce a net electron flow in a given direction.

In semiconductors, the Fermi level is located in an energy gap (ε_{gap}) between the bands where there are

---

[7] The reciprocal lattice is defined by the relation $e^{iK\cdot R} = 1$, where **R** is a vector of the direct lattice and **K** a vector of the reciprocal lattice; the reciprocal lattice is the Fourier transform of the direct lattice.

[8] Here states are the space functions $\varphi_{n,\mathbf{k}}(\mathbf{r})$, which are taken 2 times each, because of the 2 possible electron spin values.

no states available, so that at T = 0 K, when the Fermi distribution is a 0-1 step function, the lower bands are completely occupied, the higher unoccupied, and the material is actually an insulator. The higher occupied band is called valence band, the lowest unoccupied conduction band (both names being somewhat misleading, as pointed in the following).

At T > 0K some electrons are thermally promoted from the VB to the CB, allowing for non-zero conductivity: the conductivity is not only due to the electrons in the *conduction* bands, but even to those in the valence band – or in an equivalent representation - to the empty electronic states in the VB, which are called holes. The probability distribution of holes is the complement to 1 of the Fermi function.

VB and CB can be seen as the natural extension, to a very large and periodic set of atoms, of the highest-occupied and lowest-unoccupied molecular orbitals (HOMO-LUMO) that result from the LCAO method (linear combination of atomic orbitals) usually applied to molecules. In the corresponding tight binding method used for crystals, applied to III-V Zincblende compounds, *both* VB and CB of are approximately derived from the combination of the *valence s, p* atomic orbitals. Stronger bonds correspond to higher HOMO-LUMO splitting and similarly a higher cohesive energy corresponds to higher VB-CB separation. Figure A1.12 depicts qualitatively this approach. Actually, in Zincblende there are 3 similar bands that could qualify to some extent as "valence bands", so the singular is to be understood as a collective term when no differently specified.

The VB absolute maximum and the CB absolute minimum can be aligned in k-space as in Fig. A1.12 (*direct* bandgap) or correspond to different values of **k** (*indirect* bandgap). This has very important implications in relation to the optical properties of the semiconductor (section A1.9).

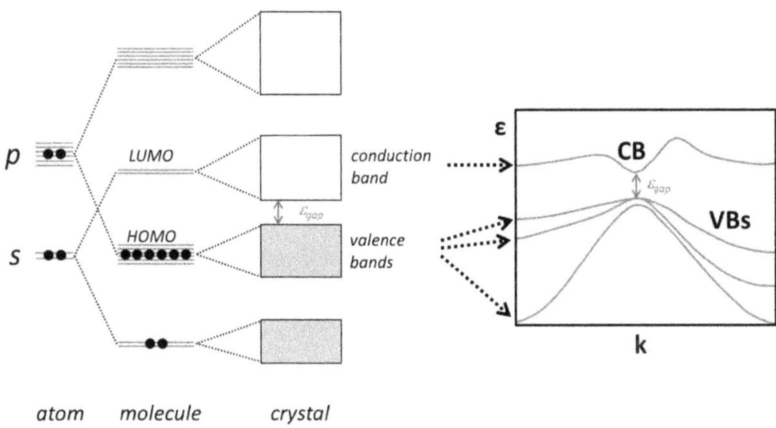

Fig. A1.12 Qualitative illustration of band formation from atomic orbitals in III-V compounds: to a first approximation, the CB originates from the s valence atomic orbitals, while three VBs originate from p valence atomic orbitals; to the right, a qualitative sketch of how the *energy* bands actually look like in k-space, in the case of direct bandgap III-V compounds.

At thermodynamic equilibrium, the bands depend only on temperature and pressure, while more in general they can be modified by other factors, e.g. the application of external electric fields. Both pressure and temperature dependences are related to the change in the internuclear distances, but the temperature dependence is even significantly caused by effects not included in the Born-Oppenheimer approximation (vibronic or "electron-phonon" interactions) [1].

Moreover, the band energies are not strictly independent from the way in which the band states are populated, because of the electron-electron interactions, which are contained explicitly in the multi-electron Hamiltonian, and in an averaged way in the effective potential *U(r)* of the one-electron Hamiltonian: in particular, increasing the carrier density in the CB and VB, the bandgap shrinks, which can be intuitively understood as a weakening of the bonds and of the crystal cohesive energy as the electrons are added to antibonding states or removed from bonding states[9].

The shape of the energy bands in k-space and their density of states can be rather complex, but for many applications it is mainly important the knowledge of the limited k-space regions around the maximum of the VB (energy ε_{VB}) and the minimum of the CB (energy ε_{CB}), because in most situations those are the regions populated by the carriers and involved in the current conduction and light emission mechanisms. To a first degree of approximation, in these regions the bands $\varepsilon_n(\mathbf{k})$ are parabolic functions of the components of **k**, and the density of states $g(\varepsilon)$ is proportional to the square root of the energy difference $\varepsilon_n(\mathbf{k}) - \varepsilon_{CB}$ for the CB and $\varepsilon_{VB} - \varepsilon_n(\mathbf{k})$ for the VB.

The integrated electron density in the CB given by E A1.8 is conventionally expressed in a non-integral form as the product of the Fermi-Dirac distribution f calculated at the CB bottom times a parameter N_C called *effective* density of states of the CB. Similarly the hole density in the VB is the product of $(1-f)$ - with f calculated at the top of the VB - times the effective density of states in the VB, N_V. This corresponds to consider all the states as concentrated at the band edges. Moreover, the Fermi-Dirac distribution is approximated with a Maxwell-Boltzmann distribution referred to the band-edge energy, and the carrier densities are given by:

$$n \approx \frac{N_C}{e^{(\varepsilon_{CB}-\varepsilon_F)/k_BT}+1} \approx N_C \exp\frac{\varepsilon_F - \varepsilon_{CB}}{k_BT} \qquad \text{E A1.9a}$$

$$p \approx \frac{N_V}{e^{(\varepsilon_F-\varepsilon_{VB})/k_BT}+1} \approx N_V \exp\frac{\varepsilon_{VB} - \varepsilon_F}{k_BT} \qquad \text{E A1.9b}$$

where n is the density of electrons in the CB, p is the density of holes in the VB. In GaAs at room temperature N_C=4.7×10^{17} cm^{-3}, N_V=9.0×10^{18} cm^{-3}. N_C, N_V can be expressed under the parabolic approximation by:

$$N_{C,V} = \frac{1}{\sqrt{2}}\left(\frac{m^*_{e,h} k_B T}{\pi\hbar^2}\right)^{3/2} \qquad \text{E A1.10}$$

where $m^*_{e,h}$ is a parameter inversely proportional to the band curvature called *effective mass* of the electrons or of the holes (better defined in section A1.8.1). It has actually the dimension of a mass, and for Zincblende III-V semiconductors is a fraction (in the range 0.01-0.9) of the electron rest mass m_0. From E A1.9 is can be seen that the carrier density is strongly temperature dependent when the Fermi energy is far from the band edge.

In using the equations E A1.9, it should be considered that The Boltzmann approximation of f is valid only for non-degenerate semiconductors, where ε_F is at least $3k_BT$ away from the band edge; above this level it starts to overestimate the carrier density. Moreover, equations E A1.9 are not valid for degenerate semiconductors *even* without using the Boltzmann approximation, because the resulting carrier density saturates at N_C or N_V as f approaches unity, underestimating the carrier density. Expressions valid in the degenerate case can be found for example in Ref. [58].

Figure A1.13 illustrates how the distribution function and the density of states combine to determine the carrier density in the VB and in the CB, considering the cases of an intrinsic (not doped) and of an n-doped semiconductor: the electrons accumulate near the bottom of the CB and the holes near the top of the VB (the top of the VB is the point of minimum energy for the holes). In an intrinsic semiconductor the carriers can be originated only by the promotion of electrons from the VB to the CB, so the electron

---

[9] The effect is also called bandgap "renormalization"; in doped semiconductors, even the presence of fixed ionized impurities does contribute to the bandgap narrowing.

density in the CB and the hole density in the VB must be identical (n = p); the Fermi level lies approximately in the center of the band gap, slightly shifted towards the band with the lower density of states. In extrinsic semiconductors, n and p can differ because ionized donor and acceptor impurities enter the charge balance: from E A1.9 it can be immediately derived that their product is subject to the so-called law of mass action:

$$n \cdot p = n_i^2 = N_C N_V e^{-\varepsilon_{gap}/k_B T} \quad \text{E A1.11}$$

where n_i is the electron (and hole) density of the intrinsic material at the selected temperature; as E A1.9, E A1.11 is strictly valid in the non-degenerate case.

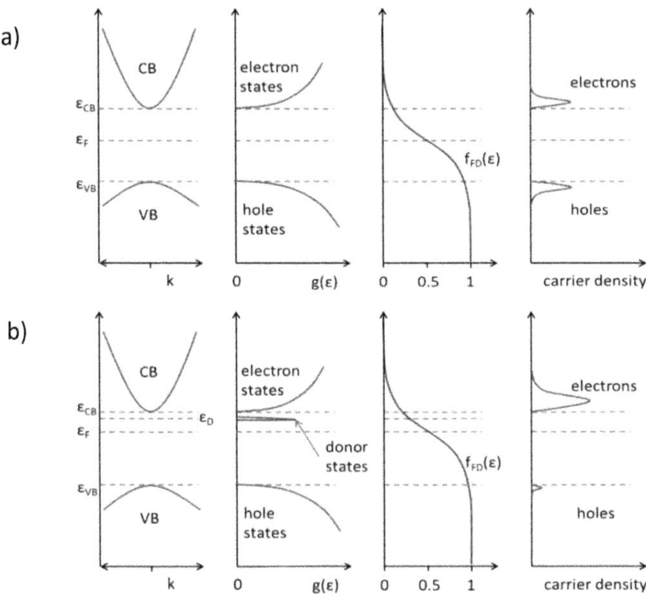

Fig. A1.13 From left to right: valence and conduction energy bands profiles, density of states including the localized donor states at energy ε_D, Fermi distribution function and carrier densities; **a)** intrinsic direct semiconductor, **b)** n-doped direct semiconductor.

The intrinsic carrier density depends strongly on temperature and band gap, examples of room-temperature values of n_i are:

1×10^{15} cm^{-3} for InAs (ε_{gap}= 0.36 eV),

2×10^{6} cm^{-3} for GaAs (ε_{gap}= 1.43 eV),

1×10^{3} cm^{-3} for $Al_{.32}Ga_{.68}As$ (ε_{gap}= 1.93 eV).

Excluding the very low bandgap compounds, these values are much smaller than the typical range of intentional doping (≈ from 1×10^{15} to 1×10^{20} cm^{-3}) and even of a typical unintentional background doping (≈1×10^{14} cm^{-3}). In a doped semiconductor at $T \approx T_{amb}$ the shallow levels are almost completely ionized, and assuming a donor concentration $N_D >> n_i$ the hole density must become very low to satisfy E A1.11: the conduction is dominated by the carriers originated by the dopants (majority carriers). In this case, the Fermi level will be very near the dopant level. In general, the value of ε_F and the equilibrium concentrations of carriers, ionized donors and acceptors can be obtained from a system of equations comprising the charge balance and equations E A1.9 (one for the CB, one for the VB and one for each donor or acceptor type).

Another consequence of E A1.11 is that, when donors and acceptors are simultaneously present, they tend to compensate each other, and when $N_D = N_A$ the carrier concentrations become as low as that of the intrinsic material. An effect equivalent to compensation can be obtained even due to the presence of deep traps: taking as example a material with shallow donors and deep electron traps in concentration $N_T \geq N_D$, the donors will be ionized but the electrons will be mostly trapped, the Fermi level will be "pinned" towards the center of the bandgap near the energy of the deep traps, and the carrier concentrations will be similar to those of an intrinsic material.

Figure A1.14 shows qualitatively how the temperature impacts the Fermi level position (left) and the carrier density (right). At low temperature the carrier concentration is determined by the incomplete ionization of the shallow levels and ε_F is near the donor or acceptor level (which are very near to the CB or VB respectively), at intermediate temperature – which with the above definition of shallow levels includes room T - the ionization is complete and the carrier concentration coincides with the donor (or acceptor) concentration, while ε_F moves towards the center of the gap, by an extent that depends on the dopant concentration. At a sufficiently high temperature (depending again on the dopant concentration) the direct promotion of electrons from the VB to the CB dominates, and ε_F reaches the middle of the bandgap.

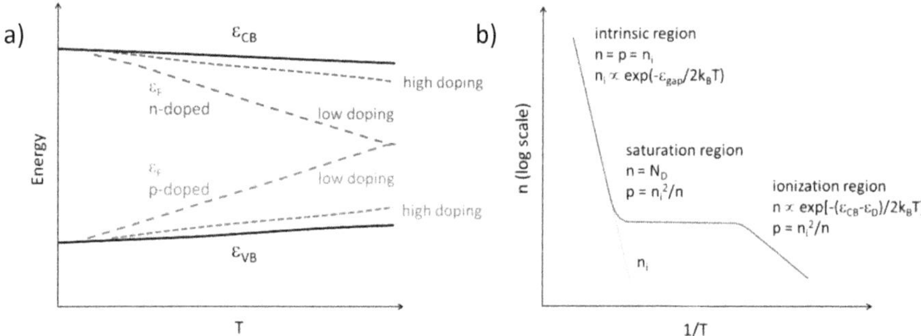

Fig. A1.14 **a)** Fermi level position at different temperatures for high and low dopant concentrations; **b)** carrier concentration as a function of reciprocal temperature for an n-doped semiconductor with N_D = donor concentration and ε_D = donor energy level [5].

A1.7.3 Interfaces and band offsets

When a system is compositionally structured in real space – as in the presence of interfaces or quantum structures - the electronic bands approach needs to be corrected to take into account non-periodic perturbations of the effective potential for the electrons $U(\boldsymbol{r})$; to a first approximation, this impacts primarily the envelope part of the electronic wavefunctions. A local perturbation can lead to the presence of localized or partially localized wavefunctions, as in the case of quantum-well structures. In most descriptions, the focus is concentrated only on the band energies, and in particular on the band energies *edges*, which are treated as well-defined real-space functions, although what is properly a real-space function is $U(\boldsymbol{r})$, since any band energy is associated to delocalized wavefunctions. In other words, CB and VB edge band energies are qualitatively described *as if* they were the potential energy for the carriers (electrons or holes), but this is only partially correct (see section A1.8). Nonetheless, in the following this simplified descriptive approach is followed.

An important characteristic for semiconductor devices fabrication is how the bands of two different semiconductors align when they are joined in a heterostructure.

When the difference between the two parts of a junction lies only in the doping type and concentration (homojunction), the band alignment is controlled by the position of the Fermi level relative to VB and CB on each side of the junction: the carriers redistribute across the interface creating an electrical potential difference (built-in potential, V_{bi}) that compensates the chemical potential difference. The extent of the transition region where the carrier density differs from the bulk value depends on the doping concentration, and is larger for smaller doping; in the case of pn junctions, the transition region is depleted from carriers with respect to bulk values. Taking the case of an abrupt pn junction in GaAs at room T, the depletion region would extend by 1.7 μm when $n = p = 1\times10^{15}$ cm^{-3}, and only by 0.02 μm when $n = p = 1\times10^{19}$ cm^{-3}.

When the compositions of the two materials are not the same with regards to the main constituent atoms (heterojunction) a complication arises from the abrupt change in the effective potential for the electrons of E A1.4 at the interface, which induces a similarly abrupt change in the bands. Focusing on the minimum of the CB and the maximum of the VB, the problem is to determine how they align with each other at the interface. A simplistic model that gives nonetheless qualitatively correct results, is to assume that the band energies of the separate materials can be determined with respect to a common reference, and that the offset between the corresponding bands is retained at the interface of the joined materials, while the bulk values adjust in order to equalize the Fermi level as in the homojunction case. Based on this picture, a transitivity of the interface offsets is expected, which is approximately confirmed by experiments. More accurate models of heterostructure band alignment must take into account the specific characteristic of the interface and the presence of strain.

Figure A1.15 depicts the different types of band alignment that can occur in a heterojunction (straddled, staggered and broken) and four examples of junctions: a pn homojunction and three straddled heterojunctions with different doping types.

For the heterojunctions, the vacuum level ε_0 is taken as common energy reference, the difference between this level and the bottom of the CB is the electron affinity ε_{EA}. In the three heterojunctions, a *spike* is formed in the CB or in the VB, because of the abruptness of the change in ε_{EA}. This kind of feature, which introduces a well followed by a barrier for the carriers, is in many cases unwanted, and can be avoided when it is possible to realize compositionally graded heterojunctions, a typical example being the system GaAs-$Al_xGa_{1-x}As$ where the x value is varied continuously, distributing the variation of ε_{EA} over the appropriate length, typically tens on nanometers.

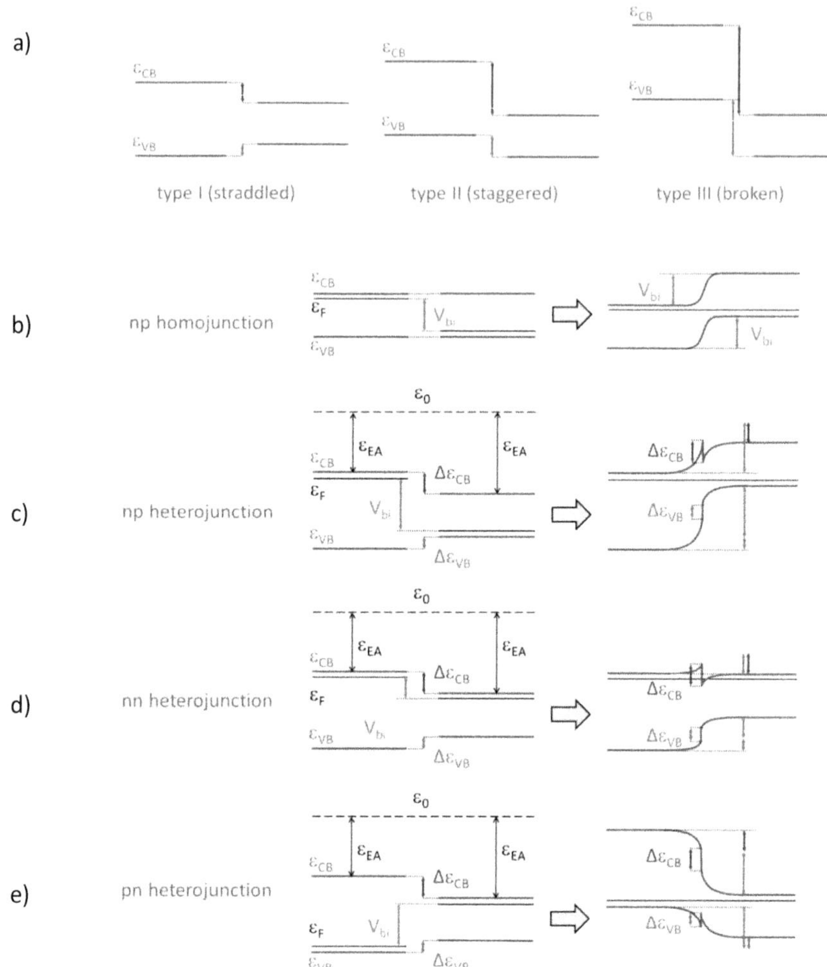

Figure A1.15 **a)** types of band alignment in heterojunctions; **b)** homojunction; c-e) type I heterojunctions; V_{bi} is the built-in potential, ε_0 the vacuum potential and ε_{EA} the electron affinity.

A1.7.4 Band structure of III-V Zincblende semiconductors

The energy bands have as domain the first BZ in the k-space and would need four dimensions to be represented, but for practical reasons they are normally drawn taking cuts along certain high-symmetry lines in the first BZ. In general, the center of the first BZ is called Γ, and other Greek letters are used to indicate lines inside the BZ, while roman letters are used for points and lines on the surface of the BZ. For the face centered cubic direct lattice (to which corresponds a body centered cubic reciprocal lattice) the points and lines are indicated in Fig. A1.16a. The directions[10] correspond to those of the lattice according to:

<100> → line Γ-X (Δ) <110> → line Γ-K (Σ) <111> → line Γ-L (Λ)

Figure A1.16b is a representation of the energy band structure of GaAs [166], where only the four bands adjacent or very near the band gap are drawn. These bands can be approximately considered as derived from the s and p valence atomic orbitals, as described in the previous section. There is a single conduction band with 3 relative minima in L, Γ and X, of which Γ is the absolute minimum, and two valence bands, degenerate in their single maximum, corresponding again to Γ. Based on the effective mass (inversely related to the curvature of the band) the VBs are called heavy hole (HH) and light hole (LH) bands. The density of states is proportional to the effective mass: that of the HH band is significantly higher than that of the LH band (e.g. $0.55m_0$ vs $0.083m_0$ in GaAs), which implies that, even near the degeneration point Γ, the HH band has a much higher carrier population. It should be added that, when the material is strained, the HH-LH degeneration is removed: a compressive strain shifts the HH maximum above the LH maximum and vice-versa a tensile strain shifts the LH maximum above the HH maximum. The correspondence of CB minimum and VB maximum in Γ makes of GaAs a direct band gap semiconductor.

The fourth band, called split-off band (SO), is similar in shape to the HH and LH bands, but has lower energy and is not so easily populated by holes as the others. The SO band would be degenerate with HH and LH in Γ in absence of spin-orbit interaction[11]: this interaction is stronger for compounds with heavy atoms, so for example the separation in Γ is 0.09 eV for GaP, 0.34 eV for GaAs and 0.87 eV for InSb.

The CB and the SO bands have spherical symmetry in Γ, while the LH and especially the HH bands are anisotropic, as can be seen comparing the <111> and <100> profiles. Moreover, it can be seen that the Γ point is unique, while there are 8 L and 6 X points in the BZ.

The other III-V compounds with Zincblende structure (binaries and higher order compounds) have band structures qualitatively very similar to that of GaAs (shifted of course to different energy values). An important difference that can arise is related to the relative energy of X, Γ and L valleys in the CB. In some compounds, including the binaries AlP, GaP, AlAs and AlSb, the X valley becomes the absolute minimum. More rarely, this can happen for the L valley. The maximum of the VB is in all cases fixed in Γ, and the compounds with CB absolute minima in X or L are *indirect* band gap semiconductors.

The effect of increasing the pressure is to increase the bandgap, approximately by 0.01-0.02 meV bar^{-1} for all the III-V Zincblende compounds [1]: this effect is due to the reduction in the lattice parameter.

The effect of an increase in temperature is to reduce the bandgap, approximately 0.3-0.5 meV K^{-1} (near room temperature) for all the III-V Zincblende compounds: this is partially due to the increase in the lattice parameter caused by thermal expansion, and partially to electron-phonon interactions [1].

The effect of doping is to reduce the bandgap; for GaAs, the narrowing is approximately given by [5]:

$\Delta\varepsilon_{gap} = -6.6 \times 10^{-5} \cdot \sqrt[3]{N_D}$ (for n-doping) $\Delta\varepsilon_{gap} = -2.4 \times 10^{-5} \cdot \sqrt[3]{N_A}$ (for p doping)

where the bandgap variation is in meV, and the doping concentration in cm^{-3}.

---

[10] In a cubic lattice the direction of direct and reciprocal lattice are the same.

[11] HH, LH and SO bands are generated from the p atomic orbitals, which have non-zero angular momentum.

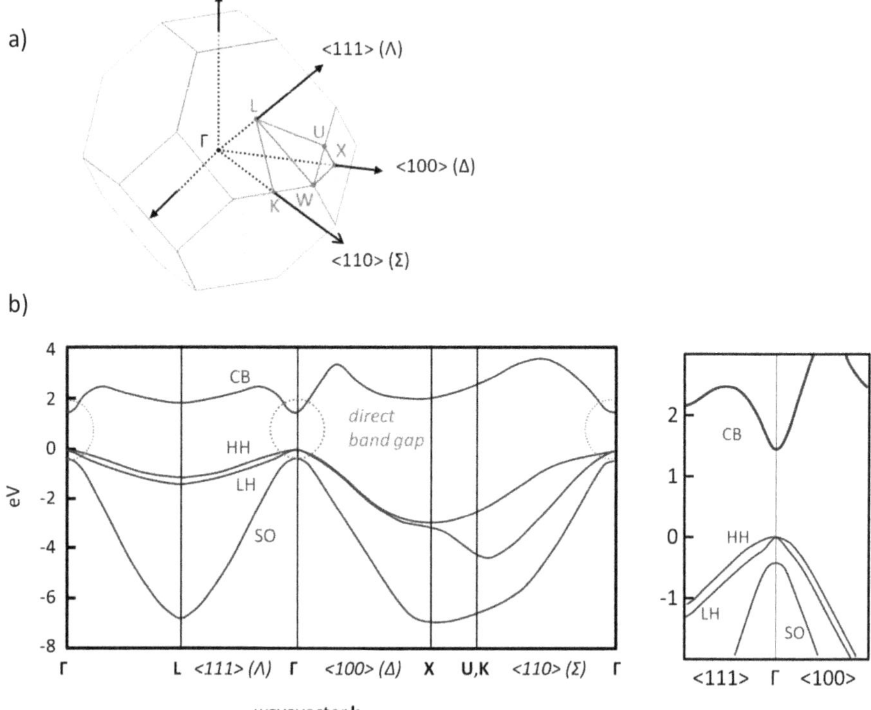

Figure A1.16 **a)** first Brillouin zone of fcc lattice with indication of lattice directions and symmetry points and lines; **b)** energy band structure of GaAs represented along high-symmetry lines, the area around direct bandgap is zoomed-in to the right.

Figure A1.17 represents the bandgap energy and the lattice parameter of several III-V compounds. The position of the binary compounds is indicated by points, while the lines connecting the binary points represent the ternary compounds (the composition of a ternary is univocally determined by its lattice parameter from Eq. E A1.1). The ternary lines are continuous for direct bandgap and dashed for indirect bandgap semiconductors. For example, the shift from direct to indirect along the line connecting GaAs with AlAs corresponds to the variation of the relative position of Γ and X valleys, which has been already shown in Fig. A1.5. Quaternary compounds are not explicitly represented; their possible positions in the diagram span the area defined by the corresponding three or four ternary lines. The lattice constant corresponding to a specific composition can be easily evaluated from linear interpolation of the binary values (section A1.4) while the bandgap requires quadratic terms in the interpolation and comparison of X, Γ and L values.

In order to avoid the detrimental effects of lattice relaxation, Zincblende multilayer epitaxial structures for optoelectronics applications are usually realized using compounds having lattice constants not too far from that of the substrate[12]. Available commercial substrates are – for technological reasons – limited to the six binary compounds of In and Ga, and the fabricated devices are often divided accordingly into groups, e.g. InP-based, GaAs-based etc. Nonetheless, use of multinary alloys leaves considerable space for "engineering" the bandgap and other properties of the materials maintaining an acceptable lattice mismatch with the substrate.

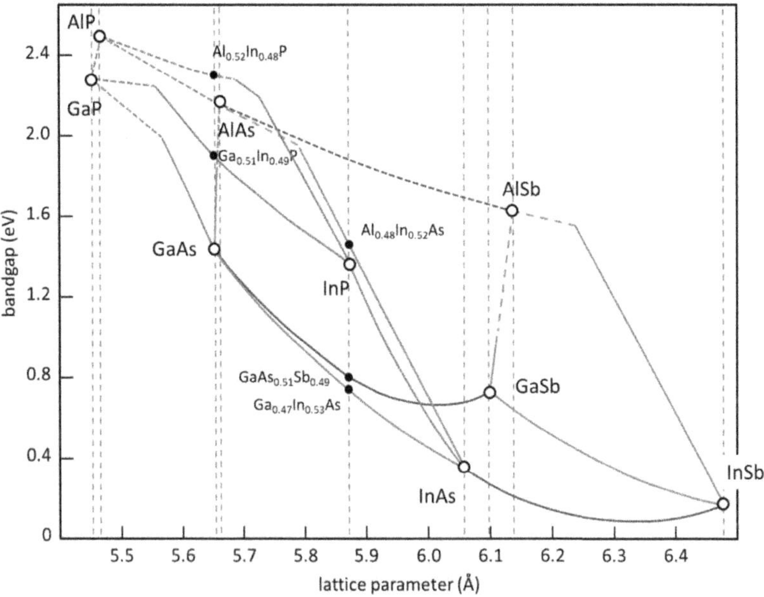

Figure A1.17. Room-temperature bandgap energy of Zincblende III-V semiconductors plotted vs lattice constant. Empty circles indicate the binaries, full circles indicate selected ternaries (lattice-matched with GaAs or InP). Lines indicate ternary compounds, continuous lines are used for direct bandgap and dashed for indirect. Data from [167, 168].

---

[12] This is not the case for nitride (Wurtzite) compounds, where high dislocation densities are comparatively less detrimental to the optoelectronic properties and lattice-matched substrate not easily available.

Band alignment parameters are less well known than bandgap values: approximate values based on the assumption of transitivity and not taking into account the effect of strain are plotted in Fig. A1.18.

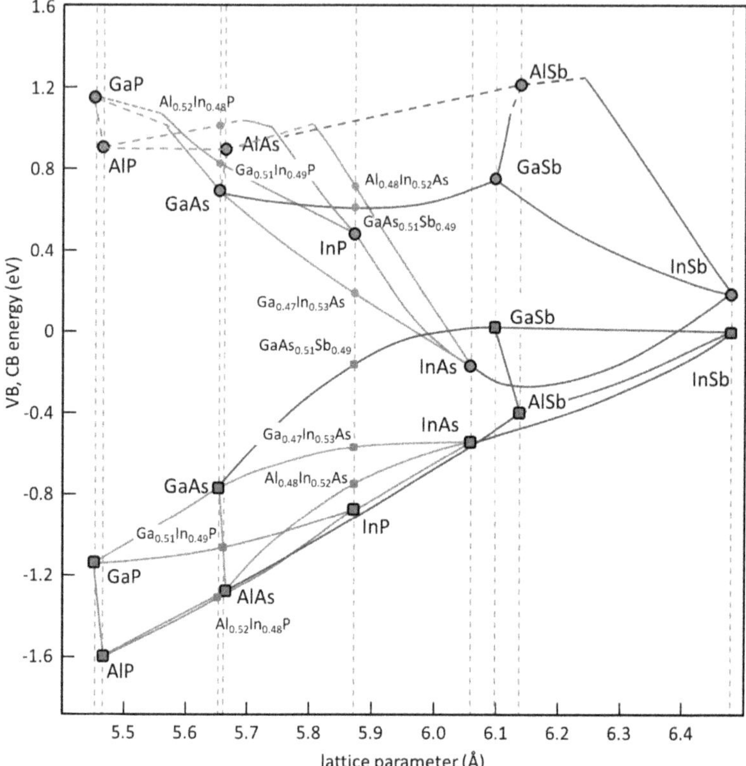

Figure A1.18 Room-temperature conduction-band edge energy (circles) and valence-band edge energy (squares) plotted vs the lattice constant: the energy values are aligned based on the transitivity property, and are evaluated for unstrained materials. Lines indicate ternary compounds, and in the CB are continuous for Γ and dashed for X or L valleys; data combined from [168, 169].

A1.8 Carrier transport [2, 146, 165]

A1.8.1 Semiclassical equations of motion – effective mass

Within the framework of the semiclassical model of electron dynamics, electrons and holes are described as wave-packets, behaving as classical particles having "momentum" $\hbar\mathbf{k}$ and moving *within a band* under externally applied electric and magnetic fields[13] according to the equations:

$$\dot{\boldsymbol{r}} = \frac{1}{\hbar}\nabla_k \varepsilon_n(\boldsymbol{k}) \qquad \text{E A1.12a}$$

$$\hbar\dot{\boldsymbol{k}} = -e[\boldsymbol{E}(\boldsymbol{r},t) + \dot{\boldsymbol{r}} \times \boldsymbol{B}(\boldsymbol{r},t)] \qquad \text{E A1.12b}$$

where $\boldsymbol{E}$ is the electric field, $\boldsymbol{B}$ the magnetic field and $\boldsymbol{k}$ is limited to the first Brillouin zone.

Equations E A1.12 correspond to classical Hamilton equations of motion, ε_n being the sum of kinetic and potential energy of the electron in the band n. The k-gradient of the energy (or the slope in one dimension) gives the particle real-space velocity, and the crystal momentum evolves in time under the external forces as a classical momentum. There are the following remarkable differences from the classical case:

- the classical relation between velocity and momentum (velocity proportional to momentum through a constant mass value) does not hold for crystal momentum;
- E A1.12b holds for the positive "particles" (holes) *without* change in the sign: the unoccupied states evolve in time as occupied states;
- when the external forces bring the crystal momentum beyond the edge of the BZ, the crystal momentum re-enters the BZ from the opposite side (Umklapp process) while the corresponding difference is transferred as real momentum to the crystal.

As long as the bands have an approximately parabolic shape in k-space, the CB with positive curvature and the VB with negative curvature, electrons in the CB and holes in the VB behave as classical particles. Qualitatively, this can be understood considering that for values of **k** near the center of the BZ the plane-wave part of the Bloch wavefunction is far from diffraction conditions, while when **k** approaches the BZ edge, the interaction becomes stronger because the BZ terminates on Bragg planes.

Requiring that – in analogy to a classical free particle – the force equals the product of a "mass" times acceleration, a quantity called *effective mass* is introduced from the equations of motion: this is not a scalar constant like the classical mass, but a function of n and **k** and in general a tensor **M** defined by:

$$[\boldsymbol{M}^{-1}]_{ij} = \frac{1}{\hbar^2} \cdot \frac{\partial^2 \varepsilon_n(k)}{\partial k_i k_j} \qquad \text{E A1.13a}$$

In the special case of spherical symmetry of the energy band in k-space, which can occur near a band minimum or maximum (e.g. in Γ for the CB), the effective mass is reduced to a simple scalar quantity:

$$m^* = \hbar^2 \left(\frac{\partial^2 \varepsilon_n(k)}{\partial k^2} \right)^{-1} \qquad \text{E A1.13b}$$

Equations E A1.13 show that the effective mass is proportional to band curvature; conventionally, the effective mass is expressed as a positive value, changing its sign in case of negative curvature of the band. Although introduced to describe dynamic properties, the effective mass can be put in relation with the density of states through the relation between density of states and energy as in E A1.10.

---

[13] To allow a description in terms of wave-packets, the wavelength of the applied field must be much larger than the lattice parameter; moreover, since the model does not include interband transitions, any applied electromagnetic field frequency ν must be such that $h\nu < \varepsilon_{gap}$, and the amplitude of the applied fields cannot be too large [146].

The effective mass is actually used in connection with the calculation of several physical quantities, especially near some band relative minimum or maximum; when it is a scalar, the form of the related expressions becomes particularly simple. In the case of anisotropic maxima or minima this wouldn't be possible, but "purpose-built" scalar effective masses can be introduced, defining them ad-hoc in order to preserve this formal simplicity (actually transferring the complexity inside the mass definition). This leads to the definition of different types of effective mass: an example is the *density of states effective mass*, as that used to calculate the density of states in E A1.10, another is the *mobility effective mass*, (used later in equations E A1.18).

A1.8.2 Intraband scattering and relaxation time

The actual description of carrier dynamics requires taking into account the fact that they are subject to elastic and inelastic scattering, due to defects and imperfections of the lattice, and to interaction with lattice vibrations (or "phonons").

In a quantum-mechanic description, lattice vibrations are described in terms of phonons: these are the elementary quanta of the normal vibrational modes of the crystal, characterized by a wavevector **k** belonging to the first BZ and an angular frequency $\omega_s(\mathbf{k})$, with the branch index *s* playing a role similar to the electron band index *n*; similarly to the case of the electrons, they possess a well-defined energy ($\hbar\omega$) and a crystal momentum $\hbar\mathbf{k}$, but they are bosons (not subject to Pauli principle) and multiple phonons can coexist with the same **k** and ω. Their energies are comparatively small, ranging for example from ≈0 to 36 meV in GaAs. When the atoms of the Bravais lattice base oscillate in phase they are called *acoustic* phonons, otherwise they are called *optical* phonons. They are further classified as transverse or longitudinal, based on their relative polarization and propagation directions.

Between two scattering events the carriers move according to equations E A1.12, while a scattering event changes their momentum, according to its own proper mechanism. A scattering process can be described as a transition of the carrier from an initially occupied state to an initially unoccupied state: in general, a scattering event can be associated to an intraband or an interband transition, but the semiclassical model of electron dynamics deals only with the first case. When only Bloch particles (electrons, phonons) are involved in a scattering process, their total crystal momentum must be conserved.

The semiclassical approach to combine scattering and equations of motion is to make use of Boltzmann's transport equation E A1.14 (below), which describes the evolution of a *non-equilibrium* carrier distribution function $f(\mathbf{r},\mathbf{k},t)$ in the **r**-**k** phase-space. It can be understood in the following way: instead of a particle, the equation describes a *fluid parcel*, i.e. an infinitesimal volume element **drdk** of the phase-space identifiable throughout its dynamic history, which contains $f(\mathbf{r},\mathbf{k},t)\mathbf{drdk}$ carriers at the time t; the parcel moves according to the equations of motion, but its content is not constant and can increase or decrease as consequence of the scattering process.

$$\frac{df}{dt} = \frac{\partial f}{\partial t} + \dot{\boldsymbol{r}} \cdot \nabla_r f + \boldsymbol{F} \cdot \nabla_k f = \left(\frac{\partial f}{\partial t}\right)_{scatt} \qquad \text{E A1.14}$$

Here **F** represents the external electrical and magnetic forces as in E A1.12b. The equation describes the effects of a concentration gradient of the carriers through the diffusion term: $\dot{\boldsymbol{r}} \cdot \nabla_r f$.

To make the scattering term explicit, several level of approximation are possible, the simplest being the assumption that a given scattering process has the effect of restoring locally the thermodynamic equilibrium, and that the scattering probability per unit time (or scattering frequency) is $1/\tau$; the parameter τ is called relaxation time and the simplified model relaxation-time approximation. The scattering term in E A1.14 becomes then:

$$\left(\frac{\partial f}{\partial t}\right)_{scatt} = \frac{f_{eq}(\boldsymbol{k}) - f(\boldsymbol{r},\boldsymbol{k},t)}{\tau} \qquad \text{E A1.15}$$

where the equilibrium distribution f_{eq} is the Fermi distribution expressed as a function of **k** and the non-equilibrium distribution f is obtained as solution of E A1.14. Within the model, the scattering process is completely defined by its relaxation time. Typical values of the relaxation time are in the picosecond range. In case of several *independent* scattering mechanisms, the total relaxation time is given by Matthiessen's rule:

$$\frac{1}{\tau_{tot}} = \sum \frac{1}{\tau_i} \qquad \text{E A1.16}$$

Based on the Boltzmann equation and the relaxation-time approximation, in absence of applied magnetic fields, the current densities for holes and electrons are described by the drift-diffusion equations (given here in one dimension):

$$j_e = -e\left(-\mu_e n E - D_e \frac{dn}{dx}\right) \quad \text{for electrons;} \qquad j_h = e\left(\mu_h p E - D_h \frac{dp}{dx}\right) \quad \text{for holes} \qquad \text{E A1.17}$$

where e is the elementary charge, E is the electric field amplitude, $\mu_{e,h}$ the mobility and $D_{e,h}$ the diffusion coefficient for electrons or holes; the product μE is the drift velocity (average velocity of the carrier under an applied electric field).

Assuming a scalar value for the effective mass, μ and D are related to the relaxation time by:

$$\mu_{e,h} = \frac{\tau e}{m^*_{e,h}} \qquad \text{E A1.18a}$$

$$D_{e,h} = \frac{\tau k_B T}{m^*_{e,h}} \qquad \text{E A1.18b}$$

Note that the effective mass is here the mobility effective mass. Both mobility and diffusion coefficient are directly proportional to the relaxation time and inversely proportional to the effective mass. Matthiessen's rule can be transferred to mobility or diffusion coefficient values.

The phonon scattering relaxation is usually divided in three separate contributions according to the specific electron-phonon interaction, which in turn is related to the kind of lattice perturbation caused by phonons: deformation potential interaction (due to both acoustic and optical phonons), piezoelectric interaction (due to acoustic phonons) and polar - or Fröhlich - interaction (due to optical phonons) [2]. The deformation potential contribution is present in any crystal structure, the piezoelectric contribution is present only in non-centrosymmetric crystals and the polar contribution is present only in crystals with polar bonds (all conditions are satisfied in III-V semiconductors). The corresponding relaxation times are strongly temperature-dependent and decrease with increasing temperature.

Conversely, the relaxation time of impurity scattering increases with increasing temperature. As it can be expected, it decreases with increasing impurity concentration; moreover, ionized impurities are much more effective than neutral impurities as scattering centers because of the longer-range interactions.

The overall result is that ionized impurity scattering dominates up to a certain temperature - dependent on the impurity concentration - beyond which phonon scattering prevails. The case of GaAs, for different temperatures and impurity concentrations, is shown in Fig. A1.18 in terms of mobility. Estimate of the corresponding electron relaxation time at 300K from the mobility values through E A1.18a gives $\tau \approx 0.3$ ps for high-purity GaAs and $\tau \approx 0.04$ ps at high n-doping. For holes, the τ values are approximately reduced by a factor of 4; the ratio of electron and hole mobility is larger than 4 because of the different effective masses.

The influence of the random perturbation of the lattice due to the compresence of different species in each sublattice (alloy scattering) has already been shown in Fig. A1.5 for the case of AlGaAs: intermediate compositions cause more scattering and correspond to lower mobility values. Besides, at high Al content the CB minimum shifts from the Γ point to the X point of the first BZ (section A1.7.2) with the consequence that the CB electrons populate the region around the X minimum, which has a higher effective mass value, leading to a lower mobility.

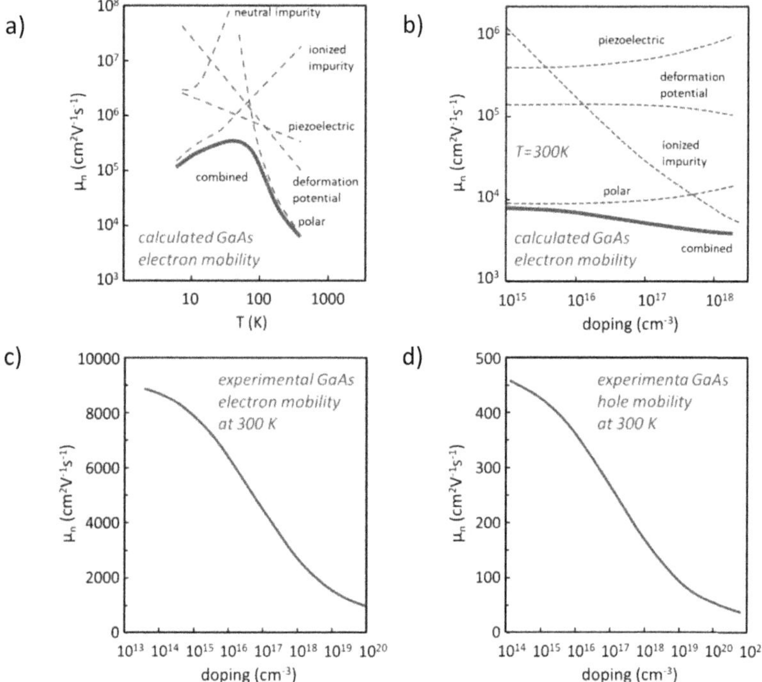

Fig. A1.18 **a)** Calculated contributions to electron mobility and combined values for low-doped (5×10^{13} cm^{-3}) n-GaAs at different temperatures; **b)** the same at 300K and for different doping concentration; c) experimental electron mobility of n-GaAs at 300K; d) experimental hole mobility of p-GaAs at 300K. Values from [170-172]

A1.9 Interband transitions

A1.9.1 Carrier populations away from equilibrium

Transition of electrons from the VB to the CB with generation of an electron-hole pair and the inverse recombination process can occur with different mechanisms. The generation process requires the absorption of an energy corresponding at least to the bandgap energy, which can be supplied for example by a photon (radiative transition), by phonons, or by other carriers; the recombination process requires the emission of energy by the same means. Along with the conservation of total energy, the total momentum must be conserved.

Although these processes can happen at thermodynamic equilibrium, most cases of interest are those when the system is subject to an external perturbation, and the population distribution of the carriers in the bands differs from the equilibrium values by an *excess* density of carriers Δp, Δn. The external perturbation can supply the energy for the pair generation, or consist in a direct injection of electrons and holes. In general the excess carrier densities can be functions of position and time, and their local values at a given position **r** and time t can be different, i.e. Δp(**r**,t) ≠ Δn(**r**,t).

When the intraband relaxation is fast in comparison to the interband recombination, and when the system can be considered in steady-state with respect to the external perturbation[14], it is possible to assume that the carriers are in thermodynamic equilibrium *within* each band, and describe separately the populations of the VB and CB with Fermi-Dirac distribution functions, substituting the Fermi level ε_F (E A1.7-A1.9) with an appropriate *quasi-Fermi level* ε_{FV} for the VB and ε_{FC} for the CB. The quasi-Fermi levels are determined by the steady-state non-equilibrium carrier density distributions in the bands, which in turn are determined by the generation, injection and recombination rates.

Under steady-state, non-equilibrium conditions, the law of mass action, E A1.11 is not more valid; the product of n and p (in the non-degenerate case) becomes:

$$n \cdot p = n_i^2 \cdot e^{(\varepsilon_{FC}-\varepsilon_{FV})/k_B T} \qquad \text{E A1.19}$$

Analogous to the relaxation time used to describe intraband transitions from local non-equilibrium to local equilibrium in the carrier distribution, a lifetime τ is used to describe the kinetic of interband electron-hole recombination. In this contest, τ can be defined in different non-equivalent ways [173], in particular:

- as the ratio of the *excess* carrier density to the *net* recombination rate R_{net}[15], expressed for electrons by:

$$\tau = -\Delta n/(d\Delta n/dt) = \Delta n/R_{net} \qquad \text{E A1.20a}$$

- as the ratio of the carrier density to the recombination rate R, expressed for electrons by:

$$\tau = -n/(dn/dt) = n/R \qquad \text{E A1.20b}$$

The first definition is most commonly applied in relation to the experimental determination of lifetime based on population decay, and implies the reversibility of the process. The second, which is applicable even to irreversible decays, corresponds to the average time a carrier spends in the CB, before being captured by a trap or recombining, and is sometimes referred to as *capture lifetime*.

Independently of definition used, since the carrier recombination processes are never properly monomolecular, τ must be generally concentration-dependent, and consequently time-dependent in non-steady state condition.

---

[14] In other words, the carrier populations must adjust to variations of the external perturbation on a timescale shorter than the timescale of the effects one wants to describe in terms of local equilibrium within each band; another common assumption is that carriers and lattice are in thermal equilibrium and a unique temperature can be defined.

[15] The net (or excess) recombination rate is the recombination rate minus the equilibrium recombination rate.

The lifetime can be measured with pump-and-probe techniques, using a level of excitation that brings both the hole and electron densities far beyond their equilibrium values, so that $n \approx \Delta n$, $p \approx \Delta p$ (*high-level excitation*): in this case, once the excitation source is switched off, the two above definitions of lifetime become practically equivalent (until the population approaches the equilibrium values).

Similarly, in the case of a doped semiconductor, if the excitation brings the minority carrier density far beyond the equilibrium value but still well below the equilibrium majority carrier density (*low-level excitation*), the two definitions become equivalent for the *minority* carrier.

The latter condition is more favorable to execute a lifetime measure, because the minority carrier population is expected to follow a simple exponential decay, as in a really monomolecular process, and the corresponding lifetime is called *minority carrier lifetime* (although the *excess* electron and hole populations decay always in the same way). Nonetheless, the recombination rates at high excitation levels can be *more* important for the operation of optoelectronic devices, in particular with regard to the active zone of laser diodes.

A1.9.2 Non-radiative recombination at deep centers (SRH)

Interband non-radiative recombination with emission of multiple phonons can satisfy energy and crystal momentum conservation; the reverse mechanism is the *thermal* generation of electron-hole pairs. Nonetheless, theoretical considerations lead to the conclusion that, when only Bloch particles are involved in a single event (electron, hole and phonons), the transition probability decreases with the number of phonons: the probability of multiphonon transitions is very low. While for intraband electron-phonon scattering it is sufficient that only one phonon is involved, for interband transitions tens of phonons would be required (e.g. at least 40 phonons are needed to reach the bandgap energy of GaAs). Experimentally, it is known that lattice defects associated with deep levels can efficiently capture the carriers and strongly promote the non-radiative recombination. The reasons behind this behavior are still object of active theoretical study. A partial explanation is simply that the number of phonons involved in the carrier capture at the defect is smaller than that necessary for the band-to-band transition. Another is that, compared to a recombination process involving only Bloch particles, where crystal momentum conservation is required, the involvement of a deep trap relaxes this constraint because momentum can be exchanged with the defect. According to the multi-phonon emission model [174], a local vibrational relaxation (atomic displacement) occurring on a picosecond scale as the carrier is captured, should be responsible of dramatically increasing the probability of multi-phonon emission. Another hypothesis, the possible existence of a "ladder" of closely-spaced electronic excited states associated to the deep defects allowing capture via cascade of single-phonon emissions, was once considered another viable explanation but has not found experimental or theoretical confirmation.

In the *Shockley-Read-Hall* (SRH) model [175] the recombination is described as a sequence of an electron (hole) capture by a deep trap, followed by a hole (electron) capture. The traps are supposed to have only two possible occupation states, differing by 1 elementary charge. According to the model, under steady-state conditions (carrier and trapped populations are described by three quasi-Fermi levels and the net rates of capture are identical for electrons and holes) the *net* recombination rate in presence of trap levels at the energy ε_t within the bandgap is given by:

$$R_{SRH} = \frac{p \cdot n - n_i^2}{\tau_{p0}(n + n_1) + \tau_{n0}(p + p_1)} \qquad \text{E A1.21a}$$

where n_1, p_1 are the electron density and hole densities, as can be calculated from E A1.9, substituting the Fermi level ε_F with the trap energy level ε_t, and τ_{n0}, τ_{p0} are the electron and hole SRH minority carrier lifetimes[16]. The equation assumes non-degenerate doping. An interesting point is that, if the trap

[16] Meaning that they are the minority carrier lifetimes in *absence* of any other recombination mechanism.

level is close to either band edge, then either n_1 or p_1 will become large, reducing the recombination rate: in other words, deep traps are expected to be more efficient than shallow traps in promoting the recombination.

The SRH minority carrier lifetimes can be expressed as:

$$\tau_{n0}{}^{-1} = N_t \sigma_n v_n \quad \text{and} \quad \tau_{p0}{}^{-1} = N_t \sigma_p v_p \qquad \text{E A1.21b}$$

where $\sigma_{p,n}$ is the carrier capture cross-section, $v_{p,n} = \sqrt{3k_BT/m^*_{h,e}}$ is the root mean square thermal velocity of the carrier and N_t the trap concentration. The SRH model itself includes no prescription to calculate the capture cross-sections; experimentally, what can usually be measured are the lifetimes (although the compresence of radiative and Auger recombination makes the evaluation of the SRH lifetimes rather difficult).

In the limit of low-level excitation, equation E A1.21a can be simplified into:

$$R_{SRH} = \frac{\Delta n}{\tau_{n0}} \quad \text{for p-type doping;} \qquad R_{SRH} = \frac{\Delta p}{\tau_{p0}} \quad \text{for n-type doping} \qquad \text{E A1.21c}$$

One can see that at low-level excitation, the rate is dominated by the minority carrier lifetime.

In the limit of high excitation: $n >> n_i, n_1; p >> p_i, p_1; n \approx p$ and equation E A1.21a can be simplified into:

$$R_{SRH} = \frac{n}{\tau_{p0} + \tau_{n0}} = \frac{n}{\tau_{high}} = An \qquad \text{E A1.21d}$$

Equation E A1.21d indicates that at high-level excitation the SRH recombination rate must be smaller than at low-level excitation, and if τ_{n0} and τ_{p0} are very different, than the longest lifetime will prevail. For a given trap, the minority lifetimes can actually be very different, in particular because of the impact on capture cross section of the charge state [176]. Considering for example a neutral trap that captures a hole (cross-section σ_p) becoming positively charged, and subsequently captures an electron (cross-section σ_n): one can expect $\sigma_n >> \sigma_p$ and consequently $\tau_{n0} << \tau_{p0}$; while at low-level excitation the lifetime would be the short τ_{n0}, at high-level excitation the lifetime would become the large τ_{p0}.

When several trap defects are simultaneously present, the recombination rate can be taken as the sum of the individual rates, and the lifetime follows Matthiessen's rule (and the shortest lifetime prevails).

To provide a quantitative example: values of the SRH recombination lifetimes for a MOVPE-grown Zn-doped GaAs (at different Zn-doping concentrations) have been evaluated in ref. [177]: al low-level excitation, τ_{n0} was in the range 5-10 ns, while at high-level excitation $1/A = \tau_{p0} + \tau_{n0}$ was in the range 500-1000 ns. In GaAs, the carrier thermal velocity at room T is $\approx 10^7$ cm/s, the deep-level capture cross sections can be as large as 10^{-14} cm^2 for ionized traps and orders of magnitude smaller for neutral traps. Hypothesizing a deep-traps concentration of 10^{15} cm^{-3} with cross-section 10^{-14} cm^2 for the minority carriers and 10^{-16} cm^2 for the majority carriers, the SRH minority carrier lifetime would be 10 ns and the high-level excitation lifetime 1000 ns, in line with the experimental results.

Note that the lifetimes in E A1.21b depend on T because both the carrier velocity and the capture cross-sections are functions of T; empirical power laws are often used to describe this dependence, of the form $\tau \propto T^m$ with $m \approx -2.5$.

Finally, it must be specified that the SRH model assumes that the relaxation time of the deep traps is negligible: at high excitation this hypothesis might become weak, and trap-saturation effect might become significant, increasing the lifetime.

A1.9.3 Surface and interface recombination

At the semiconductor surface, and to some extent at the interface between two layers of different composition, there is usually a comparatively high concentration of deep levels, which cause non-radiative recombination. The non-radiative recombination rate in a thin slab near the surface can be described in terms of the SRH model using equations E A1.21a and E A1.21b, where the carrier density values n, p and the energy differences $\varepsilon_t - \varepsilon_{CB}$, $\varepsilon_{VB} - \varepsilon_t$ (contained in n_1, p_1) are all taken in a thin slab near the surface [178, 179]. It is though more convenient to formulate a rate expression independent from the choice of slab thickness, and make use of a *surface* net recombination rate U_{SRH} (which has the dimensions of a flux, $cm^{-2}s^{-1}$) substituting the trap concentration N_t with a trap area-concentration N_{ta} (dimensions cm^{-2}):

$$U_{SRH} = \frac{p_s \cdot n_s - n_i^2}{S_{p0}^{-1}(n_s + n_1) + S_{n0}^{-1}(p_s + p_1)} \qquad \text{E A1.22a}$$

where n_s, p_s are the carrier densities at the surface and S_{p0}, S_{n0} are the minority carrier surface recombination velocities:

$$S_{n0} = N_{ta}\sigma_n v_n \quad \text{and} \quad S_{p0} = N_{ta}\sigma_p v_p \qquad \text{E A1.22b}$$

Figure A1.19 shows simulated values of carrier profiles near the surface of a p-doped semiconductor under steady-state low-level and high-level excitation [167] using values that are approximately representative of GaAs. Some simplified assumptions have been used: the excitation produces a spatially uniform carrier generation, there are only diffusion currents (no drift) and local charge neutrality is maintained. The effect of increasing the surface recombination velocity (dashed vs continuous lines) is to make the carrier density gradient near the surface steeper, while decreasing the bulk lifetime at constant excitation rate (dotted vs continuous lines) shifts downwards the carrier populations; the longest bulk lifetime is estimated assuming only radiative recombination, while the shortest assumes an the compresence of an equivalent SRH recombination process.

While the effect of surface recombination rate is negligible in the bulk, it can become extremely important near the surface/interface and of course in the case of thin layers.

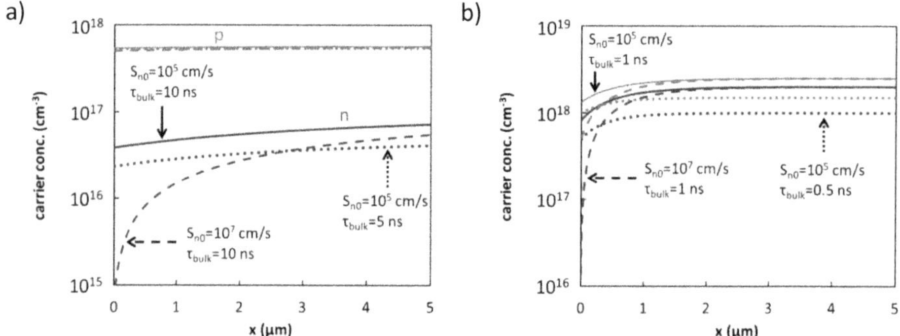

Figure A1.19 Simulated carrier concentration profiles for a p-doped semiconductor ($N_p=5\times10^{17}$ cm^{-3}) within 5 μm from the surface in x=0: **a)** under low-level excitation; **b)** under high-level excitation. The dotted lines represent the effect of a reduction of the bulk lifetime, and the dashed lines the effect of an increase of the surface recombination rate, the red lines represent the hole density and the blue lines the electron density.

The values of surface recombination velocity at a cleaved or etched semiconductor surface can be quite high, as those used in the above simulation, up to 1×10^7 cm/s at room T. On the facets of laser devices - where high non-radiative recombination is highly unwanted for reliability reasons - they are reduced by order of magnitudes with appropriate "passivating" coatings. At the heterostructure interfaces, the surface recombination velocities are usually much smaller, rather in the $10\text{-}10^3$ cm/s range, unless high contamination or significant lattice misfit are present.

A1.9.4 Auger [58]

A second fundamental non-radiative recombination mechanism is the *Auger*, where the excess energy stemming from the electron-hole recombination is released exciting a second electron to a high-energy state within the CB (eeh), or a second hole to a high-energy state within the VB (hhe). The carrier can even be excited in another band of the same type: in the case of III-V semiconductors there could be several possibilities for this, but it turns out in the analysis that only two are significant, the scattering from HH to SO, and that from HH to LH.

With a slight change of perspective, the Auger recombination can be seen as a *collision* between two electrons in the CB that *scatters* one of them to a higher energy in the same band and the other to a lower energy in the VB; a similar definition can be used for hole-hole collisions in the HH band, with scattering of one hole in the valence bands and one in the CB.

The above Auger mechanism is called "direct" or "purely collisional". The reverse mechanism - generation of electron-hole pair by a highly excited carrier - is sometimes called *impact ionization*, especially when high electric fields are applied to generate the high-energy carriers.

In direct Auger, energy and crystal momentum conservation are required. The shapes, the separation of the bands, and the fact that only the regions near the bandgap are populated by carriers, limit strongly the possibilities of satisfying these constraints, and only three types of transitions are relevant in III-V Zincblende semiconductors: they are indicated as CCCH, CHHL and CHHS and are illustrated in Fig. A1.20a; the letters in the acronyms refer to **C**B, **H**H, **L**H and **S**O bands. It can be seen that the electron-hole recombination cannot be vertical (in k-space) because crystal momentum conservation would not be possible: consequently, the recombination energy ε_{rec} must always be higher than the bandgap energy ε_{gap}.

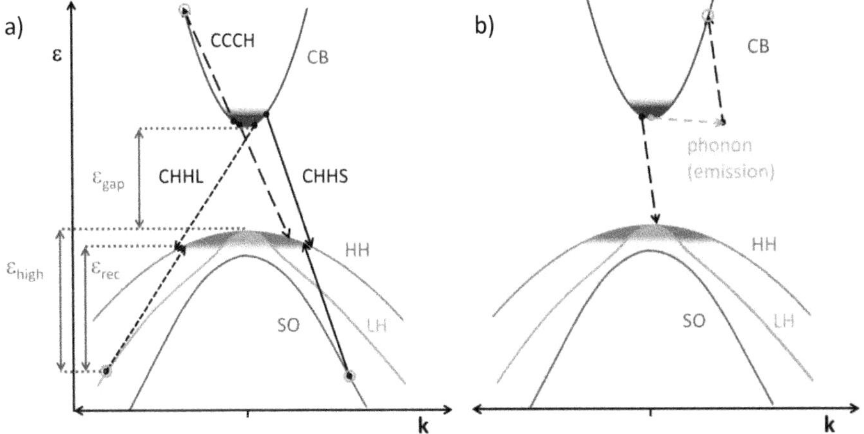

Figure A1.20 Auger recombination: **a)** the 3 possible direct recombinations in III-V Zincblende semiconductors; **b)** example of phonon-assisted Auger recombination. Shaded areas indicate the part of the bands occupied by carriers, arrows indicate electron displacements in (ε, k) space, the high-energy states involved in each transition are indicated by empty green circles.

For a transition allowed by energy and momentum conservation, the transition probability is proportional to the probability $P_{i \to f}$ that the *four* involved states have simultaneously the correct occupation. At room temperature, the high-energy state can be considered always empty, which reduces the problem to three states. Expressing $P_{i\to f}$ in terms of the carrier distribution functions in the non-degenerate case approximation (E A1.9 with quasi-Fermi levels), $P_{i\to f}$ results proportional to n^2p for the eeh process and to p^2n for hhe processes.

In the same approximation, it can be proved that $P_{i\to f}$ is proportional to $\exp(\varepsilon_{gap}-\varepsilon_{high})/k_BT$, with ε_{high} being the energy of the high-energy state relative to the band edge (indicated in Fig. A1.20a for the CCHL transition). Since $\varepsilon_{high} \geq \varepsilon_{rec} > \varepsilon_{gap}$, the argument of the exponential is always negative and the transition probability $P_{i\to f}$ increases exponentially with increasing T. The dependence of $P_{i\to f}$ on ε_{gap} depends on the shape of the bands: in the parabolic approximation it decreases exponentially with increasing ε_{gap}.

For large bandgaps, the Auger recombination does not simply disappear, because an otherwise less favorable 4-particles *phonon-assisted* Auger mechanism (Fig. A1.20b) becomes dominant [180]. The absorption or emission of a phonon provides an extra degree of freedom for crystal momentum conservation and makes the task of finding appropriate states in the bands much easier.

Based on the previous considerations[17], the net Auger recombination rate is usually expressed as:

$$R_{Auger} = C_p(np^2 - n_0p_0^2) + C_n(n^2p - n_0^2p_0) \qquad \text{E A1.23a}$$

where n_0, p_0 are the equilibrium carrier densities and C_p, C_n recombination coefficients for the hhe and eeh processes respectively.

In the limit of high excitation $p \approx n >> n_0, p_0$, and E A1.23a simplifies into:

$$R_{Auger} = (C_p + C_n)n^3 = Cn^3 \qquad \text{E A1.23b}$$

In Zincblende III-V semiconductors at room temperature, the values of the coefficient C are in the range 10^{-25}-10^{-31} cm^{6}s^{-1}, for example [58, 181-183]:

InSb (ε_{gap} = 0.17eV) $C \approx 7\times10^{-26}$ cm^{6}s^{-1}

InGaAsP (ε_{gap} = 0.8eV) $C \approx 8\times10^{-29}$ cm^{6}s^{-1}

GaAs (ε_{gap} = 1.43eV) $C \approx 7\times10^{-30}$ cm^{6}s^{-1}

GaP (ε_{gap} = 2.27eV) $C \approx 1\times10^{-30}$ cm^{6}s^{-1}

By comparison of SRH and Auger recombinations, it can be seen that the former can be expected to be the dominant non-radiative recombination process at low injection levels, and the latter at high injection levels – a condition that as already remarked is particularly relevant in the case of the active zone of laser devices.

[17] More exhaustive treatments of Auger recombination make use of the time-dependent perturbation theory, to the first order for direct and to the second order for phonon–assisted processes, and more realistic assumptions concerning the carrier population distribution and the shape of the bands [180, 184].

A1.9.5 Interband radiative transitions

Absorption of a photon of energy $h\nu \geq \varepsilon_{gap}$ can promote an electron from a state in the VB to an unoccupied state in the CB - or in terms of carriers, absorption of a photon can create an electron-hole pair. The inverse process, electron transition from the CB to an empty state in the CB (or electron-hole recombination) can happen according to two different mechanisms: spontaneous emission or stimulated emission. In the first case the photon is emitted with random phase, polarization and direction, while in the second case the emitted photon has the same phase, polarization, direction and frequency of the stimulating radiation.

Interband transitions between the VB and the CB involve photons with energy $h\nu$ comparable to the bandgap energy (in the eV range), and momentum conservation implies that only vertical transitions are allowed, i.e. between states with (almost) the same value of $\mathbf{k}$ (k-selection rule). The reason is that the photon momentum $\hbar\mathbf{k_p}$ is very small in comparison to the range of possible values of crystal momentum $\hbar\mathbf{k}$: the vacuum photon momentum module is $|\hbar\mathbf{k_p}|=h\nu/c=h/\lambda$, and assuming $\lambda \approx 1$ μm, $|\mathbf{k_p}|$ is of the order of 0.01 nm^{-1}, to be compared with a value of $|\mathbf{k}|$ on the first BZ surface of the order of 10 nm^{-1}. Non-vertical interband transitions involving phonon absorption or emission (indirect transitions) are also possible, but with much lower probability: they become important in the case of indirect semiconductors, where vertical transitions at energies near the bandgap energy are not possible. Figure A1.21 illustrates photon absorption and emission processes in direct and indirect semiconductors.

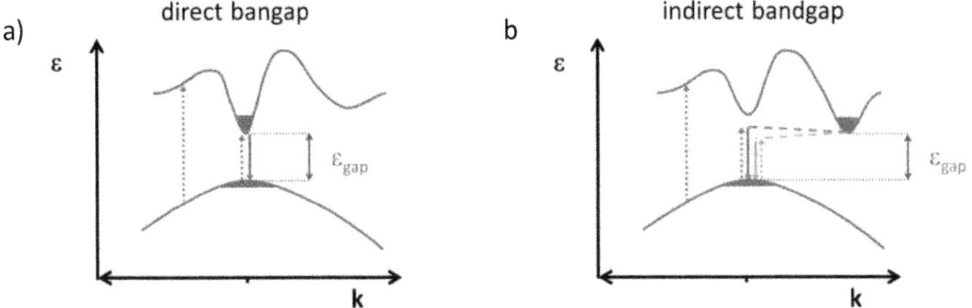

Figure A1.21 Scheme of photon absorption and emission processes, the arrows indicate the movement in ε,$\mathbf{k}$ space of the electron: **a)** direct semiconductor, absorption with transition of an electron from the VB to the CB and emission with transition of an electron from the CB to the VB; **b)** the same for an indirect bandgap semiconductor, in this case the transitions near ε_{gap} must involve either absorption or emission of a phonon.

The CB is almost completely occupied and the VB unoccupied, consequently the *absorption* beyond the fundamental edge ($h\nu = \varepsilon_{gap}$) is not limited by the occupation of the electronic states (except in close proximity of ε_{gap}), and the dependence of the absorption coefficient α on energy is primarily related to the density of states in the two bands: the shape of $\alpha(h\nu)$ around ε_{gap} is, for direct bandgap semiconductors, a very steep shoulder, while for indirect bandgap semiconductors α increases with $h\nu$ more gradually, according to the onset of successive indirect transitions involving absorption or emission of phonons. In the case of highly populated bands (due to degenerate doping, high carrier excitation or high injection) the absorption edge can shift slightly towards higher energies, because the band states near the CB minimum are filled and/or the states near the VB maximum are empty. The apparent increase of the bandgap is called Moss-Burstein effect, and is expected to be stronger for n-doping, because of the lower density of states of the CB with respect to the VB. Since at the same time high doping and/or high carrier concentration cause a real narrowing of the bandgap (renormalization), there are two opposite effects on the absorption edge. Figure A1.22a shows the measured absorption coefficient of GaAs with different doping types: with respect to the low-doped sample, the high-doped profiles are shifted: in the case of n-doping, the Moss-Burstein effect prevails (blue-shift of the absorption edge) while in the case of p-doping the bandgap narrowing prevails (red-shift).

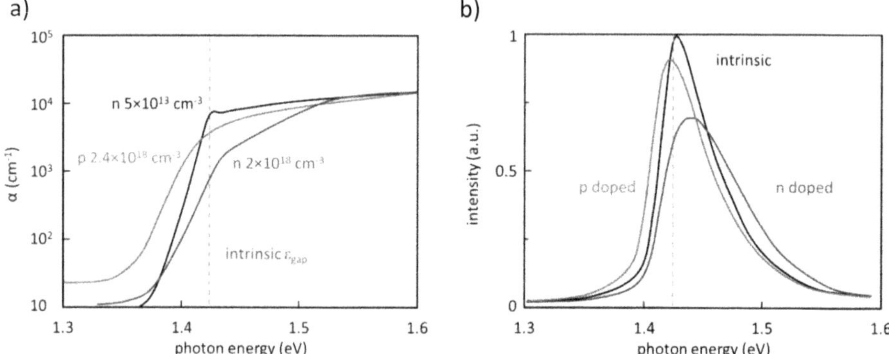

Figure A1.22 **a)** GaAs absorption coefficient at 297K for different doping types; data from [185]; **b)** GaAs luminescence at room T for different doping types (qualitative).

It can even be seen that that, in the highly-doped samples, the absorption profile is rather broadened. In general, the low-energy region immediately below the bandgap (till ≈ 0.2 eV below ε_{gap} for GaAs) is not entirely transparent, but an exponential decay of the absorption coefficient is present (Urbach tail). This is explained by several (and to some extent interconnected) effects including:

- phonon-assisted interband absorption (with phonon absorption);
- spatial fluctuations in the material composition, perturbing the states near the band edges and causing their energy to spread around the ideal band edge (impurities disorder and – in multinary compounds – alloy disorder);
- at high doping, formation of impurity bands;
- at high doping or high injection, bandgap narrowing due to carrier-carrier and carrier-ionized impurity interactions.
- at low temperature below-bandgap transitions involving defects and excitons[18] become visible; exciton-related transitions can be strong even at room temperature in quantum-confined structures.

At even lower energies ($\varepsilon < \varepsilon_{gap}-0.2$ eV for GaAs) phonon-assisted intraband transitions become the prevalent source of absorption. This effect, called free-carrier absorption, is expected to be proportional to the carrier density and to a power of the wavelength λ^m, with m in the range 1.5 to 3.5 depending on the scattering mechanism involved: 1.5 for acoustic phonons, 2.5 for optical phonons, 3 to 3.5 for ionized impurity scattering [167].

Spontaneous and stimulated emissions involve primarily the states near the band edges: this is due to the fact that intraband relaxation is normally faster (ps range) than interband recombination (ns range); the excited carriers relax towards their intra-band minima, and the populations reach equilibrium condition within the bands as discussed in section A1.9.1. Consequently, the range of emitted photon energies is limited on the lower side by the bandgap energy and on the higher side by the carrier distribution profiles: the spontaneous emission (luminescence) intensity as a function of

---

[18] Excitons are electron-hole electrostatically bounded pairs.

energy is a broad peak, as qualitatively depicted in Fig. A1.22b. The spontaneous emission is slightly broadened towards energies < ε_{gap} due to the same reasons listed above for absorption (not including the free-carrier absorption). In the case of indirect semiconductors, the rate of radiative recombination is low and non-radiative mechanisms are dominant, making the luminescence extremely weak.

Approximating the spontaneous emission process with a transition between CB minimum and VB maximum and assuming a non-degenerate carrier distribution, it can be proved that the net recombination rate is proportional to the product of electron and hole densities through a bimolecular recombination coefficient *B*[19]:

$$R_{sp} = B\left(np - n_i^2\right) \qquad \text{E A1.24a}$$

In the limit of high excitation p = n >> n_i, and eq. E A1.24a simplifies into:

$$R_{sp} = Bn^2 \qquad \text{E A1.24b}$$

B is in the range 10^{-9}-10^{-11} cm^{3}s^{-1} for direct semiconductors (and 10^{-13}-10^{-15} cm^{3}s^{-1} for indirect semiconductors, where the transition is phonon-assisted). It is ≈2×10^{-10} cm^{3}s^{-1} for GaAs at 300K and is approximately ∝ $T^{-1.5}$ (for bulk GaAs) [186]. The temperature dependence is related to the probability of finding an occupied state in the CB and an empty state in the VB with the same value of **k**, which is maximized at low temperature because the carriers accumulate towards their energy minima in Γ.

The theoretical description of interband radiative transitions rates can be done at several levels of complexity; a commonly used approach relies on a classical representation of the electromagnetic field combined with a quantum-mechanic perturbative treatment of the electronic states, although the spontaneous emission specifically requires introducing the electromagnetic field quantization [2, 58]. A noteworthy conclusion of the theory, known as Fermi's Golden Rule, is that the transition probabilities for absorption, stimulated emission and spontaneous emission are intimately connected, being in all three cases proportional to the joint density of states (density of CB and VB states with the same **k**) and to an integral of the form $\left|\langle \varphi_{n_2,\boldsymbol{k}_2} | \hat{H}' | \varphi_{n_1,\boldsymbol{k}_1} \rangle\right|^2$ which represents the superposition of the stationary state 2 with the stationary state 1, the latter perturbed by the electromagnetic field through the associated operator $\hat{H}'$.

[19] Proportionality to concentrations is strictly valid only for non-degenerate conditions.

A1.10 Optical properties in the transparency region

The macroscopic description of light propagation in a lossy, optically isotropic medium, makes use of the complex refractive index $\tilde{n}=n+i\kappa$. The real part n is the ratio of the speed of light in vacuum and the phase velocity in the medium, while the extinction coefficient κ indicates the attenuation and is related to the medium absorption coefficient α by $\alpha = 4\pi\kappa/\lambda_0$, where λ_0 is the vacuum wavelength. Using the complex refractive index, the electric field of a plane wave propagating in a medium can be written:

$$\boldsymbol{E}(x,t) = Re\left[\boldsymbol{E}_0 e^{i(\tilde{k}x-\omega t)}\right] \quad \text{E A1.25}$$

where the complex wavenumber is: $\tilde{k} = 2\pi(n + i\kappa)/\lambda_0$

An alternative description involves the complex relative dielectric function (or complex relative permittivity) $\tilde{\varepsilon} = \varepsilon_1 + i\varepsilon_2$. For non-magnetic materials $\tilde{n}^2 = \tilde{\varepsilon}$, from which $\varepsilon_1 = n^2 - \kappa^2$ and $\varepsilon_2 = 2n\kappa$.

When the complex dielectric function describes a linear response of the medium to the electric field, the real and imaginary parts of $\tilde{\varepsilon}$ are linked by a Kramers-Kronig transform, which expresses $\varepsilon_i(\omega)$ as an integral function of $\varepsilon_j(\omega)$ having $0, \infty$ as integration limits [1, 2]; a similar relation links the real and imaginary parts of $\tilde{n}$. This means that, for weak-enough electromagnetic waves, it is sufficient to determine the response of the medium in phase with the field (e.g. the extinction coefficient) to determine the out-of-phase response (e.g. the refractive index) or vice-versa. Moreover, it indicates that an effect that modifies the absorption, as the introduction of dopants, carrier injection or temperature change, will necessarily impact at the same time the refractive index.

To a first approximation, the relation between the above parameters can be interpreted classically, modelling the interaction of the electromagnetic radiation with matter using charged-oscillator models: the oscillators respond to the electric field introducing a polarization in the material, with a phase lag that causes the slow-down of the phase velocity; the energy dissipation is represented by a damping term. The electric polarization **P** can be calculated based on the oscillator model, and the relative permittivity $\tilde{\varepsilon}$ is obtained from the defining equation of the electrical displacement field $\mathbf{D}=\varepsilon_0\mathbf{E}+\mathbf{P}$ (where ε_0 is the vacuum permittivity) and assuming $\mathbf{D}=\varepsilon_0\tilde{\varepsilon}\mathbf{E}$ (applicable to linear, homogeneous, isotropic materials). The dependence of $\tilde{\varepsilon}$ and $\tilde{n}$ from frequency that can be obtained using a Lorentz damped oscillator model is qualitatively shown in Fig. A1.23.

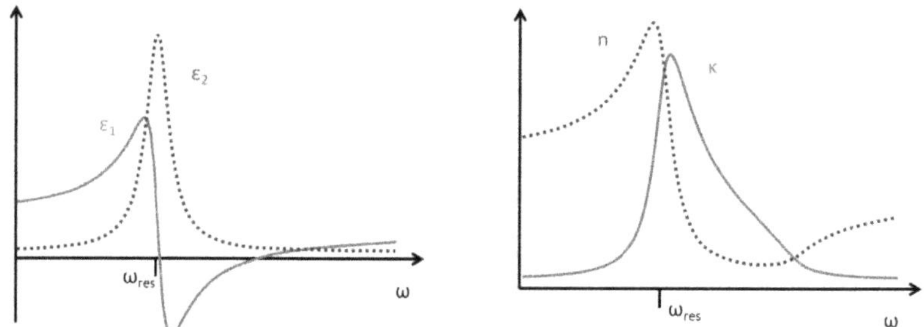

Fig. A1.23 Relative complex permittivity and complex refractive index according to Lorentz's oscillator model near the oscillator resonance frequency ω_{res}.

The experimental values of n and κ as a function of the wavelength of InP are plotted in logarithmic scale in Fig. A1.24a, in a very broad wavelength interval, encompassing the highly absorbing region where the photon energy is $>\varepsilon_{gap}$, the "Reststrahlen" region were the photon energy is comparable with the phonon energy, and the intermediate transparency region. This region can be seen as a transition between two resonances due to two different kinds of oscillators (the atoms or ions at long wavelength and the electrons in the bands at short wavelength). The trends are similar for the other zincblende III-V semiconductors, in particular for GaAs.

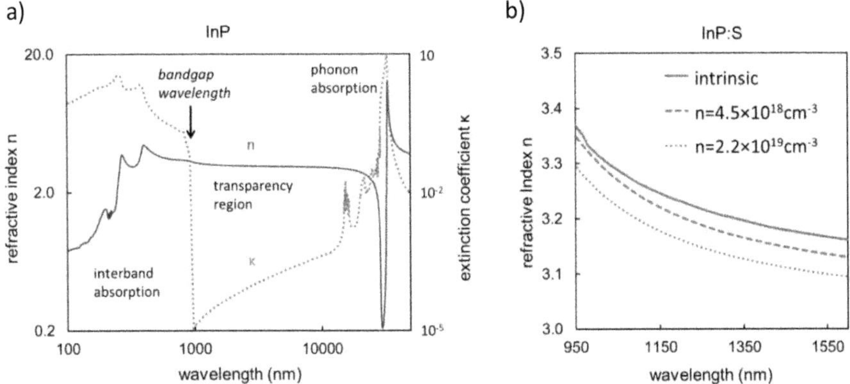

Figure A1.24 **a)** Refractive index and extinction coefficient of InP; **b)** effect of n-doping on the refractive index in the transparency region; data from [187, 188].

The effect of doping on the refractive index n in the transparency region is exemplified in Fig. A1.24b; it *decreases* with increasing doping, partially in relation to the band-filling [187] that shifts the fundamental absorption edge to lower wavelengths and partially in relation to the increase of free-carrier absorption at high wavelengths.

The effect of a temperature variation on the refractive index in the transparency region has been related principally to the bandgap shrinkage [1]; the effect is an *increase* of n with increasing temperature, with thermal coefficient dn/dT in the transparency region increasing as the wavelength approaches the bandgap value. For example, for GaAs at ≈22°C, dn/dT is 3.2×10^{-4} K^{-1} at λ=1 μm and 2.2×10^{-4} K^{-1} at λ=2μm; for InP the values are slightly lower.

A2 Some general aspects of III-V MOVPE

A2.1 Different III-V epitaxy techniques [6, 189]

Among the several existing epitaxial techniques, the most widespread for the realization of III-V semiconductor heterostructures are the liquid phase epitaxy (LPE), the molecular beam epitaxy (MBE) and the metalorganic vapor phase epitaxy (MOVPE).

In LPE the crystal is grown from a nutrient liquid phase, (e.g. an arsenic solution in liquid gallium at ≈800°C for the growth of GaAs), with conditions kept near those of thermodynamic equilibrium. High growth rates can be achieved (in the order of 1 μm/min) and high crystal perfection; drawbacks are lack of flexibility and difficulty in controlling layer thickness, alloy compositions, doping and interface smoothness; for the production of quantum-well and strained-layer structures, MOVPE and MBE techniques are generally preferred.

In MBE, the substrate is positioned on a heated rotating holder, in an ultra-high-vacuum chamber (≈10^{-9} mbar), the reagents are atoms or highly reactive molecular species, which enter the deposition chamber in the form of beams (no gas-phase interaction). For example, GaAs can be grown from Ga atoms and As_2 molecules, using a substrate temperature in the range 580°C-650°C. The reactive species are generated in external cells by thermal sublimation of solid elements, evaporation of liquid elements, or by pre-cracking of gases – as in the case for AsH_3 and PH_3, which are cracked to generate preferentially As_2 and P_2. The overall process cannot be described in terms of thermodynamic equilibrium (although a partial equilibrium between the solid and the gas phase near the surface has been assumed in certain models of MBE growth). The growth rate can reach approximately 1μm/h. MBE is much more flexible than LPE in terms of the material compositions that can be grown (including the possibility of growing thermodynamically unstable materials) and is particularly well suited for the realization of low-dimensional structures, since it is possible to control the grown thickness at sub-monolayer level; moreover, the ultra-high-vacuum allows the use of advanced in-situ diagnostic: mass spectroscopy can be used to analyze the gas-phase composition, reflection high energy electron diffraction (RHEED) is used to monitor the growth with monolayer resolution, and several surface analytic tools can be integrated, possibly using auxiliary UHV chambers. The nature of the impinging atoms or molecules is well known, and this simplifies the task of developing models of the growth mechanism. High defect density, radial disuniformity, cumbersome substrate preparation, low throughput and long maintenance downtime have been historically the main drawbacks of MBE, but these issues have largely been addressed in modern configurations.

MOVPE technique is comparable to MBE in terms of flexibility and only slightly worse in terms of control of interface abruptness (an example of MOVPE-grown MQW is shown in Fig. A2.1). The reagents are introduced in the reactor chamber at atmospheric or moderately low pressure, normally ≥ 50 mbar; gas-phase reactions and gas-phase transport phenomena play an important part in the overall growth process. Partial thermodynamic equilibrium conditions are established at the regrowth interface, a factor that can help to obtain a low defect density. The possibilities of in-situ diagnostic are more limited than in the case of MBE, the overall physical (fluid dynamical) and chemical process underlying the growth is more complicated and less accessible to direct investigation.

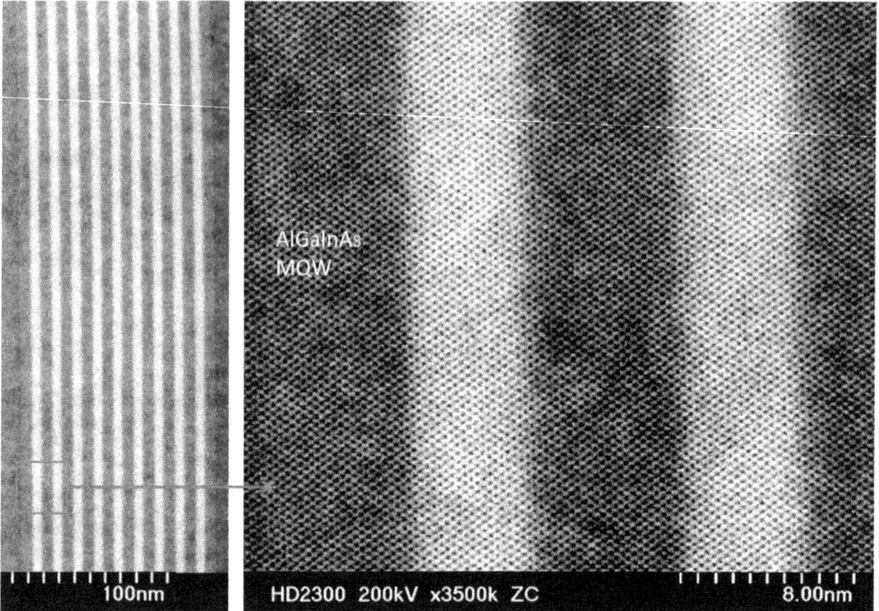

Figure A2.1 Annular-Dark-Field (Z-contrast) STEM images of an AlGaInAs MQW, grown with a MOVPE reactor. The atoms are clearly resolved, the well has higher In/Ga and In/Al ratios than the barrier and appears more bright. The interfaces are defined approximately within 2 monolayers.

A2.2 General considerations about MOVPE reactors

In section A2.2.1, growth chamber characteristics will be discussed, briefly introducing some different existing types of MOVPE tools besides of the planetary reactors already described in chapter 3; moreover, an elementary description of the overall MOVPE process, including transport, is provided.

Section A2.2.2 contains additional details on the gas-mixing system, complementing the information provided in chapter 3.

A2.2.1 MOVPE reactor chamber and overall process: a simplified description [6, 190]

The design of MOVPE reactors has evolved in time, starting from the first prototypes in the 1960', in an effort to reach higher performance in terms of layers thickness control, composition control, achievable interface abruptness, uniformity, reproducibility, efficient use of the (expensive) reagents and in general economic efficiency. Many of earlier home-made and commercial reactors used a quartz deposition chamber[1] in a **horizontal linear configuration** (Fig. A2.2); normally only 1 to 3 wafers at a time could be loaded.

[1] The deposition chamber is the "defining" part of the MOVPE tool, and is often called for short "the reactor".

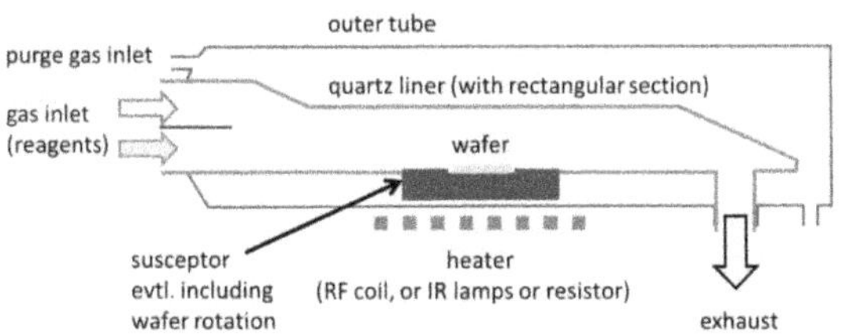

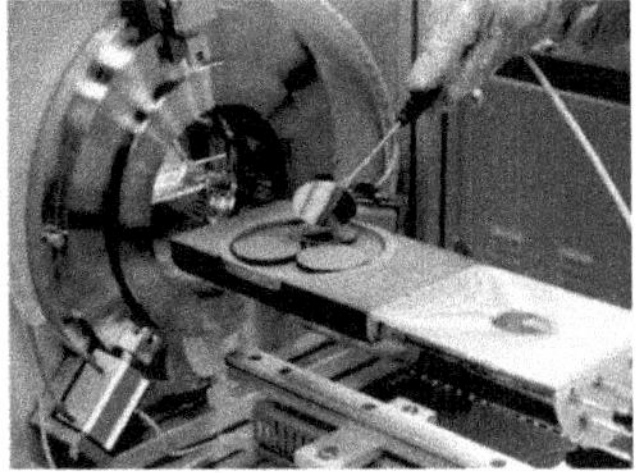

Figure A2.2 left: scheme of the growth chamber of a horizontal MOVPE reactor; right: the susceptor of an AIX-200 horizontal reactor is extracted from the growth chamber for wafer loading.

Most of the currently produced commercial reactors are larger, and can accommodate tens of wafers at a time, a solution that is cost-effective for mass production but makes research and development more expensive. On the other hand, the sophistication reached by MOVPE construction technology makes the realization of home-made reactors problematic and is anyway less attractive when a transfer from R&D to manufacturing applications is foreseen. At the time of writing, three of the most diffuse commercial reactor configurations are the Planetary Reactor®, the Close Coupled Showerhead® (both from Aixtron SE) and the "rotating disk reactor" (from several companies, including Veeco Instruments Inc. under the name TurboDisc®).

They have in common the use of a cylindrical chamber, the wafers are positioned in shallow recesses over the surface of a circular rotating susceptor, normally made of graphite, which is heated from underneath by either resistors or inductors. Chamber walls and ceiling are water and/or gas cooled to minimize surface reactions and consequent parasitic material deposition. The carrier gas and the reagents enter the chamber from an injector (of different shape and position in the three cases) located above the susceptor, and are extracted laterally all around the susceptor perimeter; the gas flow in the chamber - under ideal operative conditions - is laminar, and is sustained by a small pressure gradient across the chamber (forced convection).

The case of a large top injector, characteristic of rotating disk and showerhead reactors, is shown in Fig. A2.3. The gas carrier streamlines bend outwards as they approach the susceptor surface, so that the flow velocity becomes always horizontal in proximity of the susceptor and the vertical velocity goes to zero; only in one central point both vertical and horizontal velocity are zero (stagnation point).

A fluid-dynamics phenomenon common to all reactors is that, due to the friction at the gas-solid interface and within the gas phase, momentum is transmitted from the gas to the susceptor, with the consequence that the radial flow velocity decreases to zero at the susceptor surface; this process can be formally described as *momentum diffusion*, using as diffusion coefficient (units cm^2/s) the kinematic viscosity: $\nu = \mu/\rho$, where μ is the gas viscosity ($g \cdot cm^{-1}s^{-1}$) and ρ the gas density (g/cm^3). The region of decreasing velocity - conventionally below 99% of the "free-stream" velocity u_∞ - is called *velocity boundary layer*[2]; in the case of laminar flow over a flat surface with no upper bounds, its thickness $\delta_u(x)$ is:

$$\delta_u(x) = 4.91 \cdot \sqrt{\nu x/u_\infty} \qquad \text{E A2.1}$$

In the velocity boundary layer, transport of reagents by convection becomes increasingly less important with decreasing distance from the surface, and diffusion becomes dominant: this fact, combined with the circumstance that the ratio of momentum diffusivity ν and mass diffusivity D (called Schmidt number, ν/D) is roughly constant and ≈1, leads to the result that, if a species is consumed at the surface by a chemical reaction, a vertical concentration gradient of that species is established in a region whose

[2] The concept of "free-stream" is strictly meaningful only in the case of flow over a surface with no upper bounds.

spatial distribution approximately coincides with that of the velocity boundary layer. This region of reagent depletion is called *concentration boundary layer* and its thickness $\delta_D(x)$ is proportional to $\delta_u(x)$; in Ref. [191] Ghandhi and Field suggest that in a linear reactor the concentration boundary layer thickness can be expressed for short distances $x < h^2u/\pi D$ (with h=reactor height, u= flow velocity) as:

$$\delta_D(x) \propto \sqrt{Dx/u} \quad \text{E A2.2}$$

where the diffusion coefficient and the boundary layer thickness are referred to a specific species.

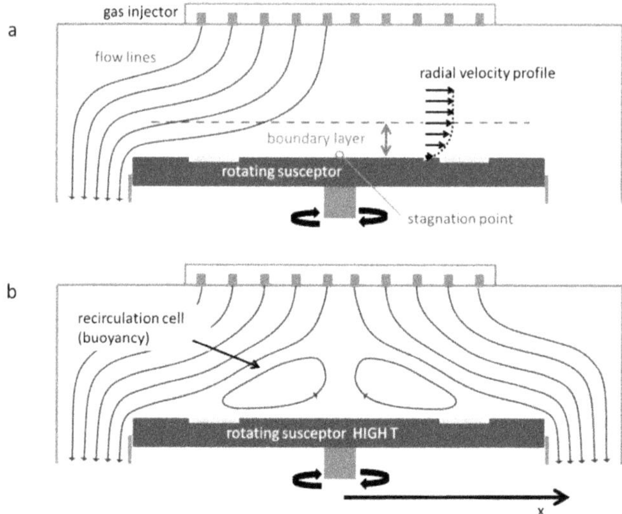

Figure A2.3: **a)** qualitative representation of flow lines and velocity boundary layer in the growth chamber of a vertical MOVPE reactor with stagnation-point flow; **b)** possible formation of recirculation cells due to gas buoyancy over the heated susceptor.

Heat transport is due to thermal convection, conduction and radiation; ignoring the latter contribution, a *temperature boundary layer* can be defined similarly to the case of the velocity boundary layer. In particular, thermal conduction can be formally described as *heat diffusion*, using as diffusion coefficient the thermal diffusivity: $\alpha_T = \kappa_T/\rho C_p$, where κ_T is the gas thermal conductivity ($Wcm^{-1}K^{-1}$) and C_p is the gas specific heat ($Jg^{-1}K^{-1}$); the values of momentum diffusivity and thermal diffusivities are very similar (Prandtl number $\nu/\alpha_T \approx 1$).

In a zeroth-order approximation (boundary layer model), the mass transport is described as due to pure convection above - and pure diffusion within - a stagnant boundary layer, although neither a sharp transition, nor a completely stagnant boundary layer do really exist. Moreover, the height of the velocity boundary layer, when defined with the 99% of free-stream velocity condition, becomes in many cases larger than the height of the MOVPE chamber: moment and mass diffusion occur to some extent in the whole chamber volume, both in the vertical and in the horizontal directions. The model can nonetheless be useful, at least as a heuristic tool and to predict rule-of-thumb trends. Within the boundary layer model, the (net) flux to the surface of the species i, $J_i(x)$, is given by:

$$J_i(x) = -D_i\Delta C_i/\delta_{Di}(x) = -D_i\Delta p_i/\left(RT\delta_{Di}(x)\right) \quad \text{E A2.3}$$

where ΔC_i is the difference between the free-stream (or input) concentration and near-surface concentration of the species *i*, re-formulated in the rightmost term using the partial pressures difference. The growth rate *G* must be proportional to the net flux of group III - or equivalently group V - atoms to the surface:

$$G(x) \propto J_{III}(x) = J_V(x) \qquad \text{E A2.4}$$

From E A2.3 and E A2.4, it can be seen that the growth rate is inversely proportional to the boundary layer thickness; combining these equations with an explicit expression of $\delta_D(x)$ as the equation E A2.2, it would be possible to predict (rather estimate) the behavior of the growth rate as a function of temperature, pressure and position.

A schematic of MOVPE overall processes is depicted in Fig. A2.4, assuming a horizontal configuration (which applies even to planetary reactors) and making use of the boundary layer approximated description.

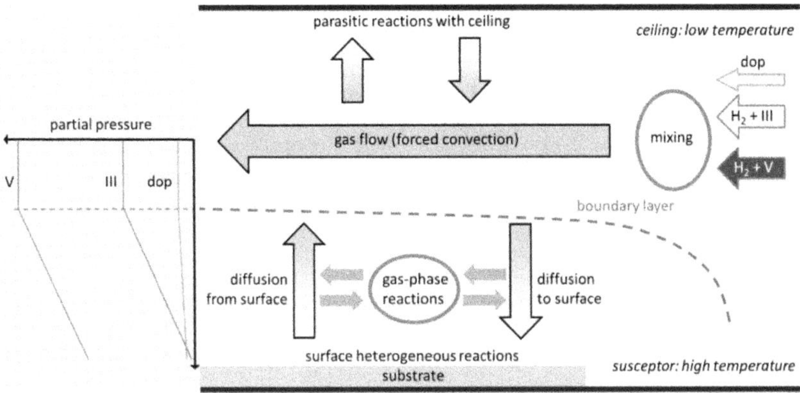

Figure A2.4: schematic representation of MOVPE transport and reactions (horizontal configuration reactor); to the left, the gas-phase concentration profiles of reagents in the boundary layer approximation.

The precursors, diluted in the carrier gas, are injected separately from the right and move towards the left under forced convection, while they intermix via diffusion. There is a vertical gradient of the horizontal flow velocity, and near the susceptor the flow velocity is reduced to zero. Because of the vertical temperature gradient, the molecules in the boundary layer are on average more energized than those in the free stream above; the high temperature promote gas-phase reactions, in particular precursor pyrolysis. When the precursors and the smaller fragments reach the hot surface of the substrate – or of the susceptor – heterogeneous reactions take place, including further pyrolysis and deposition of solid material; the volatile products of these reactions diffuse back in the gas phase, possibly further participating in gas-phase reactions.

Under usual operative conditions, once the species containing group III atoms reach the substrate, they react almost to completeness, and the group III atoms are incorporated in the crystal; simultaneously, to maintain the 1:1 stoichiometry, the same amount of group V atoms is incorporated. For both groups, a precursor concentration gradient is established in the boundary layer (qualitatively shown to the left), but since group V precursors are always introduced in excess with respect to group III, the group V species concentration does not change much – in relative terms – from the input value. Dopants can be introduced either with group III or group V, depending on the chemical compatibility; the molar flows are much smaller than those of both group V and III; their incorporation efficiency is not bound to stoichiometry and can be high or low, depending on the specific dopant, material and growth conditions. The diffusion of cracked precursors towards the ceiling leads to some amount of deposition of polycrystalline or amorphous material; an effect of this parasitic deposition is to increase ceiling reflectivity, which impacts the radiative energy transport: a small drift in the growth rate with increasing ceiling coating is typically observed.

For all the previously mentioned reactor types, the rotation contributes to average out any angular disuniformity in reagents flow and susceptor heating. The geometry and the operative condition range have been optimized in order to avoid disruptions of the laminar flow that would be detrimental for growth rate uniformity and for the possibility of obtaining abrupt interfaces. These disruptions can be regions of turbulence (chaotic behavior) - which can arise at very high gas velocity - and recirculation cells - which can arise especially because of the buoyancy effect (or "natural convection") of the gas heated by the hot susceptor, as exemplified in Fig. A2.3b. As rough guideline [192], turbulence is avoided when the Reynolds number (*Re*) is small (< 100); *Re* is defined as:

$$Re = (\rho u L)/\mu \qquad \text{E A2.5}$$

where L a characteristic length of the system (e.g. the distance between susceptor and ceiling). It can be seen that the use of a light carrier gas is beneficial to reduce Reynolds number, because of the linear dependence on density. *Re* can be reduced even lowering the pressure, the flow speed and the ceiling height, although it must be considered that in practice these parameters are to some extent interdependent: for example, at a given flow rate, reducing the pressure increases the flow speed.

Buoyancy effects can be roughly related to the thermal Rayleigh number (*Ra*) defined by:

$$Ra = \left(\beta g C_p \rho^2 L^3 \Delta T\right)/(\mu \kappa) \qquad \text{E A2.6}$$

where β is the gas thermal expansion coefficient (for an ideal gas $\beta = 1/T$), g is the gravitational acceleration and ΔT is the temperature difference between susceptor and ceiling. Large values of Rayleigh number (> 1700) are associated with the onset of natural convection. It can be seen that the use of a light carrier gas is highly beneficial to reduce Rayleigh number, because of the quadratic dependence on density. Moreover, there is a cubic dependency on the chamber height: a small distance between susceptor and ceiling is highly effective in reducing *Ra*.

The distinctive characteristic of the **rotating disk reactor** [6, 193] is that the rotation speed is quite high, in the range 500-2000 rpm (1-2 orders of magnitude higher than in the other models) and the viscous drag acts as a centrifugal pump, producing a pressure gradient that forces the gas radially outward and vertically downward: tuning of the rotation speed according to the other operative conditions (pressure gas flow, radial temperature profile) is used to obtain a streamline flow, a radially uniform reagent transport and a uniform deposition rate.

In the **close coupled showerhead reactor** [6, 194], the gas is injected uniformly over the whole susceptor surface through thousands of tightly spaced small tubes, welded inside a water-cooled showerhead which is positioned very close (10-20 mm) to the surface of the susceptor; group III and group V precursors are injected by different and interleaved families of tubes. In this configuration, the radial and vertical flow velocities are approximately decoupled; the radial velocity increases linearly with the distance x from center and the vertical velocity is independent from x: the resulting reagent supply and boundary layer thickness are uniform, which translates in a uniform diffusion rate towards the susceptor surface. A fine tuning of radial uniformity can be done controlling the radial temperature profile using separate concentric resistors.

Planetary reactors look very much alike vertical reactors, but are better described - in terms of flow configuration - as radially-isotropic horizontal reactors; they have already been described in chapter 3.

A2.2.2 Control of reagent flows [6]

In chapter 3 it has been stated that metalorganics and other liquid and solid precursors (as CBr_4) are contained in bubblers through which a bubbling gas flows and that (ideally) the bubbling gas saturates in the bubbler with the precursor vapor; the resulting net MO flow f_{MO} is given by:

$$f_{MO} = f_b \frac{P_{MO}}{P_{tot} - P_{MO}} \quad \text{E A2.7}$$

where f_b is the flow through the bubbler, P_{tot} is the bubbler pressure and P_{MO} is the partial pressure of the metalorganic. If the flows are expressed - as it is usual - in sccm (standard cubic cm per minute), the molar flow is found dividing f_{MO} by the gas molar volume at standard temperature and pressure.[3]

Ideally, the partial pressure is equal to the equilibrium vapor pressure of the MO, which in turn is a function of temperature; a commonly used approximate expression for this dependence, loosely related to the integrated form of Clausius-Clapeyron equation, is:

$$\log(P_{MO}) = A - B/T \quad \text{E A2.8}$$

where A and B are empirical parameters - normally provided by the MO source supplier. Control of the MO flow requires a tight stabilization of the temperature and the pressure in the bubbler: taking for example a TMGa bubbler kept at -5°C and 500 mbar (the conditions actually used in FBH reactors), a 1% variation of f_{MO} can be caused by a variation of either 0.15°C or 5 mbar.

The behavior of a bubbler can deviate from the ideal behavior, with the bubbler gas not reaching full saturation; this happens at high flow rates and/or when the bubbler content is low. Undersaturation can be especially significant with solid sources, as TMIn (trimethylindium), because of the formation of channels in the MO material, although in the newest bubbler models the problem has been strongly reduced. Other sources of non-ideality are the heat exchange with the carrier and the heat extraction due to the enthalpy of vaporization, resulting is a temperature difference between the precursor inside the bubbler and the thermal bath. It is possible to monitor the actual MO concentration in the exiting gas with a sensor, the standard solution being an ultrasonic cell that measures the sound velocity of the gas mixture (commercial name EPISON), but such sensors were not installed in the tools used in this work.

A gradual change in composition is obtained changing the precursor flow towards the manifold by means either of one MFC (*standard configuration*) or of three MFCs (*diluted configuration*); the two configurations are represented in Fig. A2.5.

For the hydrides, the standard configuration makes use of a *source* MFC, which determines the reagent flow, and of a *pushing* MFC, which is used to add an extra amount of carrier gas downstream of the source. This allows keeping fixed the total gas flow that reaches the manifold from this specific line, independently of the source flow, which is important in order to avoid pressure fluctuations that could ultimately introduce fluctuations in precursors supply to the growth chamber.

For MOs, in the standard configuration the source MFC is before the bubbler, the pushing MFC is after the bubbler and a subsequent PC is installed to regulate the bubbler pressure. In this case the pushing flow is beneficial even because of the dilution effect, which reduces the risk of precursor condensation along the line.

Standard mass flow controllers are characterized by a dynamic range (ratio of maximum flow to minimum controllable flow) of about 50. This means that a standard MFC should not be operated below ≈3% of its range (although some last generation MFCs should allow flow control down to 0.2% of the full scale, according to manufacturer's data). When it is necessary to control the precursor flows in a larger dynamic range, the diluted configuration can be used. In the case of hydrides, the reagent flow,

---

[3] According to the newest IUPAC convention, standard conditions (STP) are: 1 bar and 0°C, the ideal gas molar volume is 22711 cm^3, while the older STP convention was 1 atm, giving a molar volume of 22414 cm^3. The MFC must be calibrated for the specific gas, otherwise an appropriate correction factor should be applied.

exiting from the *source* MFC (used well within its dynamic range) is first diluted with carrier gas by a *dilution* MFC, then a fraction of the flow is extracted by an *injection* MFC, while the excess flow is discarded (sent to the vent line) typically through a PC. The injection flow is combined with a *pushing* flow as in the standard configuration and sent to the manifold. The resulting net hydride flow towards the reactor f_{out} is given by:

$$f_{out} = c_{hyd} \times \frac{f_{src}}{f_{src} + f_{dil}} \times f_{inj} \qquad \text{E A2.9}$$

where c_{hyd} is the original hydride concentration in the cylinder, f_{src}, f_{dil} and f_{inj} are the source, dilution and injection flows. The diluted configuration for MOs is conceptually similar to that for hydrides.

Using the diluted configuration, with MFCs dilution and injection dynamic range much larger than that of source MFC, the overall dynamic range can be extended by several orders of magnitude, at the price of wasting most of the precursor.

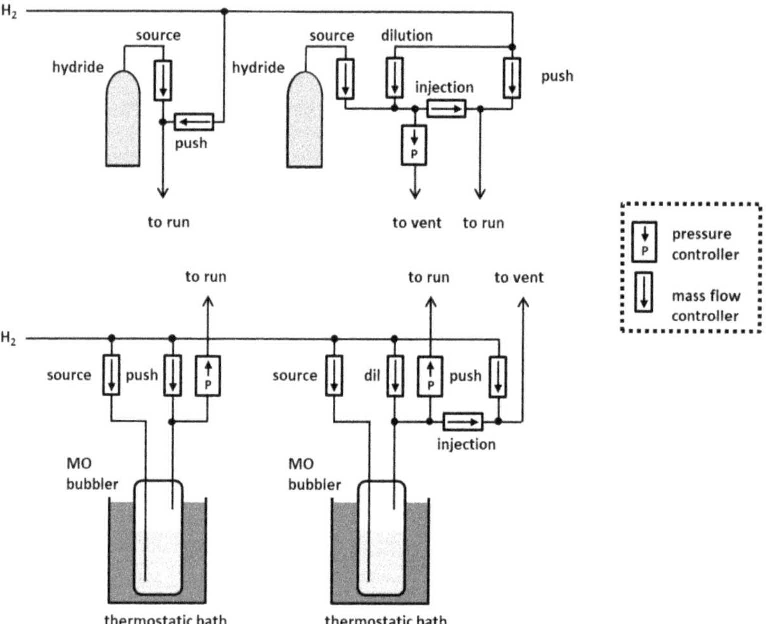

Figure A2.5: representation of standard and diluted lines of hydrides and metalorganics.

A2.3 Precursors for the growth of arsenides and phosphides

A2.3.1 General precursors requirements

Precursors for MOVPE growth must satisfy several requirements, in particular:

- high purity in terms of total metal, organic and oxygen-containing impurities, ideally at levels below parts per billion;
- in the case of liquid or solid precursors, suitable vapor pressure in the temperature range controllable with a thermostatic bath;
- sufficient stability to avoid decomposition during storage or in the gas mixing system before entering the growth chamber, approximately up to 150°C;
- they should not give premature or undesirable reactions with other reagents in the growth chamber (as formation of solid particulate in the gas phase);
- they should not be *excessively* stable, and react heterogeneously with the wafer surface at the process temperature, which is typically above 550°C, incorporating the desired elements in the solid and releasing their substituents in the gas phase as volatile species.

The last two requirements depend to some extent to the overall precursor selection, and not only on the individual chemical compounds.

A2.3.2 Molecular structure of the precursors and gas-phase diffusivity

The precursor compounds used in this work are listed Table 3.1 in chapter 3, their molecular structures are represented in Fig. A2.6. These precursors are more volatile and have higher temperature stability with respect to available alternatives which larger substituents, as for example tert-butyl-arsine $(CH_3)_3CAsH_2$ in comparison to AsH_3 and triethylgallium $Ga(C_2H_5)_3$ in comparison to $Ga(CH_3)_3$. Arsine and phosphine sources are cylinders of undiluted liquefied gas; disilane is already highly diluted in hydrogen because of the small flows required for doping. The liquid and solid compounds sources are bubblers, kept in thermostatic baths at appropriate temperatures, based on their vapor pressures (TMGa and DMZn at -5°C, the other sources at 17°C).

TMIn and TMGa have a trigonal planar structure, TMAl is a dimer at room temperature; consequently a factor of 2 must be introduced when calculating the molar flows with E A2.7; at $T > 200°C$ it is completely dissociated in the monomer, which has the same structure as TMGa and TMIn.

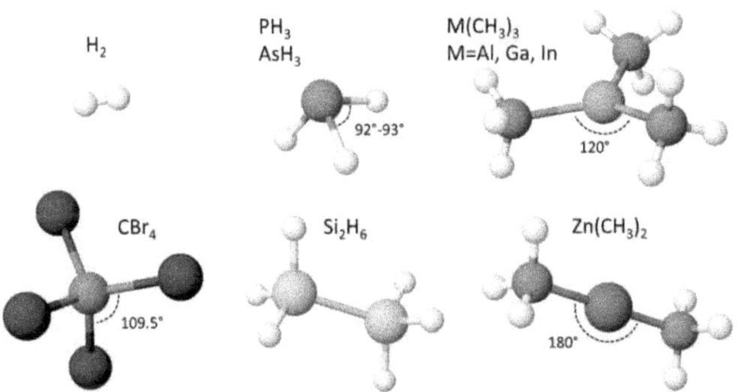

Figure A2.6 Ball & stick representation of the precursor molecules of Table 3.1.

From Tab. 3.1, it can be noted that the precursor molecular weight M_p is spread in a large interval (34-332 g/mol); M_p is always much larger than carrier (hydrogen) molecular weight M_c, and this has an impact in equalizing the precursors gas-phase diffusion coefficients, which at low pressure and high dilution can be expressed according to Chapman-Enskog model [195] by:

$$D_{p,c} = A_{p,c}(T) \cdot T^{3/2} \cdot P^{-1} \cdot \left(M_p^{-1} + M_c^{-1}\right)^{1/2} \xrightarrow{M_c \ll M_p} D_{p,c} \propto M_c^{-1/2} \quad \text{E A2.10}$$

where $D_{p,c}$ is the diffusion coefficient of the precursor, $A_{p,c}(T)$ is a function related to the range and strength of intermolecular forces, weakly dependent on T (approximately $\propto T^{0.2}$ to $T^{0.4}$). As indicated, the sum in parenthesis is dominated by the light hydrogen molecule: for example, the calculated diffusion coefficients for PH_3 and AsH_3 differ by 6% in hydrogen and by 18% in nitrogen.

A2.3.3 Pyrolysis of the precursors

Before reacting heterogeneously on the wafer surface incorporating new atoms in the crystal, the precursors undergo dissociation into smaller fragments (pyrolysis) already in the gas phase. If equilibrium conditions were reached within the gas phase, the concentrations of the species formed starting from the considered precursors, would be predictable based on their thermodynamic properties, solving a system containing mass-balance equations and chemical-equilibrium equations. The necessary input data are the standard enthalpy and entropy of formation and the heat capacity at constant pressure for each species that can (reasonably) form in the gas phase. According to calculations reported in Refs. [196, 197], in the temperature and pressure range of MOVPE growth, mixtures of arsine, phosphine and group III precursors are expected to form prevalently tetramers As_4, P_4, dimers As_2, P_2, methane, atomic indium, $GaCH_3$, GaH_2, $AlCH_3$ and AlH_3.

The actual degree of pyrolysis is in practice largely controlled by kinetic factors. The dissociation energies listed in Table 3.1 might be considered as a first indication of the relative kinetic stability of the corresponding molecules, and to the extent to which they should undergo thermal cracking at a given temperature, but this assumption is only partially true; there are at least four factors that make pyrolysis' kinetic difficult to predict: the role of energy transport, the existence of multiple monomolecular reaction paths, the presence of radical or other polymolecular reactions and the heterogeneous pyrolysis.

The simplest assumption is that precursors' pyrolysis occurs via unimolecular gas-phase reactions, whose rate can be expressed by the first order kinetic equation:

$$-\frac{d[XY]}{dt} = k[XY] \xrightarrow{(integrating)} [XY] = [XY]_0 e^{-kt} \quad \text{E A2.11}$$

The rate constant k - for any kinetic order - is usually phenomenologically expressed by the Arrhenius equation:

$$k = A \cdot e^{-E_a/RT} \quad \text{E A2.12}$$

where R is the gas constant, E_a is called activation energy and A is called pre-exponential factor (or frequency factor for unimolecular reactions). A more general, differential definition of E_a that does not assume a temperature-independent A in E A2.12 is:

$$\frac{dln(k)}{dT} = \frac{E_a}{RT^2} \quad \text{E A2.13}$$

Unimolecular dissociation reactions [198, 199] require significant activation energies and can be considered as a sequence of energization and dissociation steps:

$$XY \xrightarrow{energy} XY^* \xrightarrow{XY\ddagger} X + Y \quad \text{R A2.1}$$

In R A2.1 the energized molecule XY^* has sufficient internal energy to form the products $X + Y$ via the intermediate, metastable configuration $XY^{\ddagger}$, called transition state, which corresponds to the maximum value of Gibbs free energy that must be necessarily overcome along the reaction path. The overall reaction rate depends on:

- energy transport and XY excitation mechanisms - which determine the rate of the first step;
- the structures of the energized molecule XY^* and of the transition state $XY^{\ddagger}$ - which determine the rate of the second step.

Energy transport in the gas phase occurs primarily via intermolecular collisions (neglecting the radiative contribution), and is actually a bimolecular process; lowering the pressure reduces the energization rate, while it has no influence on the following unimolecular dissociation. Below a certain pressure value, the first step of R A2.1 becomes rate-determining (the slowest of the two) and the overall rate starts to drop with decreasing pressure; the corresponding pressure range is called "fall-off region". Vice-versa, when the energy transport is fast compared to the dissociation, the overall rate is controlled by the latter step; this condition is referred to as the "high pressure limit". The faster the dissociation of XY^*, the higher the pressure at which the transition between the two regimes occurs: larger molecules as TMGa can store energy in several internal modes besides of those possibly leading to the dissociation and consequently- once excited - are slower in dissociating, and enter the fall-off region only at very low pressures (below those used in MOVPE), while smaller molecules as AsH_3 enter the fall-off region already above atmospheric pressure [200].

In the high-pressure limit, the concentration of the energized molecules can be considered in thermal equilibrium, and the kinetic is well described by Eyring's transition-state theory (TST), according to which the rate constant – for any kinetic order - is given by:

$$k = \frac{k_B T}{h} \cdot e^{-\frac{\Delta G^{\ddagger}}{RT}} = \frac{k_B T}{h} \cdot e^{\Delta S^{\ddagger}/R} \cdot e^{-\frac{\Delta H^{\ddagger}}{RT}} = \frac{k_B T}{h} \cdot e^{\Delta S^{\ddagger}/R} \cdot e^{-\frac{P\Delta V^{\ddagger}}{RT}} \cdot e^{-\frac{\Delta E^{\ddagger}}{RT}} \qquad \text{E A2.14}$$

where $\Delta G^{\ddagger}$, $\Delta S^{\ddagger}$, $\Delta H^{\ddagger}$, $\Delta E^{\ddagger}$ and $\Delta V^{\ddagger}$ are the Gibb's energy, entropy, enthalpy, internal energy and occupied volume differences between $XY^{\ddagger}$ and the reagents.

By differentiation of E A2.14 and comparison with E A2.13 the activation energy results:

$$E_a = \Delta E^{\ddagger} + RT \qquad \text{E A2.15}$$

Considering reactions where the volume change is primarily due to a change $\Delta n^{\ddagger}$ in the number of gas-phase molecules (again between $XY^{\ddagger}$ and reagents) and assuming ideal gases, the pre-exponential term in E A2.12 becomes:

$$A = e^{1+\Delta n^{\ddagger}} \cdot \frac{k_B T}{h} \cdot e^{\Delta S^{\ddagger}/R} \qquad \text{E A2.16}$$

In the case of a simple bond dissociation as in R A2.1, E_a is equal to the dissociation energy. The frequency factor A can be theoretically derived from the molecular partition functions of $XY^{\ddagger}$ and XY and has typically values in the range 10^{15}-10^{17} s^{-1} for the first bond dissociation in simple molecules[4].

In some cases, unimolecular pyrolysis mechanisms with lower activation energies with respect to simple bond breaking are also possible: in the case of disilane, the prevailing mechanism is probably the dissociation into SiH_4 and SiH_2 with simultaneous transfer of one H atom between the Si atoms; the corresponding activation energy has been estimated 218 kJ/mol, much lower than the bond dissociation energies of Tab. 3.1 [201]. For AsH_3 and PH_3, as an alternative to dissociation into XH_2 + H (X=As or P), which has a high activation energy, dissociation into XH + H_2 has been proposed [200]; the contribution of this reaction path is uncertain.

---

[4] In the fall-off region, unimolecular reactions rates are rather described by the Rice-Ramsperger-Kassel-Marcus - or RRKM – theory, which might be considered an extension of transition-state theory that drops the hypothesis of thermal equilibrium for the energized species.

Unimolecular mechanisms can be expected to prevail over the polymolecular when the precursors are extremely diluted, the temperature is high and the carrier gas is inert; in the case of MOVPE, under normal operative conditions, bimolecular mechanisms, especially chain radical reactions, play an important role along with unimolecular dissociation. Radical reactions can involve a very high number of individual bimolecular (plus some unimolecular and trimolecular) steps, and unstable intermediate species whose properties are often not well known; as a consequence, the kinetic modeling suffers from considerable uncertainty. As prototypical example, the case of GaAs growth is presented in the following.

In Ref. [202] Mountziaris and Jensen have proposed a scheme of the gas phase process leading to the decomposition of TMGa and AsH_3 in hydrogen carrier, involving 17 reactions; the scheme was part of a broader model of MOVPE growth which included heterogeneous reaction and transport. It must be said that this model was never intended as definitive or complete, and a modified version [203] was proposed which intended to better capture the experimental results relative to carbon incorporation. According to the model, the main products of gas-phase processes are undissociated AsH_3 plus a small amount of the radical AsH_2, the $GaCH_3$ radical, methane and a small amount of C_2H_6. It can be noted that, in this analysis, several of the species that would form under equilibrium conditions, as As_4, As_2 and GaH_2, are absent; this can be justified as a consequence of kinetic restrictions related to the short residence time in the MOVPE chamber: only the species that are supposed to form in this timescale are included. The most important reactions for GaAs growth are listed and commented below (neglecting those related to carbon incorporation). In the formulas, dots are used to indicate the valence electrons not involved in bonding, A and E_a are the pre-exponential factor and the activation energy in the Arrhenius expression of the kinetic constant k of each step (several values are only estimates).

Unimolecular initiation steps (M=molecule providing energy for dissociation)

$Ga(CH_3)_3 \rightarrow \dot{G}a(CH_3)_2 + \dot{C}H_3$	A≈10^{15} s^{-1}	E_a≈250 kJ/mol	R A2.2a
$\dot{G}a(CH_3)_2 \rightarrow \ddot{G}aCH_3 + \dot{C}H_3$	A≈10^{8} s^{-1}	E_a≈150 kJ/mol	R A2.2b
$AsH_3 + M \rightarrow AsH_2 + \dot{H} + M$	A≈10^{17} $mol^{-1}cm^3s^{-1}$	E_a≈310 kJ/mol	R A2.2c

Unimolecular decomposition of $GaCH_3$ is not included, because of the expected high E_a (bond dissociation energy ≈ 340 kJ/mol). AsH_3 dissociation was included only in [203], and is written as a bimolecular process because AsH_3 cannot store the dissociation energy. The anomalous low value of A in R A2.2b is experimental and has been explained assuming a highly ordered transition state (large and negative $\Delta S^{\ddagger}$) [204].

Propagation steps

$AsH_3 + \dot{C}H_3 \longrightarrow \dot{A}sH_2 + CH_4$	A≈10^{10} $mol^{-1}cm^3s^{-1}$	E_a≈10 kJ/mol	R A2.2d
$AsH_3 + \dot{H} \longrightarrow \dot{A}sH_2 + H_2$	A≈10^{13} $mol^{-1}cm^3s^{-1}$	E_a≈10 kJ/mol	R A2.2e
$\dot{C}H_3 + H_2 \rightarrow CH_4 + \dot{H}$	A≈10^{12} $mol^{-1}cm^3s^{-1}$	E_a≈35 kJ/mol	R A2.2f
$Ga(CH_3)_3 + \dot{H} \rightarrow \dot{G}a(CH_3)_2 + CH_4$	A≈10^{13} $mol^{-1}cm^3s^{-1}$	E_a≈40 kJ/mol	R A2.2g
$\dot{G}a(CH_3)_2 + \dot{H} \rightarrow \ddot{G}aCH_3 + CH_4$	A≈10^{13} $mol^{-1}cm^3s^{-1}$	E_a≈40 kJ/mol	R A2.2h

The reaction of arsine with the hydrogen radical to form AsH_2 and H_2 R A2.2e is strongly exothermic as the reaction between arsine and methyl radical R A2.2d, but it was initially not considered, based on experimental evidence reported in Ref. [205], where cracking of AsH_3 in D_2 did show no measurable production of HD molecules; it was nonetheless added in the revised version.

Termination steps (M=molecule absorbing part of the recombination energy)

$\dot{C}H_3 + \dot{C}H_3 \longrightarrow C_2H_6$	A≈10^{13} $mol^{-1}cm^3s^{-1}$	E_a≈0	R A2.2i
$\dot{C}H_3 + \dot{H} + M \longrightarrow CH_4 + M$	A≈10^{19} $mol^{-2}\cdot cm^6s^{-1}$	E_a≈0	R A2.2j
$\dot{H} + \dot{H} + M \longrightarrow H_2 + M$	A≈10^{16} $mol^{-2}\cdot cm^6s^{-1}$	E_a≈0	R A2.2k
$\dot{A}sH_2 + \dot{C}H_3 \longrightarrow H_2AsCH_3$	A≈10^{13} $mol^{-1}cm^3s^{-1}$	E_a≈0	R A2.2l

It can be seen from this example that gas-phase reactions have probably a very complex kinetic, even when the number of original species involved is limited and not all the possibility are taken into account. To thoroughly test a specific reaction model against experimental evidence is difficult, because the detection of the intermediates is often not feasible. It is not uncommon to find "apparent" *first order* overall precursor decomposition kinetic constants reported in literature, based on data relative to the few species whose concentration can be measured, possibly fitted with the Arrhenius equation when the plot of $ln(k)$ vs $1/T$ is approximately linear.

Part of the pyrolysis occurs heterogeneously on the wafer surface and on the susceptor. Heterogeneous reactions can be unimolecular (dissociative absorption, dissociation of adsorbed molecules) or polymolecular (reaction between two absorbed molecules or between one adsorbed molecule and one gas molecule). Typically they have comparatively lower apparent activation energies but at the same time lower apparent frequency factors than the competitive gas-phase unimolecular dissociations, and become increasingly important at low temperature, when the activation energy for gas-phase mechanisms is high with respect to the thermal energy of the molecules. Low pressure and low residence times are other factors that reduce the relative importance of gas-phase pyrolysis with respect to heterogeneous.

Experimentally, it is difficult to clearly discriminate between gas-phase and heterogeneous pyrolysis, especially using experimental conditions that correspond closely to those of MOVPE growth. Moreover, for heterogeneous reactions the difference between pyrolysis and growth can become blurred, since a precursor might simultaneously release its substituents and incorporate an atom in the crystal lattice.

To summarize, the variety and complexity of reaction mechanisms, and the limited possibility of determining the concentration of gas-phase and adsorbed species during the MOVPE process, make the precise determination of the individual kinetic steps very difficult. Some guidelines are as follows:

- when group III and group V are simultaneously present, the interdependent nature of the radical chemical reactions reduces the temperatures required for precursors decomposition, the effect being stronger on the hydrides AsH_3 and PH_3;
- the presence of wafer substrates has a similar effect, promoting heterogeneous pyrolysis; the importance of heterogeneous pyrolysis is expected to increase at lower values of T, P and t_r;
- use of H_2 as carrier gas increases the decomposition rates of MOs with respect to more inert carriers as N_2, He or Ar, presumably because of participation in radical reactions propagating steps;
- PH_3 is, among the listed reagents, the one most difficult to pyrolize, requiring a temperature significantly higher than that required for AsH_3; in turn, the other precursors listed in Tab. 3.1 pyrolize more easily than AsH_3.

Assuming that the apparent rate constant is known, the amount of dissociation of a precursor in the reactor chamber can be evaluated as a function of residence time and process temperature. The rate constants that can be found in literature are unfortunately widely spread, since they are strongly dependent on the specific experimental conditions. As a semi-quantitative indication of what can be expected, the degree of gas-phase pyrolysis of each precursor in hydrogen (disappearance of the original molecule) based on averaged literature values, has been plotted in Fig. A2.7. A ×10 longer residence time would shift the curves by 40-60°C towards lower T, and the compresence of metalorganics and hydrides would have similar or stronger effects.

Heterogeneous pyrolysis occurs at significantly lower temperatures, and this is a further reason why the total degree of pyrolysis can be expected to be always higher than what shown in Fig. A2.7. This is exemplified in the case of TMGa by the dashed line in Fig. A2.7, which represents the thermal decomposition of the precursor *adsorbed* on GaAs, as evaluated in ultrahigh vacuum experiments. The contribution of this low-temperature reaction path to the overall pyrolysis depends on the rate at which the precursor molecules reach the surface (gas-phase transport, volume to surface ratio) and on precursor's adsorption and desorption rates; the dashed line can be considered an upper bound to the amount of heterogeneous pyrolysis when referred to the *total* precursor concentration. Actually, the heterogeneous pyrolysis of Ga-containing species during growth involves most probably even surface reactions with adsorbed arsenic-containing species, as detailed in section A2.4.2.

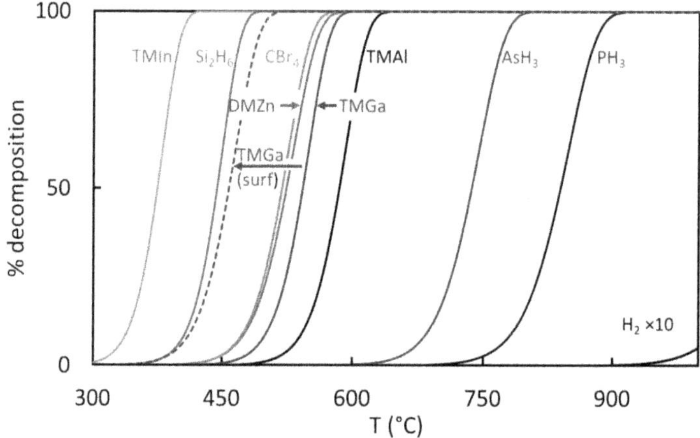

Figure A2.7 Degree of gas-phase pyrolysis in hydrogen carrier of the precursors listed in Table 3.1, based on apparent kinetic constants from Ref. [48, 200, 201, 206-212, 34ch4]; the residence time has been fixed in all cases to 1s, hydrogen values are multiplied by 10. The dashed line represents the surface pyrolysis of adsorbed TMGa, apparent kinetic constant according to Ref. [213].

A2.4 Surface processes

A2.4.1 Growth modes

Surface processes are, if possible, even more complex and less well-known than those in the gas-phase. They depend on the nature of the molecular species involved and on the composition and structure of the solid surface on which they occur (crystal planes, terraces, reconstructions). Moreover, to accurately predict the characteristics of the grown material, not only the surface chemistry but even the surface transport mechanisms should be known: unfortunately, it is experimentally difficult to access such information, especially in the case of MOVPE; more extensive information is available in the case of MBE, but this cannot be in general directly applied to MOVPE conditions, although some similarities can be expected to exist.

It is generally assumed that, when the lattice mismatch between substrate and grown material is not too high, and the strength of the bonding among the atoms in the layer is not much higher that the strength of the bonding between the atoms of the layer and those of the substrate, the growth occurs with a layer-by-layer (or Frank - Van der Merve) mechanism. This mechanism consists in the nucleation over the surface of two-dimensional islands, which then expand till a new layer is completed, as illustrated in Fig. A2.8a.

The assumption is based on various experimental evidence, including the fact that almost atomically flat interfaces can be obtained, and is justified considering that the number of bonds that an added atom (adatom) can form at the edge of an island is higher than the number of bonds that it can form when it attaches to a flat surface: this can be heuristically explained using as model a Kossel crystal [147], represented in Fig. 2.9, where the atoms are cubes that bond to each other at the facets with their 6 nearest neighbors, along the <100> directions; the total bonding energy of an adatom is higher at kink sites, intermediate at step sites and lower at surface sites.

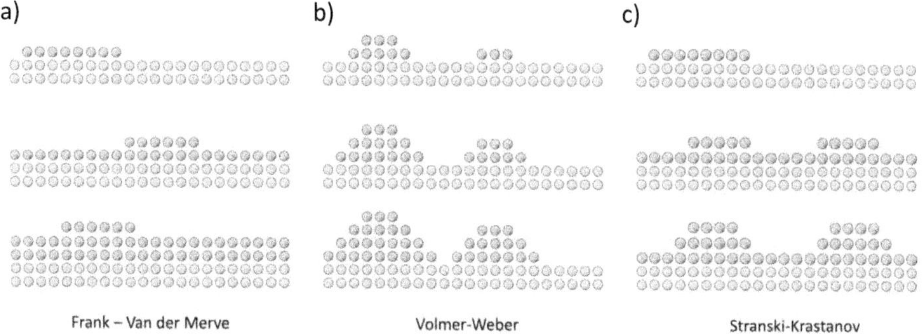

Figure A2.8 Schematic of three possible epitaxial growth modes.

The model can be refined including the 12 second neighbor and 8 third neighbor bonding interactions, respectively along the <110> and <111> directions. In the case of III-V compounds, the situation is actually more complex because of the existence of two different kinds of atoms, X and Y, which form only four X-Y bonds along the directions defined by their tetrahedral hybridization; moreover, the high-density atomic planes are the {111} and not the {100} as in a Kossel crystal. Nonetheless, if - instead of the single atoms - a cubic unit cell is considered as basis for the construction of the III-V crystal, the parallel with Kossel model becomes apparent.

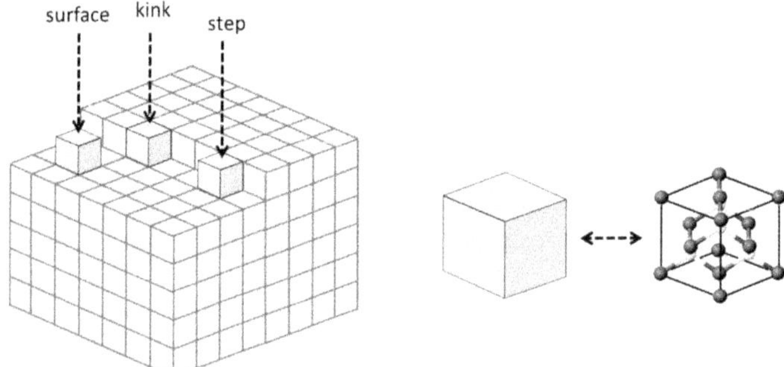

Figure A2.9 Kossel crystal, composed of cubic atoms; three adatoms (blue) are attached to a surface site (1 first neighbor), to a step site (2 first neighbors) and to a kink site (3 first neighbors). To the right: to directly apply the model to Zincblende structure, cubic atoms can be replaced with a cubic cell containing 8 atoms.

Thermodynamically, the driving force for layer-by-layer growth is the minimization of surface excess Gibbs free energy. In order to satisfy this condition, the adatoms (or the adsorbed precursor species) must be able, before being irreversibly incorporated, to move over the surface for a sufficiently long distance to reach an energetically favorable position at a step; would this not happen, they would be incorporated randomly on the surface, which would become increasingly rough during the growth. The mechanisms of surface transport in MOVPE are not well known, but they are expected to be a combination of short-range surface diffusion (the adsorbed species move on the surface without detaching) and long-range near-surface vapor-phase diffusion (species adsorbed on unfavorable sites desorb, diffuse in the stagnant layer close to the surface and are again adsorbed). Being an activated process, surface transport is reduced at low temperature. Under this aspect, when the temperature is high enough to ensure a long diffusion length of the adsorbed precursors/adatoms, the MOVPE can be considered thermodynamically controlled.

A variant of the layer-by-layer growth is the step-flow growth: in this case, starting from a surface that already contains steps, for example because of a slight misalignment with respect to a low-index plane, the nucleation of islands is not required to provide step and kink sites; this type of growth is well observable in the case of MBE.

Opposite to the layer-by-layer case, when the strength of the bonding among the atoms in the layer *is* much higher that the strength of the bonding between the atoms of the layer and those of the substrate, the growth proceed in the Volmer-Weber mode, with formation of separate tri-dimensional islands, as shown in Fig. A2.8b; this condition is similar to that of a fluid whose cohesive forces exceed the adhesive forces to a surface, resulting in high contact angle. The Volmer-Weber mode is expected to occur under thermodynamically controlled conditions in order to minimize the excess surface energy, but islands growth might result even at low temperature for kinetic reasons (low surface transport).

While a moderate mismatch is elastically accommodated, until the layer thickness reaches the critical value for dislocations to develop, a strong mismatch can lead to the Stranski-Krastanow growth mode, where at first, a Frank Van-der-Merwe continuous, highly strained "wetting" layer is grown, and then tri-dimensional islands develop, as illustrated in Fig. A2.8c. Stranski-Krastanow growth mode has found a useful application in the realization of quantum dots structures.

A2.4.2 Surface chemistry and growth rate

The case of GaAs growth is one of the most investigated and is proposed here as prototypical example, combining in the discussion some models and "classical" experimental results from literature.

In the afore-mentioned MOVPE model of Mountziaris and Jensen (M-J) [202], the growth of GaAs from TMGa and AsH_3 in H_2 carrier has been simulated considering, along with the gas-phase reactions, further 28 reactions at the heterointerface; surface transport or the effect of surface reconstruction were not included explicitly, but the different reactivity of (100), (110) and (111) crystal planes was taken into account, based on the different number of dangling bonds exposed on the surface and at the step edges in each case. Some salient features and results of the model are summarized in the following.

Surface reactions: they are divided into:

- adsorption or dissociative adsorption of the gaseous species on specific crystal sites (As or Ga atoms of the crystal)
- reactions between adsorbed species
- reactions between adsorbed and gas phase species

In the following, M^X, where X=As or Ga, indicates that the species M is absorbed on an As sublattice site or on a Ga sublattice site, forming bonds with one or more unsaturated atoms of the *other* sublattice, belonging to the underlying atomic plane, and even to the next at a step edge; a Ga (As) atom is considered incorporated once it forms an *additional* bond with an As (Ga) atom on the *growing* layer; note that in this way, atoms that are already strongly bonded to the surface, for example an As atom on (100) bonded with 2 underlying Ga atoms, are formally indicated as "adsorbed".

Non-competitive adsorption of all arsenic and gallium species is assumed.

TMGa is cracked significantly already in the gas phase; the $Ga(CH_3)_x$ species adsorb dissociatively on As atoms, to give in all cases $GaCH_3{}^{Ga}$ and releasing $CH_3(g)$; $GaCH_3{}^{Ga}$ can desorb, but with a comparatively high activation energy, estimated to be ≈190 kJ/mol on (100) planes.

Most of AsH_3 dissociation occurs heterogeneously; AsH_x species adsorb dissociatively on Ga atoms, giving in all cases AsH^{As} and releasing $H_2(g)$ or H(g). AsH^{As} desorption is difficult, requiring a high activation energy, but two AsH^{As} can react producing $As_2(g)$ and $H_2(g)$.

Methyl and hydrogen radicals can adsorb on both sites; these adsorbed radicals can react with each other or with gaseous methyl and hydrogen radicals producing the volatile species $H_2(g)$, $CH_4(g)$ and $C_2H_6(g)$. Methyl and hydrogen radicals can cause the reversion of the dissociative absorption processes, and the reaction of methyl radical with AsH^{As} leads to the formation of As^{As} adatoms and methane. The formation of Ga adatoms was considered negligible, based on experimental evidence quoted in Ref. [203].

The species responsible for GaAs(s) formation are then assumed to be the adsorbed species $GaCH_3{}^{Ga}$, AsH^{As} and As^{As} according to the reactions:

$GaCH_3{}^{Ga} + AsH^{As} \rightarrow GaAs(s) + CH_4(g)$ A≈5×10^17^ mol^-1^cm^2^s^-1^ E_a≈102 kJ/mol R A2.3a

$GaCH_3{}^{Ga} + As^{As} \rightarrow GaAs(s) + \dot{C}H_3(g)$ A≈5×10^17^ mol^-1^cm^2^s^-1^ E_a≈84 kJ/mol R A2.3b

The kinetic values are referred to (100) planes and are different for (110) and (111) planes.

Gallium carbene species as $\dot{G}a = CH_2$, are supposed to form in small amounts either in the gas phase [202] or heterogeneously [203], play a special role in the model: they adsorb and dissociate on neighboring Ga and As atoms of the crystal, incorporating C in the arsenic sublattice, where it behaves as a p dopant. Experimentally, a significant amount of carbon is actually incorporated during GaAs growth, especially at low temperature; experiments with isotopes have shown that the carbon atoms originate from the methyl ligands of TMGa, and the effect is called in MOVPE jargon *intrinsic* carbon doping. Carbon incorporates preferentially in the As sublattice, and its amount increases as AsH_3 partial pressure is reduced. The model interprets these trends, because the carbene molecule mechanism "helps" C to preferentially bond to a Ga atom, because the H-As bond is slightly stronger than the H-Ga bond. The arsine effect is explained assuming a competition between carbene incorporation and re-protonation reactions of the adsorbed carbene due to arsenic species (H is originated mostly by AsH_3 decomposition) and moreover because of competition for the same sites of the arsenic species and the carbon atom.; actually, similar arsine effects could be expected even if the molecule responsible of C incorporation was simply a CH_x radical, and the carbene hypothesis is somewhat dubious; it would be redundant in the case of AlGaAs, given the much stronger Al-C bond, and experimentally AlGaAs incorporates much higher amounts of (p-doping) carbon with respect to GaAs.

Growth rate: it is predicted in the model simulating the overall process: gas-phase transport, gas-phase reactions and heterogeneous reactions. The predictions have been be compared with the experimental results, obtained over a large range of temperatures, reported by Reep and Ghandhi in Ref. [214], partially reproduced in Fig. A2.10.

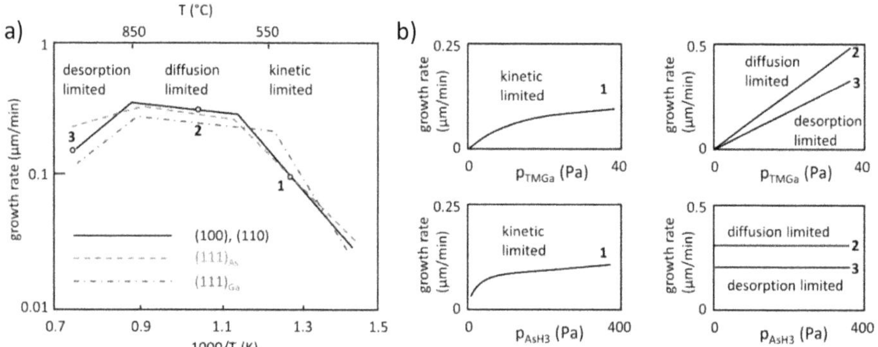

Figure A2.10 Experimental dependence of GaAs growth rate in an atmospheric-pressure MOVPE. **a)** Rate vs. T and GaAs substrate orientation; **b)** rate vs. TMGa and AsH_3 partial pressures, (100) orientation. Interpolated data adapted from Ref. [214].

Three temperature regions can be identified in relation to the growth rate: desorption limited, diffusion limited and kinetic limited.

Desorption-limited region

At high temperature (T>850°C) the experimental growth rate decreases with increasing T, increases linearly with TMGa partial pressure and is independent of AsH_3 partial pressure.

The behavior with T is *qualitatively* consistent with M-J model prediction of a rate limited by the competition between the growth reactions R A2.3 and precursors desorption, the latter becoming more strong with increasing temperature; the simulated effects of TMGa and AsH_3 partial pressures changes in this temperature range are not provided by the authors, and is not possible to tell precisely how well the model performs in that respect, but the predicted growth rates values are shown and they are largely underestimated with respect to the experimental values, up to an order of magnitude at 950°C.

The fact that the rate is "limited" by the surface processes should not be intended to exclude the simultaneous dependence on mass transport, except for the extreme case when the growth rate becomes so small that the concentration of group III at the surface equals the input value. Actually, simplified treatment of mass and heat transport conditions are the factors indicated by the authors to explain the discrepancy with the experimental data in this region.

Kinetic-limited region

At low temperature (T<550°C) the experimental growth rate strongly increases with increasing T; with increasing TMGa and AsH_3 partial pressures it increases first linearly, then sub-linearly and finally saturates. All these aspects are comparatively well described by M-J model: the temperature dependence is explained by the exponential Arrhenius form of the rates of the activated surface and gas-phase reactions (pyrolysis and growth), and the saturation effects are explained by the saturation of the adsorption sites at high partial pressures of the precursors. Even in this case the rate can be expected to depend not only on surface processes, but - simultaneously - on mass transport, except when the growth rate becomes very small and reagent depletion near the surface is negligible.

Saturation effects in the desorption-limited and kinetic-limited regions

Reep and Ghandhi suggest two simple "classical" mechanisms to explain sublinear growth rates and saturation effects at low and high temperature: Langmuir-Hinshelwood and Eley-Rideal. The Langmuir-Hinshelwood mechanism assumes a reaction between As and Ga adsorbed molecular species, whose nature is left open (they could be AsH and CH_3Ga or other pyrolysis products of AsH_3 and TMGa). The Eley-Rideal assumes a reaction between As adsorbed molecular species and Ga gas-phase species. The adsorption is assumed to be non-competitive, and is described as an *equilibrium* by Langmuir's isotherm:

$$\theta_i = \frac{K_i p_i}{1 + K_i p_i} \qquad \text{E A2.17}$$

where θ_i is the fraction of adsorption sites occupied by the ith species having partial pressure p_i and adsorption equilibrium constant K_i; θ_i increases linearly with p_i and then saturates to 1 at high p_i values. The Eley-Rideal mechanism would explain the trends at high T, provided that the surface coverage of As molecules is very high (in the saturation region of the isotherm). The Langmuir-Hinshelwood would explain the observed sublinear trends at low T, provided that the surface coverage of arsenic and gallium molecules becomes very high, so that both enter the non-linear part of Langmuir's isotherm; it could even explain the behavior at high T, provided that in this case the surface coverage of arsenic molecules is still very high while that of gallium molecules becomes low, in the linear region of the isotherm. It has been observed [11] that the assumed values of surface coverage are too high to be realistic, and that even the idea of relating the growth rate to the *average* surface coverage is not realistic; the mechanism should be probably be amended taking into account that the growth occurs preferentially at step sites, and that the adsorbed species can diffuse over the surface towards the step edges, where they encounter a perturbation in the potential energy for surface diffusion which can cause their local accumulation. The rate saturation effects might then occur even at low values of surface coverage, requiring only the filling of the step sites.

Diffusion limited region

In the intermediate temperature region, the experimental growth rate increases only weakly with T, increases linearly with TMGa partial pressure and is independent of AsH_3 partial pressure. This is explained by the fact that in this region the surface *growth* reactions are very fast in comparison to the supply rate of precursors from the gas phase, so that gas-phase transport becomes the slowest, rate determining step. The partial pressure of Ga precursors should become negligibly small near the surface, and its precise value irrelevant for the determination of the gas-phase transport rate.

The experimental moderate increase of the growth rate with increasing T is not present in M-J simulation, which predicts a temperature-independent regime; it is possibly due to the dependence on temperature of mass transport, in particular through the diffusion coefficients D (approximately $D \propto T^{1.7}$, compare E A2.10). A similar trend has been observed on FBH G3 reactor at 100 mbar, while the growth rate is almost temperature-independent in the range 575-750°C on reactor G4 at 50 mbar.

Effects of pressure and flow velocity [11, 215]

The main impact of increasing the flow velocity or of reducing the total pressure *P*, *keeping the partial pressures of the precursors constant,* is to increase the mass transport. Using the boundary-layer model

(E A2.2-A2.4) and the dependence of the diffusion coefficient on P (E A2.10) for a simple estimate, the diffusion-limited growth rate is related to flow velocity and pressures by:

$$G \propto p_{III} u^{1/2} P^{-1/2} \qquad \textit{diffusion-limited (boundary layer approximation)} \qquad \text{E A2.18}$$

When u is increased at constant pressure (increasing *all* the flows in the same proportion), even the rate should increase in proportion to $u^{1/2}$. If the pressure is increased at constant total flow (reducing the reagent to carrier flow ratio to keep the reagents partial pressures constant), being in this case the flow velocity u proportional to P^{-1} the rate should decrease in proportion to P^{-1}. The diffusion-limited region of Fig. A2.10 is expected to shrink when the flow velocity is increased or the total pressure is decreased, with the kinetic region extending towards higher T and the desorption region towards lower T; this expectation is in accord with the experimental results of Heinecke et al. in Ref. [215].

Crystal plane effects

The experimental growth rate is different on different crystal planes, especially at high and low temperature; this point is in line with what predicted by M-J model, according to which the impact of crystal planes differences should be minimal when the growth rate is not under mass-transport control. On the other hand, the growth rate order predicted by the model does not correspond exactly to the experimental one.

Experiment vs models: usefulness and limits

In the discussion above, it has been shown how both simplistic models (as the boundary-layer approximation or the Langmuir.Hinshelwood mechanism) and complex models (Mountziaris-Jensen) can provide some guidelines in interpreting the MOVPE process. Although it is impossible to verify all its assumptions, the M-J model offers at least an approximate plausible picture of the process chemistry, providing a rationalization of the observed results, and in particular of the three growth-rate regions of Fig. A2.10. At the same time, the proposed reaction scheme is probably not completely capturing all the aspects of the real process, and it is dependent on many chemical kinetic parameters that are only roughly estimated, a problem that –to the knowledge of the writer - is common to all efforts to simulate from first principles (i.e. without adjustable parameters) the MOVPE process. While the description of the MOVPE fluid-dynamic has benefited from the availability of increasingly powerful computational tools [9], the underlying chemistry is poorly understood, the kinetic and thermochemical data being scarce and uncertain; quantum-mechanical simulations, typically based on density functional theory (DFT), have been used to simulate MOVPE-related chemical reactions, an approach that is potentially very promising [216]; unfortunately these studies typically involve large clusters of heavy atoms, and the computing time and memory required scales approximately as the third power of the electron number; possibly as a consequence of this, published DFT investigations are comparatively rare (and mostly devoted to III-N and II-VI chemistry).

Selection of the growth-rate regime

In general, the growth conditions can be selected to fall in any of the three above regions, but the diffusion limited is usually preferred, not only because of the more efficient use of the reagents, but even because it often allows an overall better control on material composition, thickness and quality. In the kinetic-limited region, the effects of small temperature and surface conditions variations are stronger; moreover, the surface diffusion is low, and the morphology of the material can become poor. In the desorption-limited region, the composition of multinary alloys, especially those with different group III atoms, becomes more dependent on surface reactions and is more difficult to control. On the other hand, the temperature selection is dictated in each specific case by several considerations, including for example the control over incorporation of dopants and unwanted impurities, preservation of the shape of patterned structures on pre-processed substrates, avoidance of solid-phase inter-diffusion or dopant diffusion, avoidance of ordering effects, realization of thermodynamically unstable compositions.

A2.5 Stoichiometry, composition and impurity control in MOVPE

A2.5.1 V/III ratio, condensation phenomena and stoichiometry

The common use of an excess of group V in MOVPE is fundamentally motivated by the different equilibrium vapor pressures of the corresponding species. At the growth temperature, the III-V compounds would decompose if they were not "protected" by an overpressure of the gaseous components; experimentally, when the V/III gas-phase ratio is kept below 1, severe surface morphological problems occur, related to the presence of group III condensed phase formation, while for V/III ratios above 1, no group V condensed phase formation occurs (with the possible exception of antimony). The explanation is illustrated for the case of GaAs in Fig. A2.11, based on the simplified assumption of a system composed only of Ga and As at thermodynamic equilibrium.

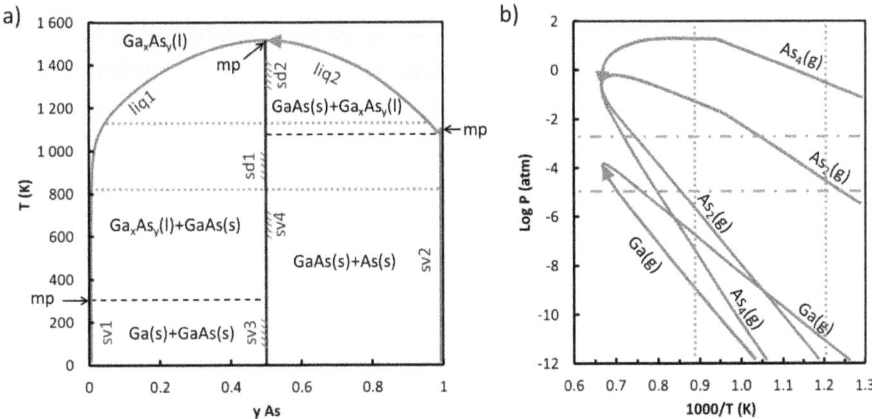

Figure A2.11 **a)** temperature-composition phase diagram of GaAs; sv=solvus, sd=solidus, liq=liquidus, mp = melting points; **b)** partial pressures-temperature phase diagram of the main gaseous species, the values correspond to the liquidus1/solvus1 (light blue) and liquidus2/solvus2 (red) curves of the first diagram. Dotted lines indicate the T interval 550-850°C, dash-dotted lines the pressure interval 1-200 Pa. Data from [11, 217]

The temperature-composition diagram shows only the condensed phases, but the gas phase is simultaneously present: a complete phase diagram would include the equilibrium partial pressures, and the total pressure as independent parameter; the shown condensed-phases diagram can be considered a section at fixed total pressure $P \approx P_{atm}$, but is approximately invariant for a moderate variation of P. In the indicated temperature range characteristic of MOVPE, 550-850°C, there are two *possible* condensed phases: for Ga-rich total composition, solid GaAs and a Ga-rich liquid phase (liquidus1); for As-rich total composition, solid GaAs and an As-rich liquid phase (liquidus2) or solid As (solvus2). The gas-phase equilibrium compositions are shown in the partial pressures-temperature diagram, taking the values along the above-mentioned liquidus and solvus lines; in this second diagram, a reasonable interval of Ga or As partial pressures (1-200 Pa) consistent with the input precursors' partial pressures applied during MOVPE growth is also indicated. For completeness, it must be added that the partial pressures of the elements *near the surface* during the growth will always be lower than what can be calculated based on the input flows and the total pressure, because of the depletion due to the incorporation in the crystal.

In the Ga-rich region, condensation of the liquid phase *can* actually occur, because the equilibrium partial pressures of arsenic and gallium species (blu lines) are below the input values, while in the As-rich region the arsenic species, especially As_4, are above the input values and condensation of a liquid phase or of solid arsenic cannot take place (although arsenic condensation would take place at lower temperatures, and can be expected on the cooler parts of the reactor chamber). Reagents depletion shifts the actual gas-phase pressures near the surface downwards, reducing to some extent the probability of condensation.

Another effect of the selection of the V/III ratio is a small variation in the crystal stoichiometry. In Fig. A2.11a, the GaAs solid phase is represented as a line, but it is actually a region of small width [218]: the stoichiometry corresponding to the left edge (sv3, sd1) is slightly more Ga-rich than the stoichiometry corresponding to the right edge (sv4, sd2). The ideal equilibrium values are not precisely known, but the difference from the 1:1 ratio is expected to be of the order of parts per million at temperatures far from the melting point, and to become larger, in the order of parts per thousand, as the temperature approaches the melting point (important for melt-growth but not a typical MOVPE condition). The excess of one element – arsenic in all practical growth conditions - is expected to correspond to the formation of vacancies (v_{Ga}) interstitials (As_i) and antisites (As_{Ga}) at the growth temperature and/or during the cooling, and possibly to the formation of arsenic precipitates.

According to the temperature-composition diagram of Fig. A2.11a, the solid stoichiometry at fixed T, P can have only two values, either on the solidus1 side or on the solidus2 side, depending on the total As fraction being either above or below 0.5; with growth conditions far from thermodynamic equilibrium as in MOVPE, the incorporation will be partially controlled by kinetic factors and the stoichiometry might vary with the V/III ratio at the surface in a continuous way.

A2.5.2 Composition control of multinary alloys

The presence of thermodynamic miscibility gaps in multinary alloys and their origin has been described in appendix 1. In this section, only the relation between the input MOVPE parameters and the resulting solid composition will be discussed, considering the cases of a ternary with two group III atoms and a ternary with two group V atoms.

For a generic ternary compound $A_xB_{1-x}C$, a distribution coefficient for the A species (Θ_A) is defined as the ratio of the fraction of A in the solid (x_S^A) and the input fraction of A in the vapor (x_v^A) phases:

$$\Theta_A = \frac{x_S^A}{x_v^A} = \frac{x_S^A}{p_A/(p_A + p_B)} \qquad \text{E A2.19}$$

where p_A, p_B are the input partial pressures of the precursors corresponding to atoms A and B (assuming that each precursor molecule contains only one A or B atom).

The distribution coefficient is experimentally a function of the growth conditions, especially the temperature and the input partial pressures. A simplified model to predict Θ_A based on a thermodynamic approach has been proposed, in particular by Stringfellow [11, 219] and by Seki and Koukitu [220]. The underlying idea is that – at least in the diffusion-limited growth regime[5] – chemical equilibrium is established between gaseous species and the solid crystal, and that the precursor molecules pyrolize completely, producing group III atoms, and that group V molecules are at their thermodynamic equilibrium concentrations (from AsH_3 mostly As_4 and from PH_3 mostly P_4). These hypothesis allow calculating the solid composition using a limited number of thermodynamic data, which are either experimentally accessible or can be calculated. The results are illustrated in Fig. A2.12, using the III-V-V ternaries AlAsP, InAsP, and the III-III-V ternary AlGaAs as examples.

---

[5] Actually, if an equilibrium model is valid for the diffusion-limited regime, it should be valid even for the desorption-limited regime, unless in the latter case enhanced parasitic pre-reactions become dominant.

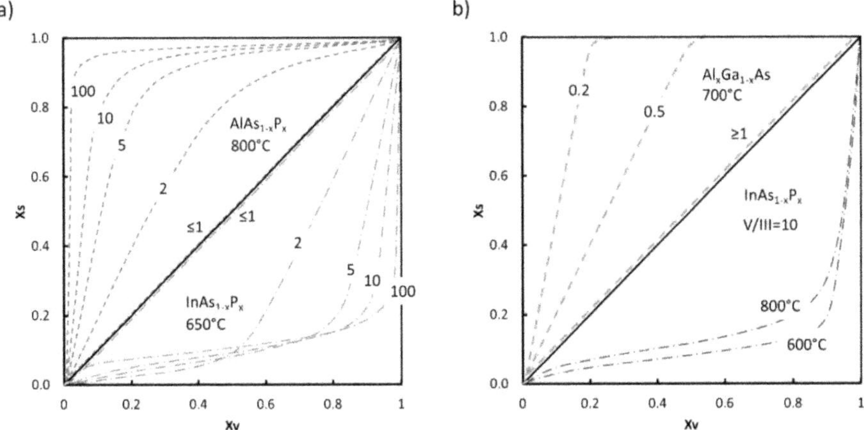

Figure A2.12 Calculated solid ternary composition vs gas-phase compositions of ternaries: **a)** AlAsP and InAsP at fixed T and different V/III ratios; **b)** AlGaAs at fixed T and different V/III ratios, InAsP at fixed V/III ratio and two different temperatures. The numbers indicate V/III ratios when not otherwise specified. Data from [220].

In the case of AlAsP (Fig. A2.12a) the relative amount of of As and P incorporated in the solid depends on the following parallel reactions:

$$Al(g) + \frac{1}{4}As_4(g) \rightleftarrows AlAs_{AlAsP}(s) \qquad \text{R A2.4a}$$

$$Al(g) + \frac{1}{4}P_4(g) \rightleftarrows AlP_{AlAsP}(s) \qquad \text{R A2.4b}$$

The corresponding equilibrium equations are:

$$K_1 = a_{AlAs}/p_{Al}p_{As_4}^{1/4} \qquad \text{E A2.20a}$$

$$K_2 = a_{AlP}/p_{Al}p_{P_4}^{1/4} \qquad \text{E A2.20b}$$

Where a_{AlAs} and a_{AlP} are the activities of the binary compounds *within* the ternary. These activities are calculated from the regular solution model, which takes approximately into account both the entropy and the enthalpy of mixing.

Observing the plot, it can be seen that, when the V/III ratio is >>1 the solid-phase composition is very different from the gas-phase composition; reducing the V/III ratio, the difference is reduced, and the distribution coefficient becomes $\Theta_P = \Theta_{As} = 1$ for V/III ≤1. This behavior is due to the fact that the equilibrium constants K_1 and K_2 are both very high and the equilibrium positions for reactions R A2.4 is strongly on the products side. The P/As ratio in the solid is given approximately by the activities ratio:

$$\frac{[AlP]}{[AlAs]} \approx \frac{a_{AlP}}{a_{AlAs}} = (K_2/K_1)\left(p_{P_4}/p_{As_4}\right)^{1/4} \qquad \text{E A2.20c}$$

When the gas-phase V/III ratio is >>1, the group III reacts completely, and P is preferentially incorporated because of a favorable enthalpy contribution (AlP having a much higher cohesive energy than AlAs) so that $K_2 > K_1$; reducing the V/III ratio to values near to 1, the gas-phase partial pressures become significantly reduced with respect to the input values, especially that of P_4, and this mitigates the thermodynamic advantage of phosphorus incorporation. At V/III ratio ≤1, group V reacts completely and the P/As ratio in the solid corresponds to the gas-phase input ratio.

The case of InAsP is explained similarly, but in this case, at high V/III ratio, As and not P is preferentially incorporated, going *against* the cohesive energies order InAs < InP: this time, it is the higher stability of the reactant P_4 over As_4 to prevail. From this, it can be already seen one critical point of the model: the selection of gaseous species that are supposed to be in equilibrium with the solid is fundamental in determining the overall result.

In Fig. A2.12b, the effect of a temperature variation on InAsP composition is shown: it can be seen that a higher T enhances the incorporation of phosphorus, consistently with the prevalence of reactant stability in determining the competition of the two parallel reactions.

In the case of the III-III-V ternary AlGaAs (Fig. A2.12b), at V/III ratio ≥1 (normal growth conditions) the distribution coefficients is fixed at $\Theta_{Al} = \Theta_{Ga} = 1$, while at V/III ratio <1 (a condition *not* used in practice) Al is expected to be preferentially incorporated because of the much higher cohesive energy of AlAs compared to GaAs, and $\Theta_{Al} > 1$.

In spite of its high level of idealization, the thermodynamic model captures the general trends, offering a simple rationalization of experimental results. Unfortunately, it does not always provide very accurate results.

Even assuming that the hypothesis of a chemical equilibrium near the surface is correct, the specific assumption that the species involved are the group III atoms and group V tetramers (or even with the possible inclusion of the dimers As_2 and P_2) is not consistent with the kinetic picture of the surface chemistry as described in section A2.4.2. In particular, the assumption that the group V species reach the equilibrium in the gas phase is definitely not correct in the case of PH_3 (unless very high temperatures are used) and this probably explains large errors ascertained in the evaluation of the As/P ratio in III-As-P compounds like GaAsP [221, 222].

A semi-empirical kinetic model for the determination of the dependence of the distribution coefficient from the input gas-phase composition is described in the following. It is assumed that the growth rate G_{ABC} of the ternary $A_xB_{1-x}C$ is composed by the sum of mutually independent binary growth rates G_{AC} and G_{BC}, each one of them proportional to the input partial pressures of the corresponding precursors: $G_{AC} = k_{AC}p_Ap_C$ and $G_{BC} = k_{BC}p_Bp_C$. The kinetic constants k_{AC}, k_{BC} are unknown functions of the growth conditions and depend on transport, gas-phase reactions and surface reactions, but are assumed independent from the input ratio of A and B species. Defining $L = k_{AC}/k_{BC}$, the fraction x_s^A of A atoms in the solid corresponding to the A input fraction x_v^A is given by:

$$x_s^A = \frac{G_{AC}}{G_{AC} + G_{BC}} = \frac{Lx_v^A}{Lx_v^A - x_v^A + 1} \qquad \text{E A2.21}$$

and the distribution coefficient for A becomes:

$$\Theta_A = \frac{L}{Lx_v^A - x_v^A + 1} \qquad \text{E A2.22}$$

The kinetic constants' ratio L is an adjustable parameter extracted from the fit of the experimental data.

In Fig. A2.13 are reported experimental values of solid vs input gas-phase compositions of $Al_xGa_{1-x}As$, $In_xGa_{1-x}As$, and $GaAs_xP_{1-x}$; all these ternaries have been grown with FBH G4 reactor on GaAs (100) substrates, from the precursors of Tab. 3.1 at P=50 mbar, using input V/III ratios >>1 (approximately constant for each material). It can be seen from these examples that the form of E A2.21 does reasonably interpolate the data; since the assumption of a constant L in the whole x_v range cannot be strictly valid, some deviations are to be expected. Note that AlGaAs has $\Theta_{Ga} < 1$, and InGaAs has $\Theta_{Ga} > 1$: this opposite behavior correlates with the cohesive energy trend of the binaries

(AlAs>GaAs>InAs)[6]. The lower indium distribution coefficient at 770°C with respect to 600°C can be explained by assuming that indium desorption rate increases with temperature more rapidly than gallium desorption rate, and the desorption-limited regime is reached earlier for indium incorporation. At very low temperature, approximately in the range 400-500°C, a completely different situation might be expected: the pyrolysis of TMIn should be still complete, but not the pyrolysis of TMGa, and the growth should enter the kinetic-limited regime for gallium incorporation but not for indium incorporation: consequently, a distribution coefficient $\Theta_{Ga} < 1$ is expected for InGaAs. This is in line with the lower than one values of Θ_{Ga} obtained at very low temperature (425-440°C) in Ref. [223].

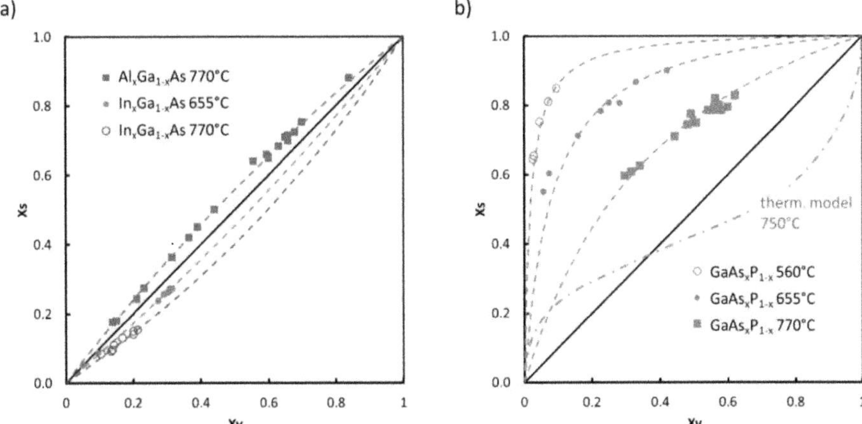

Figure A2.13 Experimental solid ternary compositions vs gas-phase compositions: **a)** AlGaAs and InGaP; **b)** GaAsP. The dashed lines are interpolations based on Eq. E A2.21, the dashed-dotted line is the prediction of the thermodynamic model of Ref. [220].

From Fig. A2.13b, it can be seen that the distribution coefficient for As in GaAsP is always >1 in the tested temperature range, and decreases with increasing temperature: the effect is most probably mainly related to the relative group V precursors stability, i.e. to the relative extent of AsH_3 and PH_3 pyrolysis. The dash-dotted line shows the prediction of the thermodynamic model of Seki and Koukitu at 750°C in similar conditions (V/III ratio, partial pressures and total pressure); the discrepancy with the experiment is in this case quite evident.

[6] As a cautionary remark, it must be said that small deviation of x_s/x_v from 1 can even be caused by a systematic error in the evaluation of the input flows: for example, the bowing of AlGaAs can be reproduced assuming a 28% overestimation of TMAl flow.

A2.5.3 Dopant and impurities incorporation

Intrinsic carbon, hydrogen and oxygen

Carbon and hydrogen are normally contained in the precursors' molecules and are always incorporated to some extent in the grown layers.

Carbon – which is tetravalent, is amphoteric in III-V compounds, and can be incorporated either in the cation or in the anion sublattice; in GaAs is incorporated preferentially in the arsenic sublattice, where it behaves as a p-dopant, and a possible kinetic mechanism involving intermediate carbene species has been briefly mentioned in section A2.4.2. Experimentally, the highest carbon concentrations are found in materials containing aluminum, as AlGaAs, where again it behaves as p-dopant as in GaAs. The same happens for InGaAs, but in InP carbon behaves prevalently as a n-dopant. Independently of the possible kinetic incorporation mechanisms, these results can be qualitatively rationalized considering the relative electronegativity and the relative strength of the bonds between carbon and the III-V constituent elements. To provide a consistent set of values, in Table A2.1 the average dissociation enthalpies of C-X bonds in $X(CH_3)_3$ molecules are reported; although these values cannot correspond exactly to those in the III-V lattice (for example because of the effect of the strain induced by the substitutional dopant), the *trends* should be roughly similar. The electronegativity differences are also reported in the same table.

Carbon has a comparatively high electronegativity, and – to a first approximation - it can be expected to be preferentially incorporated in the anion sublattice; this general forecast needs to be refined considering the relative bond strengths: it can be seen that in Al-containing compounds, the energetically more favorable situation is the formation of Al-C bonds, and consequently the incorporation in the group V sublattice. In GaAs and InGaAs the strongest bond is Ga-C, but in InP it is by far the P-C bond: the latter circumstance justifies the incorporation in the group III sublattice and the resulting n-doping. In compounds with more than one group III element, C can be expected to bond to them preferably in the order Al>Ga>In.

As already mentioned, intrinsic carbon doping can be controlled varying the group V partial pressure, probably because it competes for the same sites and because it reacts with hydrogen, which is released from the hydrides, forming methane and other volatile CH_x molecules; when p-doping is unwanted, it can be minimized using a high V/III ratio, but in compounds with high Al content it cannot be completely avoided. It is even very sensitive to the growth temperature: in GaAs the highest doping concentration is obtained at low T, probably indicating that carbon species desorption limits the incorporation. The lowest concentration has *not* been obtained at very high T but rather at intermediate values; the reason for its increased incorporation at high T is not well understood [224, 225].

element X	Al	Ga	In	P	As
X-C bond enthalpy (kJ/mol)	283	251	157	273	215
$\chi_C - \chi_X$ (Pauling)	0.94	0.74	0.77	0.36	0.37

Table A2.1 Carbon-X bond dissociation enthalpies in molecules with formula $X(CH_3)_3$, and electronegativity relative to the X elements; the uncertainty on enthalpy values is about 10%. Data from [196, 226, 227].

Hydrogen can form comparatively strong bonds with all the constituent elements of III-V compounds, especially with the group V, but at the same time, since it is monovalent, its substitutional incorporation would leave three dangling bonds in the lattice, an energetically highly unfavorable condition. Being very small, it can easily occupy interstitial positions and diffuse rapidly in the crystal; it is known to passivate the dopants, forming with them complexes from interstitial positions [228], but the electrical activity can be restored by a thermal annealing at temperatures around 500°C, in absence of sources of *atomic* hydrogen (for example an annealing under H_2 but without AsH_3 or PH_3).

Intentional oxygen incorporation in MOVPE is very rare, although possible. Oxygen is present in small amounts as contaminant in the precursors and on the wafer surface, and can enter the reactor when the chamber lid is open – since the nitrogen atmosphere in the glove-box might still contain traces of O_2 and H_2O. Moreover, it can be absorbed in periodically-changed components as for example the chamber ceiling. Similarly to the case of carbon, its incorporation is difficult to avoid in aluminum-rich materials. Opposite to other group VI elements as S, Se and Te, it never behaves as an n-dopant, and rather introduces deep levels in the bandgap. As a guideline to predict in which sublattice it is most probably incorporated, the electronegativity values and the relative strength of the bonds between oxygen and the other elements can be examined. The electronegativity differences between O and the element X, and the average dissociation enthalpies of O-X bonds in X_2O_3 oxides are listed Tab. A2.2.

element X	Al	Ga	In	P	As
X-O bond enthalpy (kJ/mol)	514	397	360	367	345
$\chi_O - \chi_X$ (Pauling)	1.83	1.63	1.66	1.25	1.26

Table A2.2 Oxygen-X bond dissociation enthalpies in compounds with formula X_2O_3 (from enthalpies of formation, assuming 6 bonds/unit), and electronegativity relative to the X elements; the uncertainty on enthalpy values is about 10%. Data from [229-231].

With respect to the constituent elements, oxygen is strongly electronegative, and it might be expected to reside on the anion sublattice; moreover, the bond energies with group III atoms are stronger or at least not less strong than those with the group V, reinforcing the conclusion. In compounds with more than one group III element, O can be expected to bond to them preferably in the order Al>Ga>In.

Oxygen has six valence electrons and no empty atomic orbitals easily accessible: consequently, rather than forming four bonds with its neighbors – which would imply a net transfer of electron density to them, it can be expected to form only two - highly polar - bonds, preferably with the more electropositive element (group III), keeping four valence electrons in localized non-bonding orbitals; this explains why, oxygen is not a donor. The heavier elements of group VI (S, Se, Te) behave as donors because they are less electronegative and because they can access low-energy empty atomic orbitals which can be employed to form the regular tetrahedral 4-bonds structure, while – assuming that they reside in the group V sublattice – the sixth electron can be allocated in a localized, non-bonding but comparatively shallow level.

As in the case of intrinsic carbon, oxygen incorporation can be minimized using high V/III ratios, probably because of the same reasons. In AlGaAs, the oxygen content is higher at low growth temperature, but it does not decrease monotonically with increasing T – again similarly to the case of carbon.

Doping from precursors; CBr_4, DMZn and Si_2H_6

As mentioned in chapter 3, the achievable p-doping in GaAs using CBr_4 can exceed 10^{20} cm^{-3} [13]. Bromine is not appreciably incorporated, but has rather an etching effect.

Zn behaves as a p dopant in all III-V Zincblende semiconductors. Based on its low electronegativity – similar or slightly smaller than that of group III elements - and the fact that it has only two valence electrons, Zn can actually be expected to reside in the group III sublattice; the easily accessible 4p atomic orbitals allow for the formation of the tetrahedral 4-bonds structure, and the electron missing to fill the bonds can be captured from the valence band, generating a hole. A consistent set of Zn-X bond energy data similar to those shown for carbon and oxygen is unfortunately not available. The achievable doping using DMZn is generally high, $\approx 10^{20}$ cm^{-3} in GaAs; it is lower in phosphide compounds, for example only $\approx 4\times10^{18}$ cm^{-3} in InP vs $\approx 4\times10^{19}$ cm^{-3} in InGaAs lattice-matched to InP [232-234], probably in relation to the small size of Zn in oxidation state II, which – within the crystal lattice - should favor the bonding with the larger As atoms over that with P atoms. A drawback of Zn is that at high temperature it can significantly diffuse in the solid material, making the doping profile more difficult to control.

Silicon – like carbon – is amphoteric in III-V compounds, but is preferentially incorporated in the cation sublattice where it behaves as n-dopant; this is compatible with an electronegativity intermediate between those of group III and group V elements, and it might be fully explained by Si-V bonds significantly stronger than Si-III bonds. Unfortunately, similar to Zn case, a consistent set of Si-X bond energies is not available. The achievable doping from disilane is comparable for arsenides and phosphides, the highest reported values being $\approx 10^{19}$ cm^{-3} in GaAs and in InGaP lattice-matched with GaAs [235, 236].

The incorporation behavior with respect to temperature and partial pressures is different for the three precursors (CBr_4, DMZn and Si_2H_6) and is summarized in Tab. A2.3 for the case of GaAs doping.

dopant	effect of dopant partial pressures	effect of TMGa partial pressures	effect of AsH_3 partial pressure	effect of increasing temperature
C from CBr_4	$\propto p_{CBr4}$	≈ independent	$\propto 1/p_{AsH3}$	decreases strongly
Zn from DMZn	$\propto p_{CDMZn}$	≈ independent	$\propto p_{AsH3}$	decreases strongly
Si from Si_2H_6	$\propto p_{Si2H6}$	$\propto 1/p_{TMGa}$	≈ independent	increases weakly

Table A2.3 Effects of precursors partial pressures and of temperature on doping concentrations in GaAs. The indicated trends are referred to doping levels below saturation, the growth temperature approximately in the range 550°C-750°C and V/III ratio >>1.

In Ref. [11] Stringfellow suggests a general interpretation for the trends of dopant incorporation based on several approximations including: boundary-layer approximation, diffusion-controlled growth rate regime and thermodynamic equilibrium between the solid and the gas-phase near the surface. In the following, a *similar* approach is outlined. The gas-phase molecules involved in the equilibrium are supposed to contain only one dopant or one constituent atom each, but for the rest their precise identity is left unspecified.

Considering either the case of a dopant *d* residing on group III sublattice (Zn, Si), or on the group V sublattice (C), the fraction of dopant atoms in the solid phase corresponds to the ratio of *d* and III (or V) species net fluxes to the surface. These fluxes are proportional to the concentration gradients via the corresponding diffusion coefficients; the concentration gradients are proportional to the difference between the "free-stream" or input partial pressures and the partial pressures near the surface, and the diffusion coefficients are approximately equal for all the species. In the case of group III, the partial pressure near the surface is much smaller than the input value, because the incorporation reaction goes almost to completeness, while in the case of group V the partial pressure near the surface is almost equal to the input value, because it is assumed a large excess of input group V. It is even assumed that - as is always the case - the molar fraction of the dopant in the solid is much smaller than 1. This implies that the net fluxes of group III and V to the wafer surface must be almost identical, because - neglecting the small contribution of the dopant - the two atoms are incorporated at the same rate to maintain the stoichiometry. Considering the solid phase as a solution of binary compounds III-V and d-V or III-d, the above hypothesis lead to the following formulae:

$$x_s^{dV} = \frac{J_d}{J_{III}} = \frac{D_d(p_d - p_d^{sf})}{D_{III}(p_{III} - p_{III}^{sf})} \approx \frac{p_d - p_d^{sf}}{p_{III}}$$ *d* on III sublattice E A2.23a

$$x_s^{IIId} = \frac{J_d}{J_V} = \frac{D_d(p_d - p_d^{sf})}{D_V(p_V - p_V^{sf})} \approx \frac{D_d(p_d - p_d^{sf})}{D_{III}(p_{III} - p_{III}^{sf})} \approx \frac{p_d - p_d^{sf}}{p_{III}}$$ *d* on V sublattice E A2.23b

The near-surface equilibrium conditions can be expressed as:

$$K_{dV} = \frac{a_s^{dV}}{p_d^{sf} \cdot p_V^{sf}} = \frac{x_s^{dV} \cdot \gamma_{dV}}{p_d^{sf} \cdot p_V^{sf}} \approx \frac{x_s^{dV} \cdot \gamma_{dV}}{p_d^{sf} \cdot p_V}$$ *d* on III sublattice E A2.24a

$$K_{IIId} = \frac{a_s^{IIId}}{p_{III}^{sf} \cdot p_d^{sf}} = \frac{x_s^{IIId} \cdot \gamma_{IIId}}{p_{III}^{sf} \cdot p_d^{sf}}$$ *d* on V sublattice E A2.24b

$$K_{IIIV} = \frac{a_s^{IIIV}}{p_{III}^{sf} \cdot p_V^{sf}} = \frac{x_s^{IIIV} \cdot \gamma_{IIIV}}{p_{III}^{sf} \cdot p_V^{sf}} \approx \frac{x_s^{IIIV} \cdot \gamma_{IIIV}}{p_{III}^{sf} \cdot p_V}$$ for the constituent elements E A2.25

where the activity coefficients of the III-V, *d*-V and III-*d* binaries *within* the ternary III-*d*-V compound have been expressed in terms of molar fractions using the activity coefficients γ.

Substituting p_d^{sf} obtained from E A.24a in E A2.23a and extracting x_s^{dV}:

$$x_s^{dV} \approx \frac{K_{dV} \cdot p_d \cdot p_V}{K_{dV} \cdot p_{III} \cdot p_V + \gamma_{dV}}$$ *d* on III sublattice E A2.26

Substituting first p_{III}^{sf} obtained from E A2.25 in E A2.24b and with the approximation $x_s^{IIIV} \approx 1$, p_d^{sf} can be expressed as:

$$p_d^{sf} \approx \frac{K_{IIIV} \cdot x_s^{IIId} \cdot \gamma_{IIId} \cdot p_V}{K_{IIId} \cdot \gamma_{IIIV}}$$ *d* on V sublattice E A2.27

Substituting p_d^{sf} obtained from E3.27 in E3.23b and extracting x_s^{IIId}:

$$x_s^{IIId} \approx \frac{K_{IIId} \cdot \gamma_{IIIV} \cdot p_d}{K_{IIId} \cdot \gamma_{IIIV} \cdot p_{III} + K_{IIIV} \cdot \gamma_{IIId} \cdot p_V}$$ *d* on V sublattice E A2.28

In the case of a dopant on III sublattice, two extreme cases are possible:

when $K_{dV} \cdot p_{III} \cdot p_V \gg \gamma_{dV}$

then $x_s^{dV} \approx p_d/p_{III}$ →type 1

when $K_{dV} \cdot p_{III} \cdot p_V \ll \gamma_{dV}$

then $x_s^{dV} \approx (K_{dV}/\gamma_{dV}) \cdot p_d \cdot p_V$ →type 2

Type 1 corresponds to a dopant that behaves like a group III element, reacting almost to completion once it reaches the surface; this situation occurs at high values of K_{dV} and is promoted by high group V and III partial pressures. The incorporation is temperature-independent and equal to the d/III input ratio. Experimentally, the behavior of silicon originating from Si_2H_6 approximates the type 1 in GaAs, although x_s^{dV} is always smaller than p_d/p_{III}, the highest value obtained on G3 reactor being for example $0.3 \cdot p_d/p_{III}$ at 760°C. The values of x_s^{dV} become slightly smaller at low temperature, possibly because of incomplete disilane pyrolysis.

Type 2 corresponds to a dopant whose incorporation is minimal, so that its concentration in the gas-phase near the surface is similar to the input value; this situation occurs at low values of K_{dV} and is promoted by low group V and III partial pressures. The incorporation is proportional to the product of the input partial pressures of dopant and group V. The temperature dependence is mostly contained in the equilibrium constant $K_{dV} = exp(-\Delta_r G/RT)$: when $\Delta_r G > 0$ (low d-V binary cohesion energy and/or highly volatile gas-phase dopant species) the incorporation decreases exponentially with increasing T. The behavior of zinc originated from DMZn approximates type 2 in GaAs.

Even in the case of a dopant on V sublattice two extreme cases are possible:

when $K_{IIId} \cdot \gamma_{IIIV} \cdot p_{III} \gg K_{IIIV} \cdot \gamma_{IIId} \cdot p_V$

then $x_s^{IIId} \approx p_d/p_{III}$ →type 3

when $K_{IIId} \cdot \gamma_{IIIV} \cdot p_{III} \ll K_{IIIV} \cdot \gamma_{IIId} \cdot p_V$

then $x_s^{IIId} \approx [K_{IIId} \cdot \gamma_{IIIV}/(K_{IIIV} \cdot \gamma_{IIId})] \cdot p_d/p_V$ →type 4

Type 3 corresponds to the case of a dopant with $\Delta_r G$ much more favorable (more negative) than the constituent group V element (because per hypothesis $p_{III} \ll p_V$); this is not a situation typical for a dopant, but it might be represented for example by oxygen in AlAs or AlGaAs: in this case, increasing the temperature or the growth rate ($> p_{III}$) should reduce oxygen incorporation (the latter *only* if O is originated by an independent source and is not contained as an impurity in the precursors).

Type 4 corresponds to a dopant with $\Delta_r G$ much less favorable than the constituent group V element, or even similar/moderately more negative, provided that the V/III ratio is sufficiently high. The dopant incorporation is proportional to the p_d/p_V ratio and decreases or increases exponentially with the temperature, depending on the difference $\Delta_r G(IIId) - \Delta_r G(IIIV)$ being negative or positive respectively. The behavior of carbon originated from CBr_4 approximates type 4 in GaAs, with incorporation decreasing as the temperature is raised, which (according to the model) implies $\Delta_r G(GaC) < \Delta_r G(GaAs)$. Lowering the V/III ratio and the temperature might shift the behavior towards type 3, but at very low V/III ratio carbon originating from the methyl groups of TMGa is simultaneously incorporated, an effect not included in the above equations. Another important point is that the hydrogen atoms generated by AsH_3 can react with CBr_4 and with C atoms adsorbed on the surface, forming CBr_xH_y and CH_x species; this effect – rather than the competition for the same sites - is often indicated as the main cause of the reduction of carbon doping with increasing arsine partial pressure [237].

As for the case of the equilibrium model used in section A2.5.2 in relation to the composition of ternary alloys, the kinetic stability of the gas-phase species involved in the assumed near-surface equilibrium is critical: if a dopant precursor is kinetically stable on the scale of the residence-time, an equilibrium-based model will unavoidably lead to partially wrong predictions. For example, this kind of model works better for disilane than for the (more stable) silane [11]. Moreover, a realistic description of near-surface processes should involve all the possible chemical reaction specific of a certain combination of reactants, as exemplified by the case of the hydrogenation of CBr_4 from AsH_3 species.

To summarize, the model offers a comparatively simple rationalization of the observed experimental trends, and can be seen as a first guide in guessing, for each dopant precursor, how the doping *might* reasonably correlate to the growth parameters, but it cannot be used as a precise predictive tool.

A3 Justification of the equations used in modeling the CBr_4+TMAl etch

Justification of E4.5

Given an arbitrary time dt, in absence of CBr_4, the thickness of the grown AlAs layer would be:

$$dx = G^{AlAs}(p_{Al}) \cdot dt \quad \text{E A3.1}$$

This thickness could be etched back by CBr_4 in the time:

$$dt' = dx/E^{AlAs}(0) \quad \text{E A3.2}$$

Combining the last two equations:

$$dt' = dt \cdot G^{AlAs}(p_{Al})/E^{AlAs}(0) \quad \text{E A3.3}$$

Growth and etch are assumed to be non-interacting processes. For low values of p_{TMAl}, $dt' < dt$ is fulfilled and under these conditions $dt - dt'$ is the part of the time dt during which GaAs is actually etched, at the rate $E^{GaAs}(0)$.

The etched GaAs thickness will then be:

$$dx' = E^{GaAs}(0) \cdot (dt - dt') \quad \text{E A3.4}$$

this, together with the E A3.3 leads to:

$$dx'/dt = E^{GaAs}(0) \cdot \left(1 - G^{AlAs}(p_{Al})/E^{AlAs}(0)\right) \quad \text{=E4.5}$$

Justification of E4.7 and E4.8

It is *assumed* that during steady-state etching of GaAs in presence of TMAl, a fraction of the surface is covered with a partially-complete monolayer of AlAs, and that growth and etching of AlAs are non-interacting processes. In terms of the fraction θ of group III sites on the surface that are occupied by Al, the chemisorption rate of Al is *assumed* to be proportional to TMAl partial pressure times the fraction of free sites, and the desorption rate (more properly: the etching rate at fixed p_{CBr4}) is *assumed* to be proportional to the fraction of occupied sites, resulting in the differential equation:

$$d\theta/dt = k_G \cdot p_{TMAl} \cdot (1 - \theta) - k_E \cdot \theta \quad \text{E A3.5}$$

Under stationary conditions $d\theta/dt = 0$, and θ is given by:

$$\theta = k_G \cdot p_{TMAl}/(k_E + k_G \cdot p_{TMAl}) \quad \text{E A3.6}$$

defining $K \equiv k_G/k_E$ E A3.6 can be rewritten:

$$\theta = K \cdot p_{TMAl}/(1 + K \cdot p_{TMAl}) \quad \text{=E4.7}$$

Note that here K is the ratio of two kinetic constants relative to *different* reactions under steady-state conditions, not to the *same* reaction proceeding in opposite directions: the chemisorption of Al is due to TMAl decomposition, the desorption to a reaction with CBr_4. Consequently K is *not* an equilibrium constant.

K must now be related to the growth rate and etch rate of AlAs: $G^{AlAs}(p_{Al})$, and $E^{AlAs}(0)$.

The right side of E A3.5 contains a positive term associated to chemisorption and a negative term associated to desorption; when $\theta = 0$ we are left only with the first term, which can be related to the growth rate through a proportionality constant d, which is the AlAs monolayer thickness:

$$G^{AlAs}(p_{Al}) = d \cdot d\theta/dt = k_G \cdot p_{TMAl} \quad \text{E A3.7}$$

When $\theta = 1$ the right side of E A3.5 is reduced to the negative desorption term, that can be related to the etch rate using again d:

$$E^{AlAs} = -d \cdot d\theta/dt = k_E \quad \text{E A3.8}$$

From the last two equations it follows that:

$$K \equiv k_G/k_E = G^{AlAs}(p_{Al})/\left(p_{TMAl} \cdot E^{AlAs}(0)\right) \quad \text{=E4.8}$$

Justification of E4.10

Equation E4.6 assumes that the surface is entirely covered by at least one complete monolayer of AlAs, which means that $\theta = 1$ and $E^{Ga(Al)As}(p_{Al}) \leq 0$ (growth prevails).

To extend its validity to the range of p_{TMAl} where $E^{Ga(Al)As}(p_{Al}) > 0$ we can write:

$$E^{Ga(Al)As}(p_{Al}) \approx E^{GaAs}(p_{Al}) + E^{AlAs}(0) \cdot \theta - G^{AlAs}(p_{Al}) \quad \text{E A3.9}$$

It has been assumed that AlAs grows with the same rate on AlAs as on GaAs, since in the considered temperature range there is no desorption of Al and there is also no indication in literature that the decomposition of the Al precursor is significantly affected by the slightly different bonding energies to these two surfaces.

Introducing E4.9 into E A3.9 gives E4.10.

Note that the fraction of the surface θ covered by AlAs is estimated with a Langmuir equation through E4.7 and E4.8, but this estimate cannot be correct when the (net) etch rate $E^{Ga(Al)As}(p_{Al})$ approaches 0 or becomes negative (growth regime): in the Langmuir equation the condition θ=1 is fulfilled only for $p_{TMAl} \to \infty$, therefore E4.9 predicts that the etching of GaAs will go to 0 only for $p_{TMAl} \to \infty$. The result is that there is a small spurious GaAs etching term that persists even when the GaAs is completely buried by the growing AlAs. In the equation E4.10 there is a second term that depends on θ and that represents the etching of AlAs: this term is slightly underestimated when $E^{Ga(Al)As}(p_{Al})$ is near 0 or becomes negative, for the same reason. This introduces two errors of opposite sign, which almost eliminate each other.

A4 Model for the calculation of α_{ip} in the implanted sections

For the derivation of an explicit expression for α_{ip} the same approach described in the appendix 5 of [58] is followed; while there it is derived for the case of a laser with a single passive section at one end, here it is extended to the case of two identical passive sections of length L_p, each one at one end of the laser cavity.

The differential quantum efficiency η_d for a laser 1 without passive sections can be expressed as:

$$\eta_{d1} = \frac{\eta_i \alpha_m}{\Gamma_z \langle g \rangle_{xy}} \qquad \text{E A4.1}$$

where α_m are the mirror losses, Γ_z is the confinement factor along the longitudinal direction and $\langle g \rangle_{xy}$ is the transverse modal gain. E4.1 can be reformulated in terms of absorption, reflection and geometric parameters using the relation $\alpha_m = L^{-1} ln(\mathcal{R}^{-1})$, where $\mathcal{R}$ is the facet power reflectivity, and the threshold condition for steady-state gain and loss: $\Gamma_z \langle g \rangle_{xy} = \alpha_i + \alpha_m$. Substituting in E4.1:

$$\eta_{d1} = \frac{\eta_i \alpha_m}{\alpha_i + \alpha_m} = \frac{\eta_i L^{-1} ln(R^{-1})}{\alpha_i + L^{-1} ln(R^{-1})} = \eta_i F_1 \qquad \text{E A4.2}$$

For a laser 2 identical to the first one in all aspects except for the presence of passive sections at the facets, the differential quantum efficiency can be similarly written:

$$\eta_{d2} = \frac{\eta_i \alpha_{m1}}{\Gamma_{z1} \langle g \rangle_{xy1}} = \eta_i F_2 \qquad \text{E A4.3}$$

It has been assumed here that η_i is the same for both lasers: this assumption implies that there is no significant current leakage in the passive sections and that the length of the passive section L_p is small with respect to the active length L_a, so that the carrier density at threshold does not change much from the case where L_p=0.

To express F_2 in terms of absorption, reflection and geometric parameters, the simple approach used in the case of F_1 cannot be used, and the more sophisticated approach of [58] has been used, giving:

$$F_2 = \frac{2\alpha_{ip} L_p - ln(R)}{\alpha_{ia} L_a + 2\alpha_{ip} L_p - ln(R)} \cdot \frac{(1-R)/\sqrt{R}}{(1 - Re^{-2\alpha_{ip} L_p})/(\sqrt{R}e^{-\alpha_{ip} L_p})} \qquad \text{E A4.4}$$

α_i in E4.2 is the same as α_{ia} in E4.4, and can be experimentally evaluated in a standard way using lasers of different lengths without passive sections. In the ratio of the slope efficiencies of the two devices, η_i cancels out, and we remain with a function F_2/F_1 of α_{ip} and known parameters:

$$\eta_2/\eta_1 = \eta_{d2}/\eta_{d1} = F_2/F_1 \qquad \text{E A4.5}$$

α_{ip} can be used as a fitting parameter to interpolate the experimental values of η through Eq. E4.5. An example of the fitting is given in Fig. A.1.

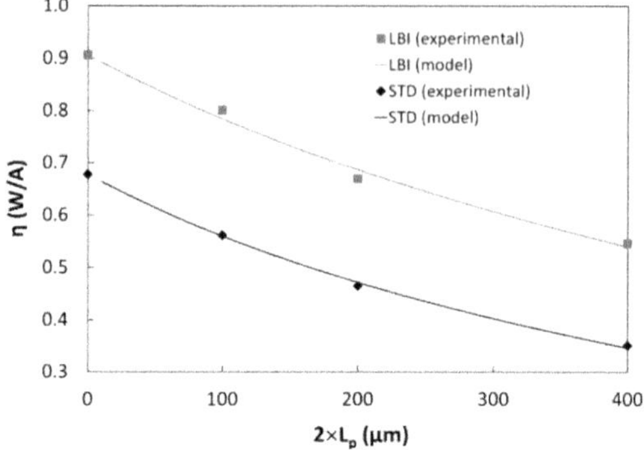

Figure A.1 Interpolation of experimental values of η for facet-implanted LBI and STD devices from sample 11, L=4 mm W=100 μm.

Glossary

Symbol or abbreviation	most common uses, meaning
$\nabla_{\boldsymbol{r}}$, $\nabla_{\boldsymbol{k}}$	Nabla operator (grad) on real space (**r**) or reciprocal space (**k**) coordinates
$\nabla_{\boldsymbol{r}}^2$	Laplacian operator (div·grad) in real space
α	- optical absorption coefficient; for lasers: α_i = internal losses; α_m = mirror losses; α_p = passive section losses - α_T = thermal diffusivity
β	thermal expansion coefficient
Γ	- confinement coefficient (of optical modes in the active zone of a laser) - origin of the reciprocal lattice
γ	- ground-state degeneracy - activity coefficient - Gibbs excess surface energy
δ	boundary layer thickness
ε	energy (electron eigenvalue energies – band-edge energies)
ε_0	vacuum permittivity
ε_F	Fermi level
$\tilde{\varepsilon}$	complex dielectric permittivity ($\tilde{\varepsilon} = \varepsilon_1 + i\varepsilon_2$)
η	- specific growth or etch rate (as defined in chapter 4) - for lasers: η = slope efficiency; η_d = differential quantum eff.; η_i = internal diff. quantum eff.; η_E = energy conversion eff.
θ	fraction of occupied surface sites
Θ	- distribution coefficient - for lasers: far-field angle
κ	extinction coefficient
κ_g	grating coupling coefficient
κ_T	thermal conductivity
λ	wavelength
Λ_B	Bragg grating period
μ	- mobility - gas viscosity
ν	- frequency - kinematic viscosity
ρ	density
σ	carrier capture cross-section

τ	relaxation time (intraband), lifetime (interband)
φ	single electron wavefunction
χ	electronegativity
ψ	multi-electron wavefunction
ω	angular frequency $\omega = 2\pi\nu$
a	lattice parameter (also the letters *b, c, u* are used)
A	- Shockley-Read-Hall recombination coefficient (high excitation) - pre-exponential factor in the Arrhenius equation
AFM	atomic force microscopy
AR	anti-reflection (optical coating)
ASE	amplified spontaneous emission
B	radiative spontaneous recombination coefficient (high excitation)
BAL	broad-area laser
BM	buried-mesa (laser, chapter 6)
BPP	beam parameter product
BZ	Brillouin Zone
C	Auger recombination coefficient (high excitation)
CB	conduction band
ccp	cubic close-packed (lattice)
CL	cathodoluminescence
COD	catastrophic optical damage
COMD	catastrophic optical mirror damage
CW	continuous-wave current (electrical current with constant value)
d	thickness
D	diffusion coefficient
DBR	distributed Bragg reflector (laser)
dc	duty cycle (of a Bragg grating, chapter 5)
DFB	distributed feedback (laser)
DFT	density-functional theory (quantum mechanics)
DIC	differential interference contrast
DMZn	dimethylzinc, C_2H_6Zn
DOP	degree of polarization (of an optical beam)
e	- elementary charge (electron in reaction equations) - Euler's number
E	etch rate
E_a , E_p	activation energy (chemical kinetics)
EDX	energy-dispersive X-ray analysis

EPD	etch pit density
f	- distribution function - gas flow
FBH	Ferdinand-Braun-Institut
fcc	face-centered cubic (Bravais lattice)
FF	far-field (of an optical beam)
FWHM	full width at half maximum
g	gravitational acceleration
g_{th}	gain per unit length; also g_0 = gain parameter
$g_n(\varepsilon)$	density of states in the band of index n
G	- Gibbs energy (thermodynamics) - growth rate (kinetics)
G3, G4	MOVPE reactors AIX2400G3 and AIX2800G4 used in this work
GRIN	graded-index (epitaxial layer)
h	- Planck constant; also: $\hbar = h/2\pi$ - reactor chamber's height
H	enthalpy
$\hat{H}$	Hamiltonian operator
HR	- high-reflection (optical coating) - high resolution in HR-XRD
hcp	hexagonal close-packed (lattice)
i	- current per unit cavity length - imaginary unit
I	current
j	current density
J_A	gas-phase flux of the species A
k	kinetic constant
K	equilibrium constant
k_B	Boltzmann constant
$\mathbf{k}$	reciprocal-space vector, wavevector
$\mathbf{K}$	reciprocal-lattice vector
LBI	lateral buried implantation (laser, chapter 7)
LIV	light-current-voltage (plot) same as PIV
LPE	liquid-phase epitaxy
m_0	electron rest mass
m^*	effective mass (of electrons or holes)
M^2	beam quality parameter
MBE	molecular-beam epitaxy

MFC	mass-flow controller
MO	metalorganic
MOCVD	metalorganic chemical vapor deposition
MOPA	master-oscillator power-amplifier
MOVPE	metalorganic vapor-phase epitaxy
MTTF	mean time to failure
L	laser cavity length
n	- refractive index; n_e = effective index, n_g = group effective index - electron density in the CB; n_i = intrinsic carrier density
$\tilde{n}$	complex refractive index ($\tilde{n} = n + i\kappa$)
n-	negative (doping, carrier type)
NAM	non-absorbing mirror
N	densities in real space:
	N(x) = ion implantation density vertical profile
	N_C (N_V) = effective density of states in the conduction (valence) band
	N_A, N_D, N_T = acceptor, donor, trap densities
NF	near-field (of an optical beam)
NIM	not-injected mirror
OM	optical microscopy
p	- partial pressure - hole density in the VB
P	- pressure - for lasers: optical power
p-	positive (doping, carrier type)
PC	pressure controller
pin, pn, pnp	positive-intrinsic-negative (junction) etc.
PIV	power-current-voltage (plot) same as LIV
PL	photoluminescence
Q	ion dose
QW	quantum well (SQW, DQW, MQW = single, double, multi QW)
QWI	quantum well intermixing
$\mathcal{R}$	power reflection coefficient (reflectance)
R	gas constant
	rate (e.g. R_{sp} = spontaneous emission rate)
R_p	projected range of ion implantation; also ΔR_p = straggle
r	real-space vector
R	direct-lattice vector

RBS	Rutherford backscattering (analysis)
RIE	reactive-ion etching
$r(I_{th})$, $r(I_L)$, $r(\eta)$	ratios of: threshold currents, leakage currents and slope efficiencies (LBI/STD lasers, chapter 7)
RTA	rapid thermal annealing
S	- entropy - surface recombination rate
SEM	scanning electron microscopy
SG-DBR	sampled-grating distributed Bragg reflector
SIMS	secondary-ion mass spectrometry
SMSR	side mode suppression ration
SRH	Shockley-Read-Hall (non-radiative recombination mechanism)
STD	standard (laser)
STM	scanning tunneling microscopy
TE	transverse-electric (electromagnetic field polarization)
TEM	transmission electron microscopy
TM	transverse-magnetic (electromagnetic field polarization)
TMAl	trimethylaluminum, C_3H_9Al
TMGa	trimethylgallium, C_3H_9Ga
TMIn	trimethylindium, C_3H_9In
T_{sp}	set-point temperature
T_w	wafer-temperature
u	flow velocity (in appendix 2)
U	- internal energy (thermodynamics) - potential energy for the electrons (quantum mechanics) - surface recombination rate
$u_{n,\boldsymbol{k}}(\boldsymbol{r})$	Bloch function
UHV	ultra-high vacuum
VB	valence band
V1, V2	variants of the lasers with buried implantation (chapter 7)
VS0, VS2	laser vertical structures with 0 and with 2 etch-stop layers (chapter 6)
W, w	width
XRD	X-ray diffraction
Z_0, Z_1	grating burst and sampling period in SG-DBR mirrors
z_R	Rayleigh distance (in a Gaussian beam)

References

[1] Sadao Adachi 2006 *Properties of Group-IV, III-V and II-VI Semiconductors* Wiley ISBN 0-470-09032-4

[2] Peter Y. Yu, Manuel Cardona 2010 *Fundamentals of Semiconductors, 4th Ed.* Springer ISBN 978-3-642-00709-5

[3] J. R. Chelikowsky and M. L. Cohen 1979 *Electronic states of the relaxed (110) surface of GaAs* Solid State Commun. **29** 267-271

[4] A. Alkauskas et al. 2016 *Tutorial: Defects in semiconductors – Combining experiment and theory* J. Appl. Phys. **119** 181101

[5] E. F. Schubert 1993 *Doping in III-V Semiconductors* Cambridge University Press ISBN 0-521-41919-0

[6] P. Capper and S. Irvine (Eds.), *Metalorganic Vapor Phase Epitaxy (MOVPE) Growth, Materials, Properties, and Applications* Wiley 2020 ISBN 978-1-119-31301-4

[7] P. M. Frijlink 1988 *A new versatile, large size MOVPE reactor* J. Cryst. Growth **93** 207-215

[8] M. Dauelsberg et al. 2000 *Modeling and experimental verification of transport and deposition behavior during MOVPE of $Ga_{1-x}In_xP$ in the Planetary Reactor* J. Cryst. Growth **208** 85-92

[9] D. Brien et al. 2007 *Modelling and simulation of MOVPE of GaAs-based compounds semiconductors in production-scaled Planetary Reactors* J. Cryst. Growth **303** 330-333

[10] S. Habermann et al. 2012 *Growth of InGaAs/AlGaAs Layers for Laser Manufacturing using 4 inch GaAs Substrate* CS-Mantech Conference 2012

[11] G. B. Stringfellow *Organometallic Vapor-Phase Epitaxy Theory and Practice* 2nd Ed. Academic Press 1999 ISBN 978-0123911742

[12] Yu-Ran Luo *Comprehensive handbook of chemical bond energies* 2007 CRC Press ISBN 978-0-8493-7366-4

[13] N. Watanabe and H. Ito 1997 *Saturation of hole concentration in carbon-doped GaAs grown by metalorganic chemical vapor deposition* J. Cryst. Growth **182** 30-36

[14] G. Hollinger et al. 1994 *Oxides on GaAs and InAs surfaces: An x-ray-photoelectron-spectroscopy study on reference compounds and thin oxide layers* Phys. Rev. B **49** (16) 11159-11167

[15] A. F. Pun et al. 2007 *Reduction of thermal oxide desorption etching on gallium arsenide* Thin Solid Films **515** 4419-4422

[16] M. Yamada and Y. Ide 1994 *Direct Observation of Species Liberated from GaAs Native Oxides during Atomic Hydrogen Cleaning* Jpn. J. Appl. Phys. **33** L671-L674

[17] M. Yamada 1996 *GaOH: Unstable Species Liberated from GaAs Surface Oxides during Atomic Hydrogen Cleaning* Jpn. J. Appl. Phys. **35** L651-L653

[18] D. A. Allwood et al. 1998 *In situ characterisation of MOVPE by surface photoabsorption I. Substrate oxide desorption* J. Cryst. Growth **195** 163-167

[19] D. A. Allwood et al. 2000 *In situ characterisation of epiready III-V substrates for MOVPE* J. Cryst. Growth **221** 160-165

[20] C. Kaspari et al. 2016 *Deoxidation of (001) III-V semiconductors in metal-organic vapour phase epitaxy* J. Appl. Phys. **120** 085701

[21] Y. Takino et al. 2012 *Improved Regrowth Interface of AlGaInAs/InP-Buried-Hererostructure Lasers by In-Situ thermal Cleaning* IEEE J. Quantum Electron. 48 (8) 971-979

[22] M. Heyen and P. Balk 1981 *Vapor phase etching of GaAs in a Chlorine System* J. Cryst. Growth **53** 558-562

[23] B. Henle et al. 1993 *In situ selective-area etching and MOVPE regrowth of GaInAs-InP on InP substrates* Semicond. Sci. Technol. **8** 994-997

[24] K. Fujii et al. 1994 Model for in-situ etching and selective epitaxy of $Al_xGa_{1-x}As$ with HCl gas by metalorganic vapor phase epitaxy J. Cryst. Growth **145** 277-282

[25] H. Q. Hou 1998 J. Cryst. Growth *In situ etching of GaAs by AsCl3 for regrowth of AlGaAs in metalorganic vapor-phase epitaxy* **195** 199-204

[26] R. Gessner et al. 2003 *Fabrication of AlGaInAs and GaInAsP buried heterostructure lasers by in situ etching* J. Cryst. Growth **248** 426-430

[27] D. Bertone et al. 1998 *Etching of InP-based MQW laser structure in a MOCVD reactor by chlorinated compounds* J. Cryst. Growth **195** 624-629

[28] T. Tsuchiya et al. 2004 *In situ Deep Etching for an InGaAlAs Buried Heterostructure by Using HCl Gas in a Metalorganic Vapor Phase Epitaxy Reactor* Jpn. J. Appl. Phys. **43** (10A) L1247-L1249

[29] H. Kizuki et al. 1997 *Time-Resolved Photoluminescence Study on a Hetero Interface Formed by Direct Regrowth of GaAs on an $Al_{0.3}Ga_{0.7}As$ Surface Prepared by an In Situ HCl Gas Etching Process* Jpn. J. Appl. Phys. **36** (10) 6290-6294

[30] Y. Inoue et al. 1994 Fabrication of $Al_xGa_{1-x}As$ buried heterostructure laser diodes by in-situ gas etching and selective-area metalorganic vapor phase epitaxy J. Cryst. Growth **145** 881-885

[31] S. Ikawa and M. Ogura 1997 *AlGaAs-GaAs Buried Heterostructure Laser with Vertically Etched Facets and Wide-Bandgap Optical Windows by in situ C_2H_5Cl Gas-Phase Etching and MOCVD Regrowth* IEEE Photon. Technol. Lett. **9** (6) 719-721

[32] A. Rebey et al. 2004 In situ reflectance monitoring of the growth and etching of AlAs/GaAs structures in MOVPE J. Cryst. Growth **261** 450-457

[33] T. Mochizuki et al. 2005 *Influence of in situ HCl gas cleaning on n/p-type GaAs and AlGaAs regrown interfaces in MOCVD* J. Cryst. Growth **273** 464-473

[34] C. Su et al. 1994 *Chemical dry etching of GaAs (100) by HCl: products, rate and a kinetic model* Surf. Sci. **312** 181-197

[35] K. Tateno et al. 1997 *Carbon doping and etching effects of CBr_4 during metalorganic chemical vapor deposition of GaAs and AlAs* J. Cryst. Growth **172** 5-12

[36] A. Maaßdorf and M. Weyers 2008 *In-situ etching of GaA/$Al_xGa_{1-x}As$ by CBr_4* J. Cryst. Growth **310** 4754-4756

[37] A. Arakawa et al. 2002 *In-situ Etching of Semiconductor with CBr_4 in Metalorganic Chemical Vapor Deposition (MOCVD) Reactor* Jpn. J. Appl. Phys. **41** 1076-1079

[38] C. Ebert et al. 2007 *Selective area etching of InP with CBr_4 in MOVPE* J. Crystal Growth **298** 94-97

[39] J. Décobert et al. 2008 *Optically in-situ monitored growth of carbon doped InAlAs by LP-MOVPE using CBr_4* J. Cryst. Growth **310** 4813-4817

[40] N. Kuznetsova et al. 2014 *Crystallographic dependent in-situ CBr4 selective nano-area etching and local regrowth of InP/InGaAs by MOVPE* J. Cryst. Growth **406** 111-115

[41] M. McEllistrem and J. M. White 1995 *Surface chemistry of gallium tetrabromide on GaAs(100)* J. Vac. Sci. Technol. A **13** (3) 1448-1454

[42] A. Maaßdorf et al. 2013 *In-situ etching of patterned GaAs/InGaP surfaces for highly efficient 975 nm DFB-BA diode lasers* J. Cryst. Growth **370** 226-229

[43] P. Della Casa et al. 2016 *CBr_4-based in-situ etching of GaAs, assisted with TMGa and TMAl* J. Cryst. Growth **434** 116-122

[44] J. P. Stagg et al. 1992 *Measurement and control of reagents concentrations in MOCVD reactor using ultrasonics* J. Cryst. Growth **120** 98-102

[45] J. L. Weyher *Defect-selective etch of III-V and wide gap semiconductors* 1st CEPHONA Workshop on Microscopic Characterisation of Materials and Structures for Photonics, Warsaw, November 24, 2003

[46] T. Bergunde et al. 2003 *Automated emissivity corrected wafer-temperature measurement in Aixtrons planetary reactors* J. Cryst. Growth **248** 235-239

[47] A. Jenichen and C. Engler 2000 *Etching of GaAs(100) Surfaces by Halogen Molecules: Density Functional Calculations on the Different Mechanisms* J. Phys. Chem. B **105** 1956-1960

[48] S. W. Benson and H. E. O'Neal *Kinetic Data on Gas Phase Unimolecular Reactions* 1970 NSRDS-NBS 21 p. 495 and p. 535

[49] A. Koukitu et al. 1995 *In situ gravimetric monitoring of arsenic desorption in GaAs atomic layer epitaxy* J. Cryst. Growth **146** 239-245

[50] Y. Matsuo et al. 2003 *Theoretical Investigation of Arsenic Desorption from GaAs (001) Surfaces under an Atmosphere of Hydrogen* Jpn. J. Appl. Phys. **42** 2578

[51] Q. Fu et al. 2000 *Mechanism of the Arsine Adsorption on the Gallium-Rich GaAs(001)-(4 × 2) Surface* J. Phys. Chem. B **104** 5595-5602

[52] W. D. Goodhue et al. 1999 *Bromine Ion-Beam-Assisted Etching of III-V Semiconductors* J. Electron. Mater. **28** (4) 364-368

[53] A. Maaßdorf, unpublished results.

[54] M. Tawfieq et al. 2017 *Concept and numerical simulations of a widely tunable GaAs-based sampled-grating diode lasers emitting at 976 nm* IET Optoelectron. **11** (2) 73-78

[55] O. Brox et al. 2017 *Realisation of a widely tuneable sampled grating DBR laser emitting around 970 nm* Electron. Lett. 53 (11) 744-746

[56] O. Brox et al. 2018 *Reflectors and tuning elements for widely-tunable GaAs-based sampled grating DBR lasers* Proc. SPIE 10553, Novel In-Plane Semiconductor Lasers XVII, 1055310

[57] M. Tawfieq et al. 2018 High-power sampled-grating-based master oscillator power amplifier system with 23.5 nm wavelength tuning around 970 nm Appl. Opt. **57** (29) 8680-8685

[58] L. A. Coldren, S. C. Corzine, M. L. Mašanović 2012 *Diode Lasers and Photonic Integrated Circuits* Wiley ISBN 978-0-470-48412-8

[59] H. Ishi et al. 1995 *Narrow Spectral Linewidth Under Wavelength Tuning in Thermally Tunable Super-Structure-Grating (SSG) DBR Lasers* J. Sel. Top. Quantum Electron. **1** (2) 401-407

[60] L. A. Coldren 2000 *Monolithic tunable diode lasers* IEEE J. Sel. Top. Quantum Electron. **6** (6) 988-999

[61] R. Phelan et al. 2008 *A Novel Two-Section Tunable Discrete Mode Fabry-Pérot Laser Exhibiting Nanosecond Wavelength Switching* IEEE J. Quantum Electron. 44 (4) 331-337

[62] J. Zhao et al. 2013 *Widely Tunable Semiconductor Lasers Based on Digital Concatenated Grating With Multiple Phase Shifts* IEEE Photonics J. 5 (5) 1502008

[63] M. C. Larson, *Widely Tunable Semiconductor Lasers* in Optical Fiber Communication Conference, OSA Technical Digest (online) (Optical Society of America, 2014), paper Tu2H.1.

[64] D. Hofstetter and H. P. Zappe 1997 *Anomalous longitudinal mode hops in GaAs/AlGaAs distributed Bragg reflector lasers* Appl. Phys. Lett. **71** 181-183

[65] M Radziunas et al. 2011 *Mode transitions in distributed Bragg reflector semiconductor lasers: experiments, simulations and analysis* J. Phys. B: At. Mol. Opt. Phys. **44** 105401

[66] F. Bugge et al. 1998 *MOVPE growth of tunable DBR laser diode emitting at 1060 nm* J. Cryst. Growth **195** 676-680

[67] F. Bugge et al. 2000 *Effect of growth conditions and strain compensation on indium incorporation for diode lasers emitting above 1050 nm* J. Cryst. Growth **221** 496-502

[68] F. Bugge et al. 2007 *MOVPE growth optimization for laser diodes with highly strained InGaAs MQWs* J. Cryst. Growth **298** 652-657

[69] F. Bugge et al. 2000 *MOVPE Growth of AlGaAs/GaInP diode lasers* J. Electron. Mater. **29** (1) 57-61

[70] F. Bugge et al. 2011 *Characterization and optimization of 2-step MOVPE growth for single-mode DFB or DBR laser diodes* J. Cryst. Growth **315** 74-77

[71] L. Hofmann et al. 2001 *(AlGa)As composition profile analysis of trenches overgrown with MOVPE* J. Cryst. Growth 222 465-470

[72] B. R. Bennett et al. 1990 *Carrier-Induced Change in Refractive Index of InP, GaAs and InGaAsP* IEEE J. Quantum Electron. 26 (1) 113-122

[73] T. G. B. Mason *InP based photonics integrated circuits* Ph.D. dissertation, University of California, Santa Barbara, 1999

[74] H. Wenzel et al. 1999 Improved theory of the refractive-index change in quantum-well lasers IEEE J. Sel. Top. Quantum Electron. **5** 637-642

[75] M. Guden and J. Piprek 1996 *Material parameters of quaternary III-V semiconductors for multilayer mirrors at 1.55 μm wavelength* Modelling Simul. Mater. Sci. Eng. **4** 349-357

[76] F. Berz 1979 *Step recovery of pin diodes* Solid-State Electron. **22** 927-932

[77] C. H. Henry et al. 1981 Spectral dependence of the change in refractive index due to carrier injection in GaAs lasers J. Appl. Phys. 52 (7) 4457-4461

[78] J-P Weber 1994 *Optimization of the Carrier-Induced Refractive-Index-Change in InGaAsP Waveguides – Application to Tunable Bragg Filters* IEEE J. Quantum Electron. 30 (8) 1801-1816

[79] P. Della Casa et al. 2017 *High-power broad-area buried-mesa lasers* Semicond. Sci. Technol. **32** 065009

[80] C. Jauregui et al. 2013 *High-power fibre lasers* Nature Photonics **7** 861-867

[81] P. Crump et al. 2013 *Efficient High-Power Laser Diodes* IEEE J. Sel. Top. Quantum Electron. **19** (4) 1501211

[82] M. Kanskar et al. 2014 *High Reliability of High Power and High Brightness Diode Lasers* Proc. SPIE 8965, High-Power Diode Laser Technology and Applications XII 896508

[83] E. Zucker et al. 2014 *Advancement in laser diode chip and packaging technologies for application in kW-class fiber laser pumping* Proc. SPIE 8965, High-Power Diode Laser Technology and Applications XII, 896507

[84] B. Kruschke et al. 2015 *Beam Combining Techniques for High-Power High-Brightness Diode Lasers* Proc. SPIE 9346, Components and Packaging for Laser Systems, 934614

[85] P. Crump et al. 2016 *Novel approaches to increasing the brightness of broad area lasers* Proc. SPIE 9767, Novel In-Plane Semiconductor Lasers XV, 97671L

[86] M. Winterfeldt et al. 2014 *Experimental investigation of factors limiting slow-axis beam quality in 9xx nm high power broad area diode lasers* J. Appl. Phys. **116**, 063103

[87] Y. Yamagata et al. 2015 *915nm high power broad area laser diodes with ultra-small optical confinement based on Asymmetric Decoupled Confinement Heterostructures (ADCH)* Proc. SPIE 9348, High-Power Diode Laser Technology and Applications XIII, 93480F

[88] P. W. Epperlein *Semiconductor Laser Engineering, Reliability and Diagnostic*, Wiley 2013; ISBN: 978-1119990338

[89] P. Ressel et al. 2005 *Novel Passivation Process for the Mirror Facets of Al-Free Active-Region High-Power Semiconductor Diode Lasers* IEEE Photon. Technol. Lett. **17** (5) 962-964

[90] M. Silver et al. 2005 *Theoretical and Experimental Study of Improved Catastrophic Optical Damage Performance in 830nm High Power Lasers with Non-absorbing Mirrors* CLEO-Conference on Lasers and Electro-Optics Technical Digest, paper CMX

[91] M. Buda et al. 1997 *Analysis of 6-nm AlGaAs SQW Low-Confinement Laser Structures for Very High-Power Operation* IEEE J. Sel. Top. Quantum Electron. **3** (2) 173-179

[92] B. S. Ryvkin and E. A. Avrutin 2005 *Asymmetric, nonbroadened large optical cavity waveguide structures for high-power long-wavelength semiconductor lasers* J. Appl. Phys. **97**, 123103

[93] J. M. G. Tijero et al. 2010 *Simulation of facet heating in high power red lasers* Proc. SPIE 7597, Physics and Simulation of Optoelectronic Devices XVIII, 75971G

[94] R.M. Lammert et al. 2006 *Advances in high brightness semiconductor lasers* Proc. SPIE 6104, High-Power Diode Laser Technology and Applications IV, 61040I

[95] R.M. Lammert et al. 2006 *High power (>10W from 100 lm aperture) high reliability 808nm InAlGaAs broad area laser diodes* Electron. Lett. Vol. **42** (9) 535-536

[96] C. L. Walker et al. 2002 *Improved Catastrophic Optical Damage Level From Laser With Nonabsorbing Mirrors* IEEE Photon. Technol. Lett. **14** (10) 1394-1396

[97] H. Taniguchi et al. 2007 25-W 915-nm *Lasers With Window Structure Fabricated by Impurity-Free Vacancy Disordering (IFVD)* IEEE J. Sel. Top. Quantum Electron. **13** (5) 1176-1179

[98] T. Morita et al. 2013 *High-Efficient and Reliable Broad-Area Laser Diodes With a Window Structure IEEE* J. Sel. Top. Quantum Electron. **19** (4) 1502104

[99] J. E. Ungar et al. 1994 *High power 980nm nonabsorbing facet lasers* Electron. Lett. **30** (21) 1766-1767

[100] A. Pietrzak et al. *Progress in efficiency-optimized high-power diode lasers* Proc. of SPIE Vol. 8898 Technologies for Optical Countermeasures X; and High-Power Lasers 2013: Technology and Systems

[101] A. Pietrzak 2011 *Realization of High Power Diode Lasers with Extremely Narrow Vertical Divergence* Dissertation zur Erlangung des akademischen Grades Doktor der Naturwissenschaften, Technischen Universität Berlin

[102] P. Crump et al. 2012 *Experimental and theoretical analysis of the dominant lateral waveguiding mechanism in 975 nm high power broad area diode lasers* Semicond. Sci. Technol. **27** 045001

[103] H. Wenzel et al. 2000 *Influence of current spreading on the transparency current density of quantum-well lasers* Semicond. Sci. Technol. **15** 557-560

[104] J. Piprek and Z. M. Simon Li *2013 On the importance of non-thermal far-field blooming in broad-area high-power laser diodes* Appl. Phys. Lett. **102**, 221110

[105] M. Winterfeldt et al. 2015 *High Beam Quality in Broad Area Lasers via Suppression of Lateral Carrier Accumulation* IEEE Photon. Technol. Lett. **27** (17) 1809-1812

[106] A. Zeghuzi et al. 2018 *Modeling of current spreading in high-power broad-area lasers and its impact on the lateral far field divergence* Proc. SPIE 10526, Physics and Simulation of Optoelectronic Devices XXVI; 105261H

[107] M. Elattar et al. 2020 *High-brightness broad-area diode lasers with enhanced self-aligned lateral structure* Semicond. Sci. Technol. **35** 095011

[108] H. Wenzel and H. -J. Wünsche 1990 A Model for the Calculation of the Threshold Current of SCH-MQ W-SAS Lasers Phys. Stat. Sol. (a) 120 (2) 661-673

[109] H. Strusny et al. 1996 *He implant-damage isolation of MOVPE grown GaAs/InGaP/InGaAsP layers* Nucl. Instr. Meth. Phys. Res. B **112** 298-300

[110] http://www.wias-berlin.de/software/tesca/

[111] J. Decker et al. 2015 *Study of lateral brightness in 20 μm to 50 μm wide narrow stripe broad area lasers* IEEE High Power Diode Lasers and Systems Conference 21-22

[112] Y. Sin et al. 2011 *Catastrophic Optical Bulk Damage (COBD) in High Power Multi-Mode InGaAs-AlGaAs Strained Quantum Well Lasers* Proc. SPIE 7918, High-Power Diode Laser Technology and Applications IX 791803

[113] D. Botez 1999 *Design considerations and analytical approximations for high continuous-wave power, broad-waveguide diode lasers* Appl. Phys. Lett. **74** (21), 3102-3104

[114] U. Bandelow et al. 2020 *Dynamics in high-power diode lasers* Proc. SPIE 11356, Semiconductor Lasers and Laser Dynamics IX, 113560W

[115] S. Y. Hu et al. 1994 *Lateral carrier diffusion and surface recombination in InGaAs/AlGaAs quantum-well ridge-waveguide lasers* J. Appl. Phys. **76** (8), 4479-4487

[116] W. B. Joyce 1980 *Current-crowded carrier confinement in double-heterostructure lasers* J. Appl. Phys. **51** (5), 2394-2401

[117] G. J. Letal et al. 1998 *Determination of Active-Region Leakage Currents in Ridge-Waveguide Strained-Layer Quantum-Well Lasers by Varying the Ridge Width* IEEE J. Quantum Electron. **34** (3) 512-518

[118] P. Della Casa et al. 2019 *High power broad-area lasers with buried implantation for current confinement* Semicond. Sci. Technol. **34** 105005

[119] D. Martin et al. 2019 Proc. SPIE 10900, High-Power Diode Laser Technology XVII, 109000M

[120] S. J. Pearton 1993 *Ion implantation in III-V semiconductor technology* Int. J. Mod. Phys. B **7** (28) 4687-4761

[121] W. Wesch 1992 *Ion implantation in III-V compounds* Nucl. Instrum. Methods Phys. Res. B **68** 342-354

[122] J. S. Williams et al. 1993 *Damage accumulation and amorphization in GaAs-AlGaAs structures* Nucl. Instr. Meth. Phys. Res. **B74** 80-83

[123] M. Skowronski 1992 in *Deep Centers in Semiconductors* 2nd ed., Gordon and Breach Science Publishers ISBN 2881245625

[124] J. Schneider et al. 1989 *Assessment of oxygen in gallium arsenide by infrared local vibrational mode spectroscopy* Appl. Phys. Lett. 54 1442-1444

[125] D. Colleoni, A. Pasquarello *Assignment of Fermi-level pinning and optical transitions to the* $(As_{Ga})_2$*-*O_{As} *center in oxygen-doped GaAs* 2013 Appl. Phys. Lett. **103** 142108

[126] M. Hata et al. 1992 *The effects of oxygen impurity in TMA on A1GaAs layers grown by MOVPE* J. Chryst. Growth **124** 427-432

[127] R. P. Bryan et al. 1989 *Compositional Disordering and the Formation of Semi-insulating Layers in AlAs-GaAs Superlattices by MeV Oxygen Implantation* J. Electron. Mater. **18** (1) 39-44

[128] H. Kakinuma et al. 1997 *Characterization of Oxygen and Carbon in Undoped AlGaAs Grown by Organometallic Vapor-Phase Epitaxy* Jpn. J. Appl. Phys. **36** 23-28

[129] S. Adachi 1988 *Si-ion implantation in GaAs and* $Al_xGa_{1-x}As$ J. Appl. Phys. **63** 64-67

[130] T. Ahlgren et al. *Concentration dependent and independent Si diffusion in ion-implanted GaAs* 1997 Phys. Rev. B **56** (8) 4597-4603

[131] R. S. Bhattacharya et al. 1983 *Si implantation in GaAs* J. Appl. Phys. **54** (5) 2329-2337

[132] R. S. Bhattacharya et al. 1983 *Epitaxial Regrowth of Si Implanted (100) and (211) GaAs* Phys. Stat. Sol. (a) **76** 131-136

[133] J. L. Tandon et al. 1978 *Si implantation in GaAs* Appl. Phys. Lett. **34** 165-167

[134] J. H. Marsh 1993 *Quantum well intermixing* Semicond. Sci. Technol. **8** 1136-1155

[135] S. Moriwaki et al. 2016 *Quantum well intermixing technique using proton implantation for carrier confinement of vertical-cavity surface-emitting lasers* Jpn. J. Appl. Phys. **55** 08RC01

[136] U. Zeimer and E. Nebauer 2000 *High-resolution x-ray diffraction investigations of He-implanted GaAs* Semicond. Sci. Technol. **15** 965-970

[137] I. Volotsenko et al. 2010 *Secondary electron doping contrast: Theory based on scanning electron microscope and Kelvin probe force microscopy measurements* J. Appl. Phys. **107** 014510

[138] http://www.wias-berlin.de/software/balaser/

[139] C. Goerke 2020 *Investigations of the Impact of Deeply Implanted Buried Current Apertures on the Electro-Optical Properties of Broad-Area Diode Lasers Emitting in the Near-Infrared* Masterarbeit im Studiengang Physik, Universität Potsdam

[140] J. C. Phillips 1973 *Bonds and Bands in Semiconductors* Academic Press ISBN 0-12-553350-0

[141] Tomonori Ito 1998 *Simple Criterion for Wurtzite-Zinc-Blende polytypism in Semiconductors* Jpn. J. Appl. Phys. **37** 1217-1220

[142] Hadis Morcoç 2008 *Handbook of Nitride Semiconductors and Devices Vol. 1* Wiley ISBN 978-3-527-40837-5

[143] A. Aresti et al. 1984 *Some cohesive energy features of tetrahedral semiconductors* J. Phys. Chem. Solids **45** (3) 361-365

[144] P. Manca 1961 *A relation between the binding energy and the band-gap energy in semiconductors of diamond or zincblende structure* J. Phys. Chem. Solids **20** (3/4) 268-273

[145] Anthony R. West 2014 *Solid State Chemistry and its Applications* Wiley ISBN 978-1-119-94294-8

[146] Neil W. Ashcroft, N. David Mermin 1976 *Solid State Physics* Saunders College Publishing ISBN 0-03-083993-9

[147] U. D. Pohl 2013 *Epitaxy of Semiconductors* Springer ISBN 978-3-642-32969-2

[148] N. Moll et al. 1996 *GaAs equilibrium crystal shape from first principles* Phys. Rev. B **54** (12) 8844-8855

[149] D. W. Shaw 1979 Morphology analysis in localized crystal growth and dissolution J. Cryst. Growth **47** 509-517

[150] M. D. Pashley 1989 *Electron counting model and its application to islands structures on molecular-beam epitaxy grown GaAs(001) and ZnSe(001)* Phys, Rev. B **40** (15) 10 481-10 487

[151] A. Ohtake et al. *2001 Surface structures of GaAs{001}A,B-(2x2)* Phys. Rev. B **64** 045318

[152] S. Adachi 2017 *III-V Ternary and Quaternary Compounds* in *Springer handbook of Electronics and Photonics Materials, 2nd Ed.(Part D)* Springer DOI 10.1007/978-3-319-48933-9_30

[153] G. B. Stringfellow 1982 *Miscibility Gaps in Quaternary III/V Alloys* J. Cryst. Growth **58** 194-202

[154] A. K. Saxena 1980 *The conduction band structure and deep levels in $Ga_{1-x}Al_xAs$ alloys from a high-pressure experiment* J. Phys. C **13** 4323-4334

[155] A. K. Saxena 1981 *Electron mobility in $Ga_{1-x}A1_xAs$ alloys* Phys. Rev. B **24** (6) 3295-3302

[156] D. C. Look et al. 1992 *Alloy scattering in p-type $Al_xGa_{1-x}As$* J. Appl. Phys. **71** 260-266

[157] T. Suzuki 2002 *Basic Aspects of Atomic Ordering in III–V Semiconductor Alloys* in: Mascarenhas A. (Ed.) *Spontaneous Ordering in Semiconductor Alloys* Springer ISBN: 978-1-4613-5167-2

[158] J. W. Matthews and A. E. Blakeslee 1974 *Defects in Epitaxial Multilayers* J. Cryst. Growth **27** 118-125

[159] Andersson et al. 1987 *Variation of the critical layer thickness with In content in strained $In_xGa_{1\_x}As$-GaAs quantum wells grown by molecular beam epitaxy* Appl. Phys. Lett. **51** 752-754

[160] R. Hull and J. C. Bean 1992 *Misfit Dislocations in Lattice-Mismatched Epitaxial Films* Critical Reviews in Solid State and Material Sciences **17** (6) 507-546

[161] H. Gottschale et al. 1978 *Stacking Fault Energy and Ionicity of Cubic III-V Compounds* Phys. Status Solidi A **45** 207-217

[162] B. T. Lee et al. 1990 *Atomic structure of twins in GaAs* Appl. Phys. Lett. **57** (4) 346-347

[163] D. Hull and D. J. Bacon 2001 *Introduction to Dislocations, 4th ed.* Butterworth-Heinemann ISBN 0-7506-4681-0

[164] J. Hornstra 1958 *Dislocations in the Diamond Lattice* J. Phys. Chem. Solids **5** 129-141

[165] P. Bhattacharya 1994 *Semiconductor Optoelectronic Devices* Prentice Hall ISBN 0-13-805748-6

[166] J. R. Chelikowsky and M. L. Cohen 1976 *Nonlocal pseudopotential calculations for the electronic structure of eleven diamond and zinc-blende semiconductors* Phys. Rev. B **14** (2) 556-582

[167] E. F. Schubert 2006 *Light-Emitting Diodes* Cambridge University Press ISBN 978-0-521-86538-8

[168] I. Vurgaftman et al. 2001 *Band parameters for III–V compound semiconductors and their alloys* J. Appl. Phys. **89** (11) 5815-5873

[169] S. Tiwari and D. J. Frank 1992 *Empirical fit to band discontinuities and barrier heights in III-V alloy systems* Appl. Phys. Lett **60** (5) 630-632

[170] K. Fletcher and P. N. Butcher 1972 *An exact solution of the linearized Boltzmann equation with applications to the Hall mobility and hall factor of n-GaAs* J. Phys. C: Solid St. Phys. **5** 212-224

[171] W. Walukievicz et al. 1979 *Electron mobility and free-carrier absorption in GaAs: Determination of the compensation ratio* J. Appl. Phys. **50** (2) 899-908

[172] M. Sotoodeh et al. 2000 *Empirical low-field mobility model for III-V compounds applicable in device simulation codes* J. Appl. Phys. **87** (6) 2890-2900

[173] R. N. Hall 1959 *Recombination Processes in Semiconductors* Proceedings of the IEE Part B **106** (17) 923-931

[174] A. Alkaukas and al. 2014 *First-principles theory of nonradiative carrier capture via multiphonon emission* Phys. Rev. B **90** 075202

[175] W. Shockley and W. T. Read Jr. 1952 *Statistics of the recombination of Holes and Electrons* Phys. Rev. 87 (5) 835-842

[176] C. H. Henry and D. V. Lang 1977 *Nonradiative capture and recombination by multiphonon emission in GaAs and GaP* Phys. Rev. B **15** (2) 989-1016

[177] M. Niemeyer et al. 2019 *Measurement of the non-radiative minority recombination lifetime and the effective radiative recombination coefficient in GaAs* AIP Advances **9** 045034

[178] D. E. Aspnes 1983 *Recombination at Semiconductor Surfaces and Interfaces* Surf. Sci. **132** 406-421

[179] R. K. Ahrenkiel et al. 1991 *Intensity-dependent minority-carrier lifetime in III-V semiconductors due to saturation of recombination centers* J. Appl. Phys. 70 (1) 225-231

[180] D. Steiauf et al. 2014 *Auger Recombination in GaAs from First principles* ACS Photonics **1** (8) 647-654

[181] S. Marchetti et al. 2001 *The InSb Auger recombination coefficient derived from the IR-FIR dynamical plasma reflectivity* J. Phys.: Condens. Matter **13** 7363-7369

[182] U. Strauss et al. 1993 *Auger recombination in intrinsic GaAs* Appl. Phys. Lett. **62** (1) 55-57

[183] Y. A. Goldberg 1999 in *Handbook of Semiconductor Parameters* Vol. 1 World Scientific ISBN 978-981-02-2934-4

[184] M .Govoni et al. 2001 *Auger recombination in Si and GaAs semiconductors:Ab initio results* Phys. Rev. B **84** 075215

[185] H. C. Casey et al. 1975 *Concentration dependence of the absorption coefficient for n- and p-type GaAs between 1.3 and 1.6 eV* J. Appl. Phys. **46** (1) 250-257

[186] G. W. 't Hooft et al. 1985 *Temperature dependence of the radiative recombination coefficient in GaAs – (Al, Ga)As quantum wells* Superlattice Microst. **1** (4) 307-310

[187] M. S. Whalen and J. Stone 1982 *Index of refraction of n-type InP at 0.633-µm and 1.15-µm wavelengths as a function of carrier concentration* J. Appl. Phys. 53 (6) 4340-4343

[188] E. D. Palik (ed.) *Handbook of Optical Constants of Solids* vol. 1 Academic Press, 1985 ISBN: 0-12-544420-0

[189] R. Pelzel *A Comparison of MOVPE and MBE growth Technologies for III-V Epitaxial Structures* CS-Mantech Conference 2013

[190] C. A. Wang 2019 *Early history of MOVPE reactor development* J. Cryst. Growth **506** 190-200

[191] S. K. Ghandhi and R. J. Field 1984 *A re-examination of boundary layer theory for a horizontal CVD reactor* J. Cryst. Growth **69** 619-622

[192] K. F. Jensen 1989 *Transport phenomena and chemical reaction issues in OMVPE of compound semiconductors* J. Cryst. Growth **98** 148-166

[193] B. Mitrovic et al. 2006 *On the flow stability in vertical rotating disc MOCVD reactor under a wide range of process parameters* J. Cryst. Growth **287** 656-663

[194] H. Li 2011 *Mass transport analysis of a showerhead MOCVD reactor* J. Semicond. **32** (3) 033006-1-033006-5

[195] B. E. Poling, J. M. Prausnitz and J. P. O'Connel *The properties of Gases and Liquids* 5th Ed. McGraw-Hill 2001 ISBN: 0639785322160

[196] M. Tirtowidjojo and R. Pollard 1986 *Equilibrium gas phase species for MOCVD of* $Al_xGa_{1-x}As$ J. Cryst. Growth **77** 200-209

[197] C. Li et al. 2003 *Thermodynamic analysis on MOVPE of* $Ga_{1-x}In_xP$ *semiconductor* J Alloy Compd. **359** 153-158

[198] K. J. Laidler *Chemical Kinetics* 2nd Ed. Tata McGraw-Hill 1965 ISBN 0070994226

[199] J. W. Moore and R. G. Pearson *Kinetics and Mechanism* 3rd Ed. Wiley 1981 ISBN 0-471-03558-0

[200] N. I. Buchan and J. M. Jasinski 1990 *Calculation of unimolecular rate constants for common metalorganic vapor phase epitaxy precursors via RRKM theory* J. Cryst. Growth **106** 227-238

[201] J. G. Martin et al. 1987 *The Decomposition Kinetics of Disilane and the Heat of Formation of Silylene* Int. J. Chem. Kinet. **19** 715-724

[202] T. J. Mountziaris and K. F. Jensen 1991 *Gas-Phase and Surface Mechanisms in MOCVD of GaAs with Trimethyl-Gallium and Arsine* J. Electrochem. Soc. **138** (8) 2426-2439

[203] M. Masi et al. 1992 *Simulation of carbon doping of GaAs during MOVPE* J. Cryst. Growth **124** 483-492

[204] C. A. Larsen et al. 1990 *Decomposition mechanisms of trimethylgallium* J. Cryst. Growth **102** 103-116

[205] C. A. Larsen et al. 1988 *Reaction mechanisms in the organometallic vapor phase epitaxial growth of GaAs* Appl. Phys. Lett. **52** (6) 480-482

[206] C. A. Larsen and G. B. Stringfellow 1986 *Decomposition kinetics of OMVPE precursors* J. Cryst. Growth **75** 247-254

[207] S. P. DenBaars et al. 1986 *Homogeneous and heterogeneous thermal decomposition rates of trimethylgallium and arsine and their relevance to the growth of GaAs by MOCVD* J. Cryst. Growth **77** 188-193

[208] M. Mashita 1990 *Reaction Mechanism in the OMVPE Growth of GaAs and AlGaAs* Jpn. J. Appl. Phys. **29** (5) 813-819

[209] M. G. Jacko and S. J. W. Price 1963 *The Pyrolysis of Trimethyl Gallium* Can. J. Chem. **41** 1560-1567

[210] S. J. W. Price in *Comprehensive Chemical Kinetics* ed. C. H Bamford and C. F. H. Tipper Elsevier 1972 vol. 4 pp. 209-214

[211] J. I. Davies et al. 1986 *In situ monitoring of chemical reactions in MOCVD growth of ZnSe* J. Cryst. Growth **79** 363-370

[212] W. C. Gardiner Jr. and G. B. Kistiakowsky 1961 *Thermal Dissociation Rate of Hydrogen* J. Chem. Phys. 35 (5) 1765-1770

[213] J. A. McCaulley et al. 1991 *Kinetics of thermal decomposition of triethylgallium, trimethylgallium, and trimethylindium adsorbed on GaAs (100)* J. Vac. Sci. Technol. A **9** (6) 2872-2886

[214] D. H. Reep and S. K. Ghandhi 1983 *Deposition of GaAs Epitaxial Layers by Organometallic CVD* J. Electrochem. Soc. **130** 675-680

[215] H. Heinecke et al. 1984 *Kinetics of GaAs growth by low pressure MO-CVD* J. Electron. Mater. **13** (5) 815-830

[216] H. Simka et al. 1997 *Computational chemistry predictions of reaction processes in organometallic vapor phase epitaxy* Prog. Crystal Growth and Charact. **35** (2-4) 117-149

[217] J. R. Arthur 1967 *Vapor pressures and phase equilibria in the Ga-As system* J. Phys. Chem. Solids **28** 2257-2267

[218] D. T. J. Hurle 2010 *A thermodynamic analysis of native point defect and dopant solubilities in zinc-blende III-V semiconductors* J. Appl. Phys. **107** 121301

[219] G. B. Stringfellow and M. J. Cherng 1983 *OMVPE growth of* $GaAs_{1-x}Sb_x$*: solid composition* J. Cryst. Growth **64** 413-415

[220] H. Seki and A. Koukitu 1986 *Thermodynamic analysis of metalorganic vapor phase epitaxy of III-V alloy semiconductors* J. Cryst. Growth **74** 172-180

[221] E. T. J. M. Smeets 1987 *Solid composition of* $GaAs_{1-x}P_x$ *grown by organometallic vapour phase epitaxy* J. Cryst. Growth **82** 385-395

[222] A. D. Maksimov et al. 2017 *A model for calculating the composition of* $GaAs_xP_{1-x}$ *solid solutions under metalorganic vapor phase epitaxy conditions* Inorg. Mater. **53** (4) 369-375

[223] L. Watanabe et al. 1994 *Low Temperature Metalorganic Chemical Vapor Deposition Growth of InGaAs for a Non-Alloyed Ohmic Contact to n-GaAs* Jpn. J. Appl. Phys. **33** L271-L274

[224] J. van Deelen et al. 2004 *Parameter study of intrinsic carbon doping of $Al_xGa_{1-x}As$ by MOCVD* J. Cryst. Growth **271** 267-284

[225] M. C. Hanna 1992 *Intrinsic carbon incorporation in very high purity MOVPE GaAs* J. Cryst. Growth **124** 443-448

[226] L. H. Long and J. F. Sackman 1957 *The heat of formation of trimethylphosphine* J. Chem. Soc. Faraday Trans. **53** 1606-1611

[227] NIST Standard Reference Database Number 69 https://doi.org/10.18434/t4d303

[228] W. Ulrici 2004 *Hydrogen-impurity complexes in III-V semiconductors* Rep. Prog. Phys. **67** 2233

[229] S. B. Hartley and J. C. McCoubrey 1963 *Enthalpy of Formation of Phosphorus Oxide* Nature **198** 476

[230] K. R. Henke, D. A. Atwood and L. Blue *Arsenic* Wiley 2009 ISBN 978-0-470-02758-5

[231] D. R. Lide (Ed.) *CRC Handbook of Chemistry and Physics, 85th ed.* CRC Press 2004 ISBN 978-0849332043

[232] R. W. Glew 1984 *Zn doping of MOCVD GaAs* J. Cryst. Growth **68** 44-47

[233] S. ChiChibu et al. 1990 *High concentration Zn doping in InP grown by low-pressure metalorganic chemical vapor deposition* J. Appl. Phys **68** 859-861

[234] K. Kurishima et al. 1993 *High-performance Zn-doped-base InP/InGaAs double heterojunction bipolar transistors grown by metalorganic vapor phase epitaxy* Appl. Phys. Lett. **64** 1111-1113

[235] P. R. Hageman et al. 1992 *Pressure and temperature dependence of silicon doping of GaAs using Si2H6 in metalorganic chemical vapor deposition* J. Cryst. Growth **116** 169-177

[236] S. Minagawa et al. 1995 *Heavy doping of silicon into $Ga_{0.5}In_{0.5}P$ at low temperatures in organometallic vapor phase epitaxy* J. Cryst. Growth **152** 251-255

[237] N. Watanabe and H. Ito 1997 *Mechanism of carbon incorporation from carbon tetrabromide in GaAs grown from metalorganic chemical vapor deposition* J. Cryst. Growth **178** 213-219

Innovationen mit Mikrowellen und Licht
Forschungsberichte aus dem Ferdinand-Braun-Institut, Leibniz-Institut für Höchstfrequenztechnik

Herausgeber: Prof. Dr. G. Tränkle, Prof. Dr.-Ing. W. Heinrich

Band 1: **Thorsten Tischler**
Die Perfectly-Matched-Layer-Randbedingung in der Finite-Differenzen-Methode im Frequenzbereich: Implementierung und Einsatzbereiche
ISBN: 3-86537-113-2, 19,00 EUR, 144 Seiten

Band 2: **Friedrich Lenk**
Monolithische GaAs FET- und HBT-Oszillatoren mit verbesserter Transistormodellierung
ISBN: 3-86537-107-8, 19,00 EUR, 140 Seiten

Band 3: **R. Doerner, M. Rudolph (eds.)**
Selected Topics on Microwave Measurements, Noise in Devices and Circuits, and Transistor Modeling
ISBN: 3-86537-328-3, 19,00 EUR, 130 Seiten

Band 4: **Matthias Schott**
Methoden zur Phasenrauschverbesserung von monolithischen Millimeterwellen-Oszillatoren
ISBN: 978-3-86727-774-0, 19,00 EUR, 134 Seiten

Band 5: **Katrin Paschke**
Hochleistungsdiodenlaser hoher spektraler Strahldichte mit geneigtem Bragg-Gitter als Modenfilter (α-DFB-Laser)
ISBN: 978-3-86727-775-7, 19,00 EUR, 128 Seiten

Band 6: **Andre Maaßdorf**
Entwicklung von GaAs-basierten Heterostruktur-Bipolartransistoren (HBTs) für Mikrowellenleistungszellen
ISBN: 978-3-86727-743-3, 23,00 EUR, 154 Seiten

Band 7: **Prodyut Kumar Talukder**
Finite-Difference-Frequency-Domain Simulation of Electrically Large Microwave Structures using PML and Internal Ports
ISBN: 978-3-86955-067-1, 19,00 EUR, 138 Seiten

Band 8: **Ibrahim Khalil**
Intermodulation Distortion in GaN HEMT
ISBN: 978-3-86955-188-3, 23,00 EUR, 158 Seiten

Band 9: **Martin Maiwald**
Halbleiterlaser basierte Mikrosystemlichtquellen für die Raman-Spektroskopie
ISBN: 978-3-86955-184-5, 19,00 EUR, 134 Seiten

Band 10: **Jens Flucke**
Mikrowellen-Schaltverstärker in GaN- und GaAs-Technologie Designgrundlagen und Komponenten
ISBN: 978-3-86955-304-7, 21,00 EUR, 122 Seiten

Cuvillier Verlag
Internationaler wissenschaftlicher Fachverlag

Innovationen mit Mikrowellen und Licht
Forschungsberichte aus dem Ferdinand-Braun-Institut, Leibniz-Institut für Höchstfrequenztechnik

Herausgeber: Prof. Dr. G. Tränkle, Prof. Dr.-Ing. W. Heinrich

Band 11: **Harald Klockenhoff**
Optimiertes Design von Mikrowellen-Leistungstransistoren und Verstärkern im X-Band
ISBN: 978-3-86955-391-7, 26,75 EUR, 130 Seiten

Band 12: **Reza Pazirandeh**
Monolithische GaAs FET- und HBT-Oszillatoren mit verbesserter Transistormodellierung
ISBN: 978-3-86955-107-8, 19,00 EUR, 140 Seiten

Band 13: **Tomas Krämer**
High-Speed InP Heterojunction Bipolar Transistors and Integrated Circuits in Transferred Substrate Technology
ISBN: 978-3-86955-393-1, 21,70 EUR, 140 Seiten

Band 14: **Phuong Thanh Nguyen**
Investigation of spectral characteristics of solitary diode lasers with integrated grating resonator
ISBN: 978-3-86955-651-2, 24,00 EUR, 156 Seiten

Band 15: **Sina Riecke**
Flexible Generation of Picosecond Laser Pulses in the Infrared and Green Spectral Range by Gain-Switching of Semiconductor Lasers
ISBN: 978-3-86955-652-9, 22,60 EUR, 136 Seiten

Band 16: **Christian Hennig**
Hydrid-Gasphasenepitaxie von versetzungsarmen und freistehenden GaN-Schichten
ISBN: 978-3-86955-822-6, 27,00 EUR, 162 Seiten

Band 17: **Tim Wernicke**
Wachstum von nicht- und semipolaren InAlGaN-Heterostrukturen für hocheffiziente Licht-Emitter
ISBN: 978-3-86955-881-3, 23,40 EUR, 138 Seiten

Band 18: **Andreas Wentzel**
Klasse-S Mikrowellen-Leistungsverstärker mit GaN-Transistoren
ISBN: 978-3-86955-897-4, 29,65 EUR, 172 Seiten

Band 19: **Veit Hoffmann**
MOVPE growth and characterization of (In,Ga)N quantum structures for laser diodes emitting at 440 nm
ISBN: 978-3-86955-989-6, 18,00 EUR, 118 Seiten

Band 20: **Ahmad Ibrahim Bawamia**
Improvement of the beam quality of high-power broad area semiconductor diode lasers by means of an external resonator
ISBN: 978-3-95404-065-0, 21,00 EUR, 126 Seiten

Cuvillier Verlag
Internationaler wissenschaftlicher Fachverlag

Innovationen mit Mikrowellen und Licht
Forschungsberichte aus dem Ferdinand-Braun-Institut, Leibniz-Institut für Höchstfrequenztechnik

Herausgeber: Prof. Dr. G. Tränkle, Prof. Dr.-Ing. W. Heinrich

Band 21: **Agnietzka Pietrzak**
Realization of High Power Diode Lasers with Extremely Narrow Vertical Divergence
ISBN: 978-3-95404-066-7, 27,40 EUR, 144 Seiten

Band 22: **Eldad Bahat-Treidel**
GaN-based HEMTs for High Voltage Operation
Design, Technology and Characterization
ISBN: 978-3-95404-094-0, 41,10 EUR, 220 Seiten

Band 23: **Ponky Ivo**
AlGaN/GaN HEMTs Reliability:
Degradation Modes and Anslysis
ISBN: 978-3-95404-259-3, 23,55 EUR, 132 Seiten

Band 24: **Stefan Spießberger**
Compact Semiconductor-Based Laser Sources
with Narrow Linewidth and High Output Power
ISBN: 978-3-95404-261-6, 24,15 EUR, 140 Seiten

Band 25: **Silvio Kühn**
Mikrowellenoszillatoren für die Erzeugung von atmosphärischen Mikroplasmen
ISBN: 978-3-95404-378-1, 21,85 EUR, 112 Seiten

Band 26: **Sven Schwertfeger**
Experimentelle Untersuchung der Modensynchronisation in Multisegment-Laserdioden zur Erzeugung kurzer optischer Pulse bei einer Wellenlänge von 920 nm
ISBN: 978-3-95404-471-9, 29,45 EUR, 150 Seiten

Band 27: **Christoph Matthias Schultz**
Analysis and mitigation of the factors limiting the effiency of high power distributed feedback diode lasers
ISBN: 978-3-95404-521-1, 68,40 EUR, 388 Seiten

Band 28: **Luca Redaelli**
Design and fabrication of GaN-based laser diodes for single-mode
and narrow-linewidth applications
ISBN: 978-3-95404-586-0, 29,70 EUR, 176 Seiten

Band 29: **Martin Spreemann**
Resonatorkonzepte für Hochleistungs-Diodenlaser
mit ausgedehnten lateralen Dimensionen
ISBN: 978-3-95404-628-7, 25,15 EUR, 128 Seiten

Cuvillier Verlag
Internationaler wissenschaftlicher Fachverlag

Innovationen mit Mikrowellen und Licht
Forschungsberichte aus dem Ferdinand-Braun-Institut, Leibniz-Institut für Höchstfrequenztechnik

Herausgeber: Prof. Dr. G. Tränkle, Prof. Dr.-Ing. W. Heinrich

Band 30:	**Christian Fiebig** Diodenlaser mit Trapezstruktur und hoher Brillanz für die Realisierung einer Frequenzkonversion auf einer mikro-optischen Bank ISBN: 978-3-95404-690-4, 26,30 EUR, 140 Seiten
Band 31:	**Viola Küller** Versetzungsreduzierte AIN- und AlGaN-Schichten als Basis für UV LEDs ISBN: 978-3-95404-741-3, 34,40 EUR, 164 Seiten
Band 32:	**Daniel Jedrzejczyk** Efficient frequency doubling of near-infrared diode lasers using quasi phase-matched waveguides ISBN: 978-3-95404-958-5, 27,90 EUR, 134 Seiten
Band 33:	**Sylvia Hagedorn** Hybrid-Gasphasenepitaxie zur Herstellung von Aluminiumgalliumnitrid ISBN: 978-3-95404-985-1, 38,00 EUR, 176 Seiten
Band 34:	**Alexander Kravets** Advanced Silicon MMICs for mm-Wave Automotive Radar Front-Ends ISBN: 978-3-95404-986-8, 31,90 EUR, 156 Seiten
Band 35:	**David Feise** Longitudinale Modenfilter für Kantanemitter im roten Spektralbereich ISBN: 978-3-7369-9116-3, 39,20 EUR, 168 Seiten
Band 36:	**Ksenia Nosaeva** Indium phosphide HBT in thermally optimized periphery for applications up to 300GHZ ISBN: 978-3-7369-287-0, 42,00 EUR, 154 Seiten
Band 37:	**Muhammad Maruf Hossain** Signal Generation for Millimeter Wave and THZ Applications in InP-DHBT and InP-on-BiCMOS Technologies ISBN: 978-3-7369-9335-8, 35,60 EUR, 136 Seiten
Band 38:	**Sirinpa Monayakul** Development of Sub-mm Wave Flip-Chip Interconnect ISBN: 978-3-7369-9410-2, 44,00 EUR, 146 Seiten
Band 39:	**Moritz Brendel** Charakterisierung und Optimierung von (Al, Ga) N-basierten UV-Photodetektoren ISBN: 978-3-7369-9465-2, 49,90 EUR, 196 Seiten
Band 40:	**Erdenetsetseg Luvsandamdin** Development of micro-integrated diode lasers for precision quantum optics experiments in space ISBN: 978-3-7369-9479-9, 39,00 EUR, 126 Seiten

Internationaler wissenschaftlicher Fachverlag

Innovationen mit Mikrowellen und Licht

Forschungsberichte aus dem Ferdinand-Braun-Institut, Leibniz-Institut für Höchstfrequenztechnik

Herausgeber: Prof. Dr. G. Tränkle, Prof. Dr.-Ing. W. Heinrich

Band 41: **Thi Nghiem Vu**
Development and analysis of diode laser ns-MOPA systems for high peak power application
ISBN: 978-3-7369-9480-5, 38,80 EUR, 138 Seiten

Band 42: **Christian Bansleben**
Differentieller Mikrowellen-Leistungsoszillator für die Realisierung ultrakompakter Plasmaquellen in Matrixanordnung
ISBN: 978-3-7369-9530-7, 34,90 EUR, 136 Seiten

Band 43: **Martin Winterfeldt**
Investigation of slow-axis beam quality degradation in high-power broad area diode lasers
ISBN: 978-3-7369-9733-2, 39,90 EUR, 158 Seiten

Band 44: **Jonathan Decker**
Investigation of monolithically integrated spectral stabilization in high-brightness broad area diode lasers
ISBN: 978-3-7369-9798-1, 49,50 EUR, 174 Seiten

Band 45: **Andreea Cristina Andrei**
Untersuchung und Optimierung robuster und hochlinearer rauscharmer Verstärker in GaN-Technologie
ISBN: 978-3-7369-9810-0, 39,90 EUR, 148 Seiten

Band 46: **Peng Luo**
GaN HEMT Modeling Including Trapping Effects Based on Chalmers Model and Pulsed S-Parameter Measurements
ISBN: 978-3-7369-9906-0, 48,00 EUR, 160 Seiten

Band 47: **Jörg Jeschke**
Entwicklung von optisch pumpbaren UVC-Lasern auf AlGaN-Basis
Chalmers Model and Pulsed S-Parameter Measurements
ISBN: 978-3-7369-9918-3, 44,90 EUR, 176 Seiten

Band 48: **Nikolai Wolff**
Wideband GaN Microwave Power Amplifiers with Class-G Supply Modulation
ISBN: 978-3-7369-9931-2, 44,90 EUR, 170 Seiten

Band 49: **Carlo Frevert**
Optimization of broad-area GaAs diode lasers for high powers and high efficiences in the temperature range 200-220 K
ISBN: 978-3-7369-9944-2, 44,90 EUR, 174 Seiten

Band 50: **Bassem Arar**
GaAs-based components for photonic integrated circuits
ISBN: 978-3-7369-9976-3, 43,60 EUR, 152 Seiten

Cuvillier Verlag
Internationaler wissenschaftlicher Fachverlag

Innovationen mit Mikrowellen und Licht
Forschungsberichte aus dem Ferdinand-Braun-Institut, Leibniz-Institut für Höchstfrequenztechnik

Herausgeber: Prof. Dr. G. Tränkle, Prof. Dr.-Ing. W. Heinrich

Band 51: **Mahmoud Tawfieq**
Development and characterisation of a diode laser based tunable high-power MOPA system
ISBN: 978-3-7369-9983-1, 44,90 EUR, 170 Seiten

Band 52: **Sebastian Preis**
Hocheffiziente frequenzagile Mikrowellen-Leistungsverstärker auf Basis von Verbindungshalbleitern und Ferroelektrika
ISBN: 978-3-7369-7004-5, 54,00 EUR, 136 Seiten

Band 53: **Simon Fleischmann**
Materialaspekte der Hydridgasphasenepitaxie von Aluminiumgalliumnitrid
ISBN: 978-3-7369-7029-8, 41,80 EUR, 160 Seiten

Band 54: **Erhan Ersoy**
Optimierung von koplanaren GaN-MMIC-Leistungsverstärkern im X-Band
ISBN: 978-3-7369-7157-8, 39,90 EUR, 154 Seiten

Band 55: **Simon Rauch**
Lateral emission characteristics of high-power broad-area lasers subject to external optical feedback
ISBN: 978-3-7369-7168-5, 29,90 EUR, 112 Seiten

Band 56: **Frank Dittmar**
Untersuchung der Strahlgüte von brillanten Hochleistungs-Trapezlasern für den Wellenlängenbereich bei 808 nm
ISBN: 978-3-7369-7167-7, 44,90 EUR, 176 Seiten

Band 57: **Florian Hühn**
Flexibler Modulator und Digitalverstärker-MMIC für den energieeffizienten Betrieb einer digitalen Sendekette im GHz-Bereich
ISBN: 978-3-7369-7190-5, 29,90 EUR, 162 Seiten

Band 58: **Dimitri Stoppel**
Interconnection development for InP-HBT terahertz circuits
ISBN: 978-3-7369-7204-9, 44,90 EUR, 156 Seiten

Band 59: **Anissa Zeghuzi**
Analysis of Spatio-Temporal Phenomena in High-Brightness Diode Lasers using Numerical Simulations
ISBN: 978-3-7369-7289-6, 49,90 EUR, 176 Seiten

Band 60: **Tasmim Alam**
Spectroscopic Applications of Terahertz Quantum-Cascade Lasers
ISBN: 978-3-7369-7297-1, 39,90 EUR, 132 Seiten

Cuvillier Verlag
Internationaler wissenschaftlicher Fachverlag

Innovationen mit Mikrowellen und Licht
Forschungsberichte aus dem Ferdinand-Braun-Institut, Leibniz-Institut für Höchstfrequenztechnik

Herausgeber: Prof. Dr. G. Tränkle, Prof. Dr.-Ing. W. Heinrich

Band 61: **Max Schiemangk**
Ein Lasersystem für Experimente mit Quantengasen unter Schwerelosigkeit
ISBN: 978-3-7369-7293-3, 49,90 EUR, 166 Seiten

Band 62: **Thorben Kaul**
Epitaxial Design Optimizations for Increased Efficiency in GaAs-Based High Power Diode Lasers
ISBN: 978-3-7369-7396-1, 38,90 EUR, 136 Seiten

Cuvillier Verlag
Internationaler wissenschaftlicher Fachverlag

www.ingramcontent.com/pod-product-compliance
Ingram Content Group UK Ltd.
Pitfield, Milton Keynes, MK11 3LW, UK
UKHW022001190726
13853UKWH00004B/1668

9 783736 973978